Resi... Cabling Manual

A Telecommunications Association

McGraw-Hill

New York   Chicago   San Francisco   Lisbon
London   Madrid   Mexico City   Milan
New Delhi   San Juan   Seoul
Singapore   Sydney   Toronto

McGraw-Hill

A Division of The McGraw·Hill Companies

Copyright © 2002 by BICSI. All rights reserved. Printed in the United States of America. Except as permitted under the United States Copyright Act of 1976, no part of this publication may be reproduced or distributed in any form or by any means, or stored in a data base or retrieval system, without the prior written permission of the publisher.

1 2 3 4 5 6 7 8 9 0 DOC/DOC 0 7 6 5 4 3 2

ISBN 0-07-138211-9

The sponsoring editor for this book was Marjorie Spencer and the production supervisor was Sherri Souffrance. It was set in New Century Schoolbook by Patricia Wallenburg.

Printed and bound by R.R. Donnelley & Sons Company.

 This book was printed on recycled, acid-free paper containing a minimum of 50% recycled, de-inked fiber.

McGraw-Hill books are available at special quantity discounts to use as premiums and sales promotions, or for use in corporate training programs. For more information, please write to the Director of Special Sales, Professional Publishing, McGraw-Hill, Two Penn Plaza, New York, NY 10121-2298. Or contact your local bookstore.

Contents

Preface

Acknowledgments

The BICSI officers and membership wish to thank the following members of the BICSI Technical Information and Methods Committee Panel 500 (Residential Publications) who contributed to the development of the *Residential Network Cabling Manual*, and provided important feedback through their reviews of this manual.

Chair:

Robert Jensen, RCDD, *Fluke Networks, Inc.*

Editorial Review Team (also Panel 500 Members):

Rick Akins, *FirstPoint Residential Communications*
Sharon Ballas, *BICSI*
Steven Elmore, RCDD, *CommScope*
Leah Fox, *BICSI*
Joan Hersh, *BICSI*
Nelda Hills, *BICSI*
RJ Hirshkind, *ADI*
Arron Hoffer, *SW Online*
John Jarmacz, *Pass & Seymour/Legrand*
Michael Jennison, *Greyfox Systems*
Rob Jewson, RCDD, *Essex Group*
Mel Lesperance, RCDD, *Ortronics, Inc.*
Todd McWhirter, *University of Washington*
Steve Munter, RCDD/LAN Specialist, *Teknon Corporation*
Arlyn Powell, *Independent Contractor*
John Pryma, RCDD/LAN Specialist, *Genesis Cable Systems*
John Seger, *Leviton Voice & Data Division*
Ron Shaver, RCDD/LAN/OSP Specialist, *BICSI*
Jonathan Werner, RCDD, *RSL COM U.S.A., Inc.*

Panel 500 Members:

Alan Amato, *Times Fiber Communications*
Marcy Antiuk, RCDD, *Leviton Integrated Networks Group*
Donna Ballast, RCDD, *dbi*
Edmund Baulsir, *Design Systems, Inc.*
Jeffrey Beavers, RCDD, *University of Kansas*
Roger Billings, *Wide Band Corporation*
Bill Black, *Copper Development Association*
Ian C.R. Blackman, RCDD, *solus communications Ltd*
Jerry Bowman, RCDD/LAN Specialist, *Superior Systems Technologies*
Doug Coleman, *Corning Cable Systems*
Ray Craig, RCDD/LAN Specialist, *ComNet Communications, Inc. (CCI)*
James Darnell, RCDD/OSP Specialist, *Technology Dynamics*
Jack Dix
Edward J. Donelan, RCDD/LAN Specialist, *Telecom Infrastructure Corp.*
David Dormer, RCDD, *CDM*
Jon Dreymann, *Technical Engineering Specialists*
Richard Dunfee, RCDD/OSP Specialist, *BICSI*
Robert Faber, RCDD/LAN Specialist, *The Siemon Company*
Al Feaster, RCDD, *ADC*
Robert Futch, RCDD, *Southern Voice & Data*
David Gibson, *Ortronics, Inc.*
David Graney, *Porta Systems Corporation*
Allan Hawkes, RCDD, *Cap Gemini UK, PLC*
Mark Johnston, RCDD, *Microtest, Inc.*
Joe Jones, *BICSI*
Greg Keys, RCDD, *Media Integrators*
Trevor Kleinert, RCDD, *Krone Australia*
Steve Lampen, RCDD, *Belden Electronics Division*
Denis Leduc, *FIPOE*
Tom Lyga, RCDD, *Pass & Seymour/Legrand*
Chas MacKenzie, RCDD/LAN Specialist, *NetAC*
Matthew Masi, RCDD, *Interface Network Systems*
Chris McPherson, RCDD, *Southwire Cyber Technologies*
Bill Minnis, RCDD, *InfoNet Technologies*
Frank Murawski, *FTM Consulting*
Don Nelson, RCDD, *Nelson Consulting Associates (NCA)*
Steve Oates, RCDD, *Verizon Connected Solutions*
Bruce Olson, RCDD, *Best Buy World Headquarters*
Jason Piazza, *V.91 Technologies*
Ned Sigmon, RCDD, *Tyco Electronics Corporation*
Richard Tobier, RCDD/LAN Specialist, *Proactive Solutions, Inc.*
Jimmy Underwood, RCDD, *CHR Solutions*
George Ward, RCDD/LAN Specialist, *Atlantic Pacific Communications*
Jay Warmke, *BICSI*
Kenneth White, RCDD, *KLK Technologies*

About BICSI

BICSI, a not-for-profit telecommunications association, was founded in 1974 to serve and support the professionals responsible for the design and distribution of telecommunications wiring in commercial and multi-family buildings.

BICSI has grown dramatically since those early days and is now recognized worldwide as an educational resource for the cabling infrastructure industry. Our membership spans the globe and our services cover the broad spectrum of voice, data, and video technologies. BICSI offers training, conferences, publications, and registration programs for cabling distribution designers, as well as commercial, and most recently, residential installers.

BICSI Member Benefits and Opportunities

BICSI members receive substantial discounts on quality education—design courses, conferences, and manuals. Members also gain access to valuable telecommunications information with the *BICSI News*, *Region News*, *District News*, standards and regulatory updates, and BICSI's Web site.

Membership offers ample opportunities for professional networking, and career development and advancement.

Members may pursue and obtain prestigious credentials— RCDD, RCDD/LAN Specialist, RCDD/OSP Specialist, Registered Installer and Technician, and Registered Residential Installer.

BICSI Training and Registration Programs for Commercial and Residential Installers

BICSI's commercial installation program

Even the most properly designed telecommunications cabling systems cannot function optimally unless they are properly installed. In 1996, BICSI created a comprehensive training and testing program to help commercial telecommunications cabling installers learn how to properly install a high-performance cabling system.

The Telecommunications Cabling Installation Training and Registration Program became an instant success. As of late 2001, 140 licensed training centers worldwide have produced more than 14,000 BICSI Registered Installers (Levels 1 and 2) and Technicians. The broad-based, comprehensive, vendor-neutral, career-building program produces highly competent and knowledgeable installers, in a minimal amount of time at a reasonable cost.

BICSI Residential
Cabling Program

BICSI's residential program

The residential network cabling market is growing rapidly. Residential structured cabling systems (the key enabling technology for home connectivity) are now included in nearly one-third of home starts, according to the National Association of Home Builders Research Center. And the number is growing.

Today's homeowners want properly installed cabling, at a minimal cost, to allow them to enjoy all the benefits of the connected home—whole-house Internet access and networking, intercom and paging, home theaters and automation, security, and more. Qualified installers are currently in demand to meet the increasing needs of today's electronic home.

Using the successful commercial installation program as a model, BICSI recently designed a comprehensive training and testing program for residential cabling installers to help ensure a properly installed residential structured cabling system.

The Residential Network Cabling Training and Registration Program includes instruction, followed by written and hands-on exams, then structured on-the-job training. The written exam is based on this manual. Those passing the exams receive the designation BICSI Registered Residential Installer. Graduates are well-equipped to design and install a variety of residential network systems, including new construction and retrofits.

The five-day, 40-hour course, **RES150: Residential Network Cabling: Theory and Hands-On Training**, is followed by a written and hands-on exam. Individuals not needing hands-on experience and BICSI Registered Installers and Technicians may take **RES100: Residential Network Cabling: Theory**, a $2^1/_2$-day, 20-hour course, plus separate written exam.

The theory component features modules on voice, data, and video distribution based on ANSI/TIA/EIA-570-A, *Residential Telecommunications Cabling Standard*, and the *National Electrical Code*®. The outline covers topics found in Chapters 1–12 and 17 of this manual.

The hands-on component includes new construction and retrofit rough-in skills for locating and mounting distribution devices and outlets. Students will practice properly pulling, securing, and terminating copper, coaxial, and fiber cables. Trim-out finish techniques will be covered, as well as system testing, documentation, and clean up (as described in Chapters 13–16 of this manual).

For Further Information and to Obtain Errata

For a complete packet of information on BICSI's wide range of products and services, including the Residential Network Cabling Training and Registration Program, please contact:

BICSI World Headquarters
8610 Hidden River Parkway
Tampa, FL 33637-1000 USA
813-979-1991 or
800-242-7405 (USA/Canada toll free)
fax: 813-971-4311
e-mail: bicsi@bicsi.org
Web site: www.bicsi.org

BICSI strives to provide up-to-date and accurate information. If errors are found in this manual, corrections will be posted on the BICSI Web site. Visit www.bicsi.org, click on Publications, and go to *Residential Network Cabling Manual*. Comments about the manual, including possible errors, may be e-mailed to publications@bicsi.org.

BICSI Policy for Numeric Representation of Units of Measure

International System of Units (SI)

BICSI technical manuals primarily follow the modern metric system, known as the International System of Units (SI). The SI is intended as a basis for worldwide standardization of measurement units. All units of measure in this manual are expressed in SI terms, followed by an equivalent empirical (U.S. customary) unit of measure in parentheses (see exceptions listed below).

Style guidelines

- In general, SI units of measure are converted to an empirical unit of measure and placed in parentheses. Exception: When the reference material from which the value is pulled is provided in empirical units only, the empirical unit is the benchmark.

- In general, soft (approximate) conversions are used in this manual. Soft conversions are considered reasonable and practicable; they are not precise equivalents. In some instances, precise equivalents (hard conversions) may be used when it is a:

  - Manufacturer requirement for a product.

  - Standard or code requirement.

  - Safety factor.

- For metric conversion practices, refer to ANSI/IEEE/ASTM SI 10-1997, Standard for Use of the International System of Units (SI): The Modern Metric System.

- Trade size is approximated for both SI and empirical purposes. Example: 103 mm (4 trade size).

- American wire gauge (AWG) and plywood are not assigned dual designation SI units. Dimensions shown in association with AWGs represent the equivalent solid conductor diameter. When used in association with flexible wires, AWG is used to represent stranded constructions whose cross-sectional area (circular mils) is approximately equivalent to the solid wire dimensions provided.

- In some instances (e.g., optical fiber media specifications), the physical dimensions and operating wavelengths are designated.

- When Celsius temperatures are used, an equivalent Fahrenheit temperature is placed in parentheses.

Background

Residential Environment

Introduction

Today's homeowner expects more from their house than just a living space. Now, more than ever, we must look to the future and the electronic house, which may soon become yours:

- Office
- School
- Shopping center
- Home theater
- Secure living space

Ask yourself:

- Can your house support these services today?
- What can your house do for you today, and what about tomorrow?
- How do you get there?

The residential environment for the single- and multi-family dwelling is in a constant state of change. Technological demands on the existing residential telecommunications cabling in these dwellings have driven the development and deployment of advanced products and services that must meet these demands. This change in the residential environment requires a higher degree of focus on the networks and the residential cabling that supports it.

Electronic Home of the Future

With the proliferation of the Internet, satellite, and community antenna television (CATV), homeowners are networked in an environment dominated by two-way, real-time telecommunications. In the future, the electronic home will require a more demanding telecommunications infrastructure to provide the homeowner with a truly interactive lifestyle.

Lighting control will be as natural and as expected in a new home as air conditioning and heat are today. Access to high-speed Internet data communications will be required virtually everywhere in the home. In November 2000, more than 50 percent of all Americans accessed the Internet from their homes—only five or six years after the Internet reached mass distribution.

The average home is built for a functional life of 50 years or more. What applications will then become standard? What will happen to the property values of homes that lack today's most advanced offerings?

This manual provides information to the homeowner, telecommunications distribution designer, builder, and property manager who will identify the key design, installation, and maintenance issues surrounding the implementation of residential structured cabling systems (RSCSs) in the residential market.

Homes that are properly cabled today will provide a wide range of benefits to the homeowner at minimal costs. Homes without RSCSS may simply become obsolete. Tomorrow's homeowners will expect to find the following applications, among others, available within their interactive homes.

TABLE 1.1   Top 20 uses for a residential telecommunications system

Services and Applications	Structured Cabling			
	Voice, Data, Video	Entertainment	Security	Home Automation
Internet access	✔	✔	✔	✔
Home office	✔		✔	✔
Home schooling and continuing education	✔	✔	✔	
Multiple telephone and facsimile (fax) lines	✔		✔	
Networking computers, printers, and devices	✔	✔	✔	✔
Intercom and paging	✔		✔	
Home health monitoring	✔		✔	
Theater quality sound and video	✔	✔		
Delivery and distribution of satellite and CATV, pay-per-view	✔	✔		
Distribution from internal video sources	✔	✔	✔	
Multi-zone whole home audio		✔	✔	
Improved use of floor space		✔		✔
Securing the family and belongings against crime	✔		✔	

continued on next page

TABLE 1.1   Top 20 uses for a residential telecommunications system (continued)

Services and Applications	Structured Cabling			
	Voice, Data, Video	Entertainment	Security	Home Automation
Remote control of home entry for children after school	✔		✔	✔
Monitoring against flooding, carbon monoxide, fire, smoke, and power outages	✔		✔	✔
Closed-circuit security cameras and remote baby monitoring	✔		✔	
Smart appliances	✔	✔		✔
Preprogrammed lighting scenes				✔
Automated control of sprinklers and other subsystems				✔
Power control for increased efficiency				✔

Reprinted with permission of FirstPoint, *The Interactive Home* © 2001.

Why Upgrade Traditional Residential Cabling?

Today, the basic residential cabling that is installed in most homes is based on 25-year-old technologies. Many homes have "flat" or "quad" four-conductor cable (or low-grade twisted-pair cabling commonly referred to as station wire) used for providing plain old telephone service (POTS). Most homes also have 75 Ω coaxial cable that is used to provide connection to the incoming CATV service.

Neither of these cables is recommended for providing access to today's high-speed data telecommunications systems. The four-conductor telephone cable cannot reliably transmit data at rates that are increasing well beyond the megabit per second (Mb/s) range, and often experiences trouble connecting to low-speed dial-up service. Digital subscriber line (DSL) and cable modem technologies will place still more demands on this residential cabling infrastructure. Further, the traditional coaxial cable used in homes is not adequate for the two gigahertz (GHz) plus frequency range needed to transport today's satellite television and communications service.

The proper choice of material is not the only issue. The design of the residential cabling system historically has also been inadequate. Many homes have been designed with daisy-chained cabling run in series from one telecommunications outlet/connector to another. Daisy-chaining cable between outlets limits the system's effectiveness to a significant degree and does not conform to the standards requirements of ANSI/TIA/EIA-570-A, *Residential Telecommunications Cabling Standard*, which specifies that each cable from each outlet location needs to be run back to the home's central cabling distri-

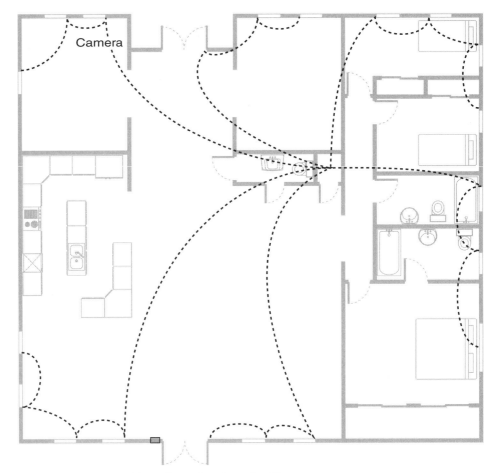

Figure 1.1   Example of daisy-chained residential cabling.

bution points (DPs), also known as distribution devices (DDs) (see Chapter 2: Residential Structured Cabling Systems). The lack of a centralized telecommunications cabling DD strictly limits the functionality and manageability of the home's various subsystems. The lack of proper testing and labeling limits the effectiveness of the systems upon move-in and decreases the manageability and usability over the life of the house.

Standards-Based Structured Cabling

To prepare for and meet the requirements of current and future telecommunications technologies and residential networking applications, a standards-based RSCS design and installation procedure should be followed. These standards allow the residential structured cabling professional to provide the following benefits:

- A proper cabling infrastructure
- Quality components
- Consistent installation techniques
- Appropriate testing and labeling

Knowledge of the applicable telecommunications cabling standards, particularly ANSI/TIA/EIA-570-A, is critical. Understanding the requirements of the various telecommunications networking standards is also of significant importance.

Experience in the particular requirements of the residential cabling installation environment is a key factor in the final cost and success of an installation. This is of greater importance in retrofit applications, in order to limit the potential for damage being done to the existing house.

Types of Services

The residential customer requires up to four primary low-voltage services in the home—telecommunications, entertainment, security, and automation. Because most individual homes represent a relatively small revenue opportunity, builders and owners often desire a single-source subcontractor for all integrated solutions.

Structured residential telecommunications cabling installers in this market may be capable of providing the following services:

- Telecommunications—High performance structured cabling (twisted-pair, coaxial, and optical fiber) for voice, data, and video (VDV) applications, networking and telephony hubs, residential gateways, and video distribution devices.

- Entertainment—In-wall and in-ceiling speakers, home theaters, whole-house audio systems, and satellite television and communications. The ability to deliver audio and video to any point in the home is becoming increasingly important.

- Security—Security and monitoring systems, closed-circuit surveillance cameras, and intercom/paging systems. Control and monitoring of the home and family can protect both family members and family belongings from a wide range of hazards.

- Automation—Lighting control, heating, ventilating, and air conditioning (HVAC) control, "smart" appliances, and other systems. These systems increase fuel efficiency and provide everyday conveniences to the homeowner.

The demand for these services revolves around work, learning, and lifestyle.

In terms of work, a large percentage of the American workforce now calls their home their office (small office, home office [SOHO]), while millions more work at least part-time (telecommute) from home. Many of the telecommunications applications taken for granted in the office are likely missing in the home. However, telecommunications services are of even more importance to the telecommuter—whose only contact with fellow workers and customers is through the telecommunications system—than for the typical office worker.

In terms of learning, many school-aged children are being home schooled, driving the demand for Internet access as a primary teaching tool. For those in college, or returning to college part-time while in the workforce, the home has become the lecture hall, group study portal, and classroom rolled into one.

Many homeowners want their homes to provide convenience, safety, and entertainment. Convenience is provided by lighting and HVAC control (with the added benefit of energy efficiency). Safety is provided through monitoring homes against intrusion, fire, carbon monoxide (CO), and other hazards. Entertainment requirements are met through preinstalled audio and home theater systems.

Integrated Service Provider

Vendors, residential cabling installers, and other professionals are focused on the development of the residential cabling market. The response from the home-buying community has been strong and the market is growing at a healthy rate.

For the past five years, marketing experts have waited for the one application that would spark a revolution in residential telecommunications systems. However, the growth in demand has not resulted from any single application.

Instead, the market has grown because there now exists a broad spectrum of mature applications consumers want access to in the home. Individuals and families are spending more on their homes, demanding:

- A safe and secure environment
- A number of technological conveniences
- The ability to entertain themselves and their guests
- Access to efficient telecommunications with the outside world
- Reliability

Homeowners want these telecommunications systems to integrate with one another for improved control and efficiency.

These market changes have opened the door for companies looking to integrate services, as shown in Figure 1.2. Integrated services companies are single-source solutions providers who can implement systems during the time of construction. Installation during construction limits the aggravations, time consumption, and cost previously required from the homeowner trying to obtain these services from vendors after the residence is constructed.

As with any market, there will be room for growth in businesses that specialize in one of the telecommunications technologies. Whether the home-

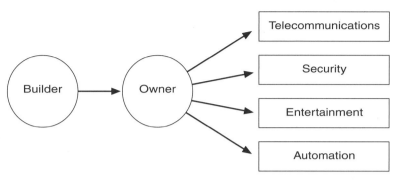

Figure 1.2   Homeowner securing services in aftermarket from multiple vendors.

builder or homebuyer works with an integrated service provider (SP) or a number of specialists (see Figure 1.2), the benefits of properly integrated telecommunications solutions are critical, as is the ability to install these systems during the initial construction of the house.

Projected Size of the Residential Market

There are a number of factors driving the growth of the residential cabling and home networking markets. These include the growth in the number of homes with multiple personal computers (PCs) and other devices that need to be networked together and share Internet services. The number of home-based businesses has increased dramatically, as has the number of people telecommuting from home on a part-time basis.

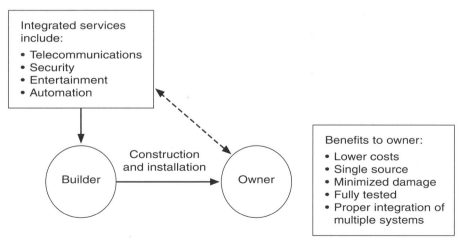

Figure 1.3   Homeowner secures services from a builder working with a single integrated services company or multiple specialists.

By 2004, the estimated size of the U.S. residential structured cabling business for VDV applications is estimated to top $4 billion. By that time, structured cabling systems will be installed into nearly 50 percent of all new U.S. homes and a steadily growing percentage of existing homes. The residential cabling market is much broader in scope than the commercial field. Whereas the commercial structured cabling installer can focus on the cabling system alone, or perhaps with an eye on integrating the networking equipment, the residential cabling intaller often must implement a variety of other telecommunications technologies.

Typically, residential installers install cabling systems and components for VDV applications and for home entertainment systems, home security, and home automation. The homebuilder or homeowner may also require various telecommunications applications to the property, including satellite television and communications, broadband Internet, and security/monitoring.

The market for the entire integrated services package in the U.S. residential market may approach $50 billion annually by 2007—a much larger market than is currently available in the commercial structured cabling market.

Entering the Market

For many residential cabling professionals, the residential market is unexplored territory. Ultimately, the quality of work and longevity of a residential telecommunications cabling company is largely based on the proper understanding of the issues to be faced within the home environment.

The average amount of revenue generated by a residential cabling installation from a single-family residence is typically smaller than for a commercial business application. This makes the expense of marketing and selling to individual homeowners more significant and burdensome. To overcome this obstacle, residential structured cabling installers can elect to market through homebuilders or architects, rather than directly to the homeowner. This strategy allows the cabling installation professional to efficiently market to more homeowners per year at lower costs.

The residential cabling professional may have to meet each customer to explain and sell optional upgrades. While this provides the full-service integrator the opportunity to deliver potentially $3,000–20,000 of products and services to each home, there is no guarantee that this up-front work will increase revenue. In addition, many builders require a margin for services sold to their homeowners. This represents an additional cost to the subcontractor or customer.

Within the residential cabling market, dealing with logistics is also quite difficult. While working in the commercial networking environment, there are often critical scheduling requirements. In the home market, however, there are often two phases—"rough-in" and "trim out"—particularly when dealing with new construction (see Figure 1.4).

During the rough-in phase, residential cabling is installed within the walls, and various mounting fixtures are added to the house. The window of time

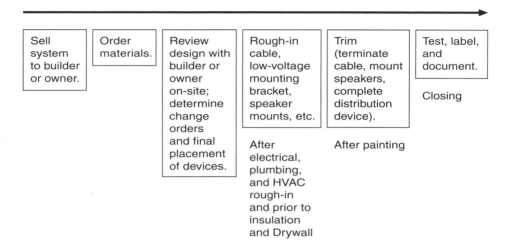

Sell system to builder or owner.	Order materials.	Review design with builder or owner on-site; determine change orders and final placement of devices.	Rough-in cable, low-voltage mounting bracket, speaker mounts, etc.	Trim (terminate cable, mount speakers, complete distribution device).	Test, label, and document.
			After electrical, plumbing, and HVAC rough-in and prior to insulation and Drywall	After painting	Closing

HVAC = Heating, ventilating, and air conditioning

Figure 1.4 Typical residential cabling installation time line.

available to complete this work may be only a few days, or perhaps only a day in terms of working with larger production-style builders. The work should begin after the completion of the rough-in work of the electrician, plumber, and HVAC professionals and should be completed prior to the start of the installation of insulation and Drywall.

Once the cable and components are roughed in, several weeks may pass before the second time-critical event—the termination, testing, and labeling of the cabling system. Since this final trim phase of the RSCS is one of the last tasks to be completed during the preparation of a new house, the deadline may literally be the home's closing date—the day the homeowner takes possession.

Additional logistical difficulties may rise from the nature of the residential construction market. First, the business may be geographically dispersed—even when working with a single homebuilder. Each house may differ in terms of design (and installation requirements), as opposed to the commercial telecommunications cabling environment in which drop ceilings present consistent design and installation challenges. On-site determination of the location for various systems may be required, often requiring an interface with both the homebuilder and the homebuyer. Meeting the homeowner often requires meeting after hours or on the weekend.

Potential Problems

Some homes can be roughed-in (or trimmed out) in less than half a day—leaving one or more technicians out of work or in a potentially lengthy transit to another job site. Dealing with consumer-oriented products opens up the integration company to a potentially heavy load of service calls.

Residential cabling installation work is difficult and messy. Some factors that must be dealt with include:

- Working around other trades
- Asbestos
- Fiberglass insulation
- Concrete and plaster walls

There is also the possibility of having the preinstalled cabling and components damaged during the construction phase of the house (prior to the final trim). Inclement weather, including rain and snow, may also interfere with the progress of work. Often, job sites are without power during the rough-in period.

In some situations, tooling costs are higher. Ladders must be purchased to handle cathedral ceilings and other locations that are not the standard drop-ceiling height found in the commercial building environment. Drill bits must be obtained to penetrate a number of materials. Residential cabling installers must be provided the tooling and training to work with a variety of cable types. Some of the tools needed by the residential cabling installer are:

- Cable reel holders
- Cable ties
- Cable/wire cutters
- Coaxial cable termination tools
- Cordless drills
- Field test instruments (e.g., optical fiber loss test set, Category 5 or 5e test set, multimeter)
- Fish tape
- Flexible drill bits
- Labeling equipment
- Labeling tape
- Levels
- Long drill bits
- Optical fiber cable termination tools
- Portable heater
- Power drill with masonry bits
- Power generator
- Pull string

- Staple guns
- Stepladders (1 m [3 ft], 1.8 m [6 ft], and 2.4 m [8 ft])
- Strippers for cables up to 14 AWG (1.6 mm [0.063 in])
- Stud finder
- Twisted-pair cable termination tools

Single-Family versus Multi-Dwelling

In general, the residential cabling market is divided into single-family residences and multi-dwelling units (MDUs). In many ways, the MDU system design and residential cabling installation practices approximate the commercial building environment once it enters the dwelling unit. Important considerations pertaining to MDUs are covered in this manual. The main focus of this manual, however, is the single-family residence, an area that presents the greatest differences from the commercial structured cabling environment.

If the decision is to provide services to the residential cabling market, a number of secondary decisions must be considered. These decisions include whether to focus on single-family residences, or MDUs, or both.

The residential cabling system designed and installed within each individual unit of an MDU (typically townhouses, condominiums, apartment complexes, or assisted living facilities) should follow the practices detailed in ANSI/TIA/EIA-570-A. The individual units have similar cabling needs as a single-family residence. The provision of incoming telecommunications services—voice, data, and video—however, will require a substantial backbone telecommunications cabling system (see Chapter 18: Multi-Dwelling Unit Structured Cabling Installation).

In other cases, the decision may be to cable from a centralized point within the MDU directly to the voice and video outlets in the individual units. This design does not follow industry standards but has been the design of choice for many years due to its reduced cost structure. In this case, some of the design criteria concerning standard RSCSs may be bypassed altogether, and many of the benefits of this type of system will be lost.

Geography (e.g., large city vs. suburbs, etc.) and the general housing conditions of the local market will influence the decision of a contractor to focus on the single-family or multi-dwelling environment, or both. A commercial structured cabling company may focus on the MDU market, which in many aspects works the same as a commercial building. Tooling and training are similar. The sales and marketing techniques are also similar since more of the design and purchasing decisions for an MDU are determined ahead of time by the builder.

Some of the potential residential telecommunications services also have a diminished role in the MDU environment (e.g., home security). A commercially oriented structured cabling company will not need to invest as much time and resources in these areas to remain competitive. However, other products and services, such as in-wall speaker systems, may be of more interest to owners of space-constrained MDUs.

Due to the potential of greater sales per unit, a residential-only integrated services company may choose to focus on the single-family market. Homeowners in the single-family market are often more likely to pay extra for adding functionality to their own homes. Meanwhile, a major goal of an MDU building owner is to minimize up-front residential cabling installation costs.

> **NOTE:** Many multi-dwelling properties do not have an owner but are managed through a condominium or homeowner association.

The cost structure and pricing for an individual home can be better controlled than in the MDU environment. A small miscalculation per unit (when spread over hundreds of properties within an MDU) can result in a large loss for the residential cabling installer.

The market for retrofitting existing MDU properties may be limited due to the construction of the building (e.g., concrete slab). However, the retrofit market represents an enormous and potentially profitable opportunity in the single-family environment.

New Construction versus Retrofit Systems

The next key decision is whether to focus on the new construction market or the retrofit market. The cabling contractor should consider the geographic region in which the company is located and the general state of the new construction market. If the new construction housing market is strong in the integrator's region, the company can build a business around providing services to, and through, homebuilders. However, if the number of new homes built in the immediate area is low, a marketing approach directed through remodelers and architects into retrofit applications may be more productive.

Installation costs are much lower when working with the open walls of a new house. The cost of the systems can often be rolled into the home's mortgage. In addition, new construction customers are a captured market and can be led by the homebuilder's sales professional through the options available through the low-voltage integration company.

The risk of experiencing situations that are both unexpected and time consuming are significantly larger during retrofit projects. Market prices are higher, or will be based on time and material billing. If the jobs are run consistently well, profit margins can be significant. The retrofit market may also become much larger than new construction. While there are over one million new homes built each year in the United States, there are some 90 to 100 million existing homes. Many of these properties will need a technological upgrade in the coming years.

Working with homebuilders during new construction often implies the inclusion of a margin for the builder, which is added to the total charged to the homeowner. This may not be the case in the retrofit market.

Consider the sales and marketing costs associated with generating name recognition for the integrated services firm in the general homeowner market versus the low costs associated with marketing to a small number of new homebuilders in a geographic area. The production homebuilder may be an attractive customer due to the large number of homes being built each year. However, this also makes that builder a prospect for other companies already on site offering competitive low-voltage services.

Residential Cabling Installer

Integrated telecommunications services are ultimately purchased by the homeowner, whether the property is new construction or an existing home. However, in the new home environment, the builder becomes an invaluable part of the process. Often, the homeowner is not known until after the work of the structured cabling company is complete. Therefore, in new construction, the needs of the homebuilder must be met consistently to create a strong long-term working relationship. The requirements include:

- Training the builder's sales staff and providing marketing tools
- Answering homeowner's questions that cannot be answered by the builder's sales staff
- Providing a profit margin on upgrades
- Providing cost-effective standard packages
- Meeting all construction deadlines
- Working effectively with the builder's billing and payment processes
- Providing a warranty

In the retrofit market, it is important to work with architects and remodelers. One requirement in dealing with architects is the ability to aesthetically blend the connectivity solutions into the overall design of the home. It is important to determine the final decision-making authority (architect, homeowner, or general contractor) up front. Most importantly, the integrator must retain the flexibility of continually redesigning a system to meet the changing requirements of the homeowner.

Manual Overview

This manual provides information that will identify the key design, installation, and maintenance issues surrounding the implementation of structured cabling systems in the residential market.

Chapter 1 provides an introduction to the future of residential telecommunications cabling, today's broadly accepted technologies, and various issues affecting the contractors, homeowners, and homebuilders working in this area.

Chapter 2 covers the details of the ANSI/TIA/EIA-570-A standard. Proper design considerations are covered, focusing on each major subcomponent of the cabling system. A useful sample design of outlet and service locations for a home is offered, providing insight into the future networking and telecommunications requirements within the home.

Chapter 3 provides a listing of other resources available to the reader. This includes a brief description of the most important standards and codes governing the residential cabling market.

Chapters 4 and 5 detail proper component requirements, first in terms of cabling media, and then with a discussion of connectors and other devices.

Chapter 6 provides information regarding the integration of consumer components. These range from cable modems and DSLs to residential gateways and computer networking devices.

Chapter 7 describes the technical requirements of a properly functioning low-voltage cabling system. Transmission fundamentals are detailed to identify potential hazards to avoid when designing and installing a residential cabling system.

Chapter 8 provides critical information regarding electrical protection systems, with a particular focus on proper electrical grounding techniques.

Chapter 9 details safety issues for both the homeowner and the cabling professional.

Chapter 10 reviews alternative choices for transmission media within the home, including wireless, telephone line, and power line technologies. This may be of particular interest to the owner of an existing home.

Chapter 11 provides an overview for other major residential telecommunications technologies, including direct broadcast satellite, intercom/paging systems, home theater, home security, and more.

Chapter 12 brings the discussion back to single-family residences. The process of planning and implementing of the network cabling system is covered. Alternative building constructions are also addressed. The materials developed before construction, and those provided to the builder or owner after the work is completed, are described.

Chapter 13 reviews the installation of a residential network cabling system. This chapter describes the "rough-in" phase of a project, when the cable is run within the walls to the final locations.

Chapter 14 describes the completion of the installation for a retrofit (i.e., existing home) project.

Chapter 15 describes the final "trim-out" of a cabling project for a new home (beginning after insulation and Drywall have been installed).

Chapter 16 provides critical information concerning the proper testing of a residential network cabling system, ensuring proper functionality from the first day of use. Troubleshooting information is also provided.

Chapter 17 provides an overview of special design and installation requirements when dealing with an MDU. Backbone cabling systems, not typically found in single-family residences, are described.

Chapter 18 covers the multi-dwelling environment, focusing on the unique demands of an MDU.

The glossary consists of common definitions and frequently used acronyms from this manual. Terms and definitions marked with an asterisk are reprinted with permission of Telecommunications Industry Association.

The bibliography provides the reader with a summary of other industry resources that will provide in-depth analysis of specific issues.

2

Residential Structured Cabling Systems

Overview

This chapter provides an overview of residential telecommunications premises cabling infrastructure.

Residential construction has traditionally differed from commercial construction. Commercial buildings have evolved to incorporate cabling systems to address networking capabilities, while most single-family and multi-dwelling unit (MDU) construction has had residential cabling installed primarily to support telephones and televisions. However, beyond being living spaces with limited telecommunications needs, homes have become small offices and home offices. These homes are incorporating sophisticated automation systems and are using bandwidth-intensive multimedia applications.

In this chapter, residential services are correlated to grades of cabling and are aligned with the principles and requirements of ANSI/TIA/EIA-570-A, *Residential Telecommunications Cabling Standard*. Preparing homes for these services with the proper grade of telecommunications cabling is the focus of this chapter.

This discussion of residential structured cabling system (RSCS) infrastructure clarifies the use of structured cabling in the home and describes what systems fall within this category.

Grades of Residential Premises Telecommunications Cabling

There are two defined grades of residential premises telecommunications cabling for the home, as defined in ANSI/TIA/EIA-570-A. These grades have been established based on services that are expected to be supported within each residential unit and to assist in the selection of the cabling infrastructure.

"For each cabled location…Grade 1 provides a generic cabling system that meets the minimum requirements for telecommunications services. As an example, this grade provides for telephone, satellite, community antenna television (CATV) and data services. Grade 1 specifies twisted-pair cable and coaxial cable placed in a star topology. Grade 1 cabling minimum requirements consist of one 4-pair UTP (unshielded twisted-pair) cable and associated connectors that meets or exceeds the requirements for Category 3 cabling…and one 75-ohm coaxial cable…. Installation of Category 5 cable in place of Category 3 cable is recommended, to facilitate future upgrading to Grade 2."

"For each cabled location…Grade 2 provides a generic cabling system that meets the requirements for basic, advanced, and multimedia telecommunications services. This grade provides for both current and developing telecommunications services. Grade 2 specifies twisted-pair cable, coaxial cable, and optionally optical fiber cable, all placed in a star topology. Grade 2 cabling minimum requirements consist of two 4-pair UTP cables and associated connectors that meet or exceed the requirements for Category 5 cabling…and optionally, 2-fiber optical fiber cabling…. Installation of Category 5e cabling…in place of Category 5 cabling is recommended."

> **NOTE:** It is recommended that Category 5e twisted-pair cable be used for both Grade 1 and Grade 2 installations. It is also recommended that Series 6 coaxial cable be used for both Grade 1 and Grade 2 quad-shielded.

TABLE 2.1   Typical residential services supported by grade

Service	Grade 1	Grade 2
Telephone	√	√
Television	√	√
Data	√ (limited)	√
Multimedia	—	√

TABLE 2.2   Recognized residential cabling by grade

Cabling	Grade 1	Grade 2
4-pair UTP	√ (Category 3; Category 5 cable recommended)	√ (Category 5; Category 5e cable recommended)
75 Ω coaxial cable	√	√
Optical fiber	—	√ (Optional)

> **NOTES:** It is highly recommended that each living area be cabled with a minimum of one outlet. A living area is defined as any room in which voice/data/video (VDV) applications may be used. This

includes bedrooms, kitchens, den/study/home offices, and living/family rooms.

Multimedia is defined as a collection of technologies designed to create and view content made up of any combination of text, graphics, still images, audio, video, and animation.

It is also highly recommended that multiple outlets be installed in each living area to allow for flexible furniture configurations and to reduce the use of long patch cords.

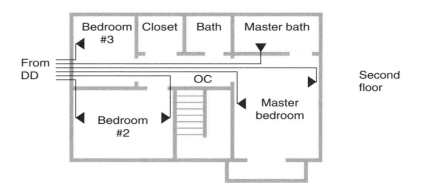

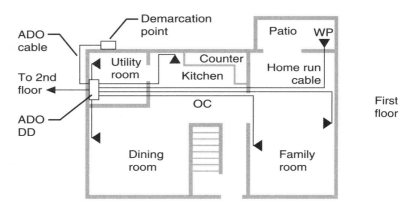

▼ = Telecommunications outlet/connector
ADO = Auxiliary disconnect outlet
DD = Distribution device
OC = Outlet cable
WP = Waterproof outlet box

Figure 2.1   Example of home run cabling.

All cabling must be home run directly from individual outlets to the distribution device (DD) (i.e., without any other connection point along the cabling).

Additional information about CATV cabling is covered in Chapter 22: Private CATV Distribution Systems, of the *BICSI Telecommunications Distribution Methods Manual,* 9th edition.

Typical Residential Premises Cabling System

A typical residential premises cabling system appears in Figure 2.1. The components are shown as though they perform only one function, but they can perform multiple functions. Note that cabling should be home run from individual telecommunications outlets/connectors to the DD (Figure 2.2).

Figure 2.3 illustrates an example of a residential cabling layout.

Residential Structured Cabling Systems (RSCSs) and Cabling Topologies

An RSCS is defined as the complete collective configuration of cabling and associated hardware at a given site that has been installed to provide a com-

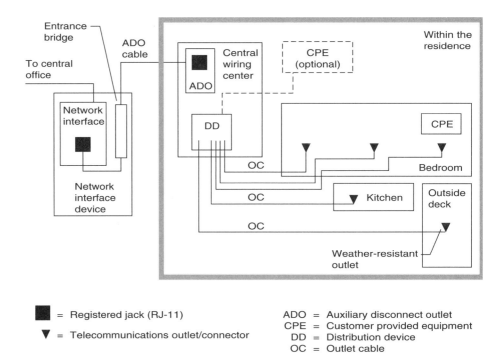

Figure 2.2 Typical residential premises cabling system.

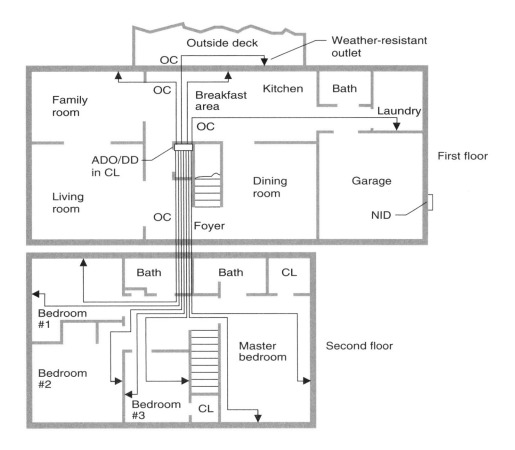

Figure 2.3   Residential cabling layout.

prehensive telecommunications infrastructure. This infrastructure is intend-ed to serve a wide range of uses, from providing basic voice service to advanced entertainment services, and should not be device- or application-dependent.

In this manual, an RSCS is further defined in terms of ownership. The RSCS begins at the point at which the access provider (AP) terminates its infrastructure. This point is also known as the demarcation point (DP) and may be evidenced by a network interface device (NID).

In a telephone system installation, the AP furnishes one or more service lines dependent on the customer's requirements and connects them to the

NID. Everything beyond the DP, including the cabling system installation, maintenance, and the equipment hooked to it is the responsibility of the home or building owner.

There are many possible configurations of an RSCS. This is due to variations in the:

- Architectural structure of the building that houses the residential cabling installation
- Cable and connection products
- Function of the residential cabling installation
- Types and characteristics of equipment that the residential cabling installation will support, both present and future
- Configuration of any existing residential cabling installation (e.g., upgrades and retrofits)
- Customer requirements and actual needs
- Warranties offered by manufacturers

Standardization of residential cabling installation

Although the specifics of any installation may be unique, the overall components of an RSCS and the methods used to complete and maintain the installation are relatively standard. The standardization of residential cabling installations is necessary to ensure acceptable performance in increasingly complex systems. The cabling industry in the United States accepts the Telecommunications Industry Association as the responsible organization for providing and maintaining standards and practices within the cabling profession. In Canada, the Canadian Standards Association is responsible for providing and maintaining standards.

Standards-compliant installations provide the following benefits:

- Consistency of design and installation
- Conformance to physical infrastructure and information transmission requirements
- A basis for examining proposed system expansions and other changes
- Uniform documentation
- Ease of maintenance
- Reliability

Cabling topologies

An item's topology refers to its features, shape, or physical appearance. For example, a topographical map represents the physical appearance of the geographic area shown. Network topologies can have physical or logical configurations.

- The physical appearance of the network is known as its physical topology.

- The logical topology is determined by how the messages are transmitted from device to device, by how the network actually functions.

There are instances where a network has a certain physical appearance but logically transmits its messages in a different topology. For this reason, it is necessary to make the distinction between the physical and logical topologies of a network. This section illustrates the physical appearance of the network.

There are two fundamental RSCS topologies—star and bus.

Star topology.   In a star topology, the DD is placed in the physical, as well as logical, center of the network. The remaining devices are connected to this central hub like the points on a star. See Figure 2.4.

Each device has its own direct, dedicated line to the DD. This is known as a home run. Star topology advantages are:

- Cabling is easier to install and maintain.

- If one device is disabled or isolated from the DD, it is the only device affected.

- Faults are easier to locate and isolate.

- It provides a central location for managing the network.

Bus topology.   A bus topology is a linear configuration. It places all of the devices on one length of cable. See Figure 2.5.

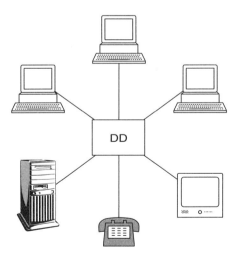

DD = Distribution device

Figure 2.4   Star topology.

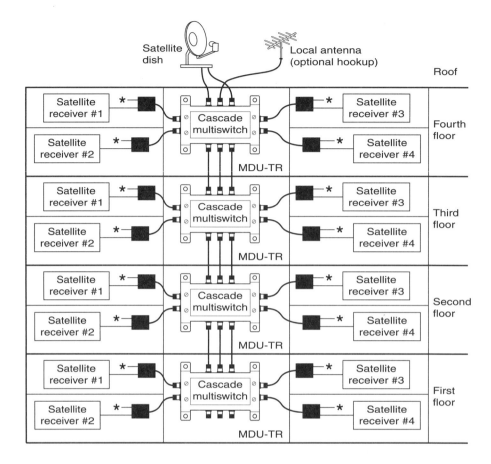

Satellite dish

Local antenna (optional hookup)

Roof

Fourth floor

Third floor

Second floor

First floor

■ = Diplexer, if hooking up local antenna to multiswitch

★ = To local antenna feed on TV or receiver

MDU-TR = Multi-dwelling unit telecommunications room

Figure 2.5 Bus topology.

The ends of the transmission channel in this arrangement are not connected to network devices. Problems occur when the transmitted signal is sent along the cable and it reaches either of the ends. For this reason, each end of the cable is connected to a terminator.

The most common bus topology residential applications include:

- MDU video/direct broadcast satellite (DBS) distribution
- Intercom
- Fire alarm

Components of a Residential Structured Cabling System (RSCS)

This section describes the basic components of an RSCS designed for telecommunications.

Demarcation Point (DP)

The DP is the point of interface between APs and the home. For single-family homes, the DP is usually located on the outside of an exterior building wall adjacent to the power meter. APs are responsible for maintaining service only to this point. The property owner is responsible for all inside cabling beyond the DP.

To comply with local regulations and building codes, the residential cabling installer should contact the AP to determine the proper DP location. Where the total length of cabling from the DP to the farthest outlet in the home exceeds 150 m (492 ft), the AP should be notified at the design stage to ensure that transmission requirements can be met. However, the AP is not necessarily responsible to accommodate this situation and additional DD equipment may be required.

The DP may be evidenced by a NID that is provided and installed by the AP, and may contain an entrance bridge. The DP is meant to serve only as an interface between the AP and the customer, and it is not to be used as a DD.

It is recommended that some documentation be provided at every AP's DP to instruct the AP of the existence of an RSCS in the home. This documentation instructs the AP on the basic cabling infrastructure within the home and any special notes for service hookups.

> **NOTE:** Consider that the DP for multiple service providers (SPs) is usually in the same place but not always (e.g., DBS service and a multi-channel multi-point distribution system can be considered a type of DP and requires a separate DP and auxiliary disconnect outlet [ADO] cable).

The following are considerations for locating and installing the DP:

- Locate the DP on the exterior building wall near the electrical power meter.

> **NOTE:** The DP should be within 6 m (20 ft) of the electrical service power meter so that a connection point to electrical ground is available. Otherwise, a ground rod will have to be driven and a bonding conductor extended between the ground rod and the electrical service entrance point.

- Consult the APs for proper location.
- Comply with local regulations and building codes.

- Place the DP within 150 m (492 ft) of the farthest outlet in the home.
- The DP should be readily accessible to both the APs and the residential cabling installer.
- The DP should provide documentation for the AP installer.

Auxiliary Disconnect Outlet (ADO) cable

ADO cables extend access lines from the DP to the ADO. Where a single residential unit is part of a multi-residential building, the ADO cable may extend from the MDU-telecommunications room (TR) to the ADO in the tenant space.

The ADO cable that is used will ultimately be determined by the media and signal type provided by the AP at the DP and by the tenant's needs. The ADO cable should be of a type and specification suitable to adequately carry the AP service (and future anticipated services from that AP) to the ADO/DD.

It is recommended that the ADO cables be installed inside of an appropriately sized duct. For more details, refer to Chapter 11: Residential Cabling, of the *Telecommunications Distribution Methods Manual (TDMM)*, 9th edition.

Twisted-pair.   Twisted-pair ADO cable should:

- Use 24 AWG (0.51 mm [0.020 in]) paired wires, with a minimum of Category 5e cable recommended.
- Have a minimum of eight pairs (two 4-pair cables) per residential unit to provide for:

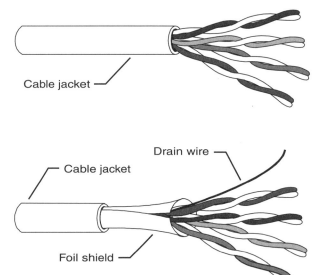

Figure 2.6   Twisted-pair cable.

– Telephone lines

– Data (e.g., modems)

– Security

– Separation of services, if needed

The tenant's needs will ultimately determine the proper number of pairs required for the ADO cable.

Coaxial.   Coaxial ADO cable should:

▪ Use a quad-shielded Series 6 or Series 11 or both types of coaxial cable.

▪ Accommodate transmission of both baseband and broadband signals.

▪ Supply a minimum of two coaxial cables per AP to each residential unit to provide for:

– Telephone lines

– Data

– Community antenna television (CATV)

– Security

– Multimedia service

Coaxial patch cords.   Equipment and outlets serviced by coaxial cable requires coaxial patch and equipment cords. The cords typically provided with consumer audio/video equipment are only suitable for use as equipment cords.

▪ Equipment cords are used to connect equipment to the outlets and may be Series 6 or Series 59 cabling and should be factory terminated with an F connector. Threaded and nonthreaded (push-on) connectors are acceptable.

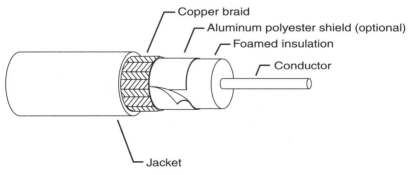

Figure 2.7   Coaxial cable.

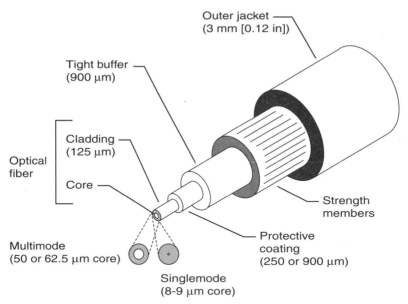

Figure 2.8   Optical fiber cable.

- Patch cords are used to connect outlet cabling to the DD and must be the same cable that is used in the outlet cabling. These cords can be either factory or field terminated with F connectors.

Optical fiber (optional).   Optical fiber ADO cable should:

- Use 50/125 μm or 62.5/125 μm multimode, singlemode, or all types of fiber within the cable.
- Have a minimum of four fibers per residential unit to provide for:
  - Telephone lines
  - Data
  - Community antenna television (CATV)
  - Security
  - Multimedia service

Auxiliary Disconnect Outlet (ADO)

The ADO provides the means for the tenant to disconnect from an AP. The ADO provides terminations for connection to the customer premises equipment (CPE), premises cabling, or both.

According to ANSI/TIA/EIA-570-A, "An auxiliary disconnect outlet (ADO) provides the means for the tenant to disconnect from an access provider. In a

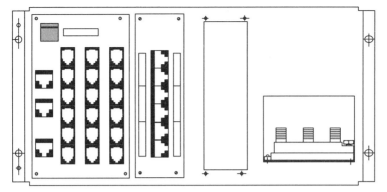

Figure 2.9   Auxiliary disconnect outlet.

single-residential unit, an ADO shall be installed where a means of disconnecting is not otherwise provided, or if the location of that disconnection point
is not easily accessible by the tenant. It is desirable to co-locate the ADO with
the distribution device (DD). The ADO and DD shall be located indoors and be
readily accessible."

ADOs should be designed for:

- A minimum of four twisted-pairs (the other four twisted-pairs may not be ter-
minated)
- Two coaxial cables
- Four fiber strands

The ADO and DD are usually colocated and can be combined in a single unit.
Install the ADO in the end user's or tenant's premises at a readily accessible
location. Consider allowing for growth by leaving space for the addition of
more ADOs in the future.

If an ADO is created within the DD, patch panels and patch cords are recommended for this purpose.

Consider the following when locating and installing ADO cable and ADO:

- Connects network interface device (NID) at distribution point to distribution
device (DD).
- Consists of twisted-pair, coaxial, and optionally, fiber optic cable.
- ADO cable is installed in an appropriately sized duct.
- ADO should accommodate a minimum of four twisted-pairs, two coaxial
cables, and optionally, four fiber strands.
- ADO may be colocated with the DD.

- ADO cable and ADO should be readily accessible to the homeowner and the residential cabling installer.

- Consider leaving space in the DD for additional ADOs to accommodate growth.

Distribution Device (DD)

The DD is a cross-connect facility used for terminating and connecting outlet cables, DD cords, equipment cords, and ADO cables (see Figure 2.10). It is the center of the residential cabling network, equivalent to the telecommunications room (TR) or equipment room (ER) in a larger commercial network.

At a minimum, the DD should provide a single point of consolidation for incoming access lines from the DP/ADO and outlet cabling within the home. The DD should provide for flexible cabling distribution using cable management. It is essential that the design anticipate flexible and user-friendly moves, adds, or changes (MACs) on both the AP side and the telecommunications outlet/connector side.

Space should be provided within or adjacent to the DD for the installation of a surge protection device for each conductive cable entering or leaving the building. Access to the building's electrical ground must be provided within 1.5 m (5 ft) of the DD, and in accordance with applicable electrical and building codes.

The DD may consist of a passive cross-connect facility, or an active cross-connect, or both. An example of an active cross-connect facility is a residential gateway.

The DD should be engineered and tested with other components as set forth in these specifications to operate and perform as a complete system.

The DD must be of sufficient capacity and functionality for the:

- Quantity and types of incoming services

- Distribution of services

- Application-specific outlets

All incoming services are delivered directly to the DD using the ADO cabling. These incoming services are then distributed throughout the home from the DD using outlet cabling.

The DD should:

- Be an enclosed cabinet with a removable or fully accessible cover.

- Be recessed within the wall or surface-mounted on the wall.

- Have adequate cable entry holes and knockouts in the top, bottom, and sides to accommodate the quantity of cables required for full capacity.

- Be equipped with plastic grommets and plugs to protect cabling and cover unused holes.

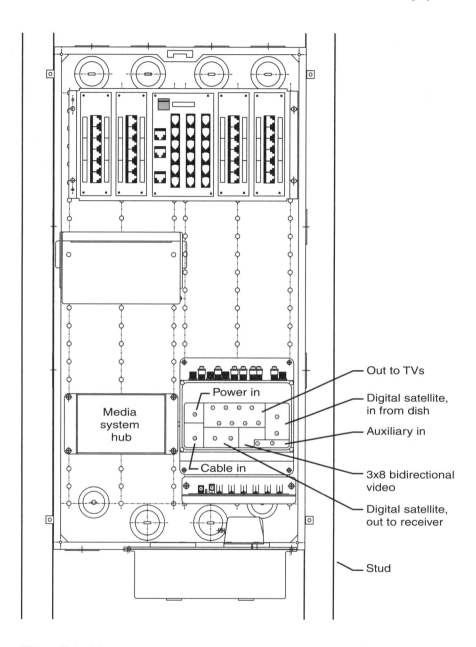

TV = Television

Figure 2.10   Distribution device.

A dedicated, nonswitched 15 ampere (120 volt alternate current [Vac]) duplex power outlet is recommended for a Grade 1 installation and is required for a Grade 2 installation. The outlet shall be within 1.5 m (5 ft) of the DD. Consider placing the outlet within the DD (lower right corner) fed from the bottom of the recessed panel, or from the rear of a surface-mounted panel.

The electrical outlet should be installed within a dedicated, approved, electrical enclosure, providing a barrier between low-voltage and electrical supply voltage cabling. The circuit shall be terminated in a duplex electrical outlet.

In larger homes, an additional DD may be needed to fully accommodate the quantity of cables and components being installed. If an additional DD is installed, it is recommended that one or multiple pathways be installed between the DDs to maximize flexibility. When multiple DDs are used, consider the benefits of aggregating each service's cabling and components to only one of the DDs, as opposed to spreading out services to each DD.

Consider including a room dedicated to acting as a DD in larger homes. In this case, components and modules may be installed onto an adequately sized backboard or dedicated equipment rack that acts as the DD.

Location requirements for the distribution device (DD). A DD must be installed inside each tenant space in a location that is accessible for cabling maintenance. The location should be centralized within the tenant space to minimize the length of outlet cables.

When deciding the placement of the DD, take into account local environmental factors that may affect the performance of the components to be placed within the device. There must be adequate climate control for the DD to allow for a proper environment for nonstructured cabling components to be used (e.g., video amplifiers or data switches and routers).

The DD should be installed where the normal ambient temperature range is 0 ºC (32 ºF) to 44 ºC (110 ºF), and where humidity levels will not readily condense moisture on the enclosure or its contents. The DD should not be located in a garage, attic, or exterior storage area. The DD must not be within the same enclosure as an exterior SP demarcation.

Wall space allocation for a distribution device (DD) and associated equipment. Space allocation for the DD is determined by the grade of service and number of telecommunications outlet/connectors to be installed in the home. Table 2.3 provides guidance for planning the wall space that should be allocated for the DD and associated equipment.

To provide for future growth, equip the DD to add 25 to 50 percent more cable runs than the original installed number of runs.

More than one DD may be used if the number of outlets requires it. In such a case, for instance, one DD may be used for voice/data and one for video.

TABLE 2.3   Space allocation guidelines for the distribution device and associated equipment

Number of Outlet/Connectors	Grade 1	Grade 2
1 to 8	410 mm (16 in) wide 610 mm (24 in) high	815 mm (32 in) wide 915 mm (36 in) high
9 to 16	410 mm (16 in) wide 910 mm (36 in) high	815 mm (32 in) wide 915 mm (36 in) high
17 to 24	410 mm (16 in) wide 1220 mm (48 in) high	815 mm (32 in) wide 1220 mm (48 in) high
More than 24	410 mm (16 in) wide 1525 mm (60 in) high	815 mm (32 in) wide 1525 mm (60 in) high

> **NOTE:** Most commercially available DDs in a residential system are 368 mm (14.5 in) wide (because the space between the studs is 368 mm [14.5 in]) to allow flush mounting in a standard 410 mm (16 in) on-center wall stud bay. If more space is required, multiple, interconnected panels will serve the user better. This is important when considering the aesthetics of the installed system.

Electrical power.   A dedicated, nonswitched 15 ampere (120 Vac) duplex electrical outlet must be provided adjacent to or within 1.5 m (5 ft) of the DD. The height of the electrical outlet should be suitable for the DD and associated equipment being installed, and shall be in compliance with applicable codes. Consider installing this electrical outlet within the DD to eliminate cable clutter.

Modules inside the distribution device (DD).   Modules should be available in varying capacities to accommodate different-sized homes. The modules should be easily removable from the DD to provide for flexible configurations (combinations of modules) and ease of replacement for repair or upgrade.

Cabling slack loops for DD modules may optionally be stored in the nearest, most accessible space for future MACs of modules within the DD.

All cables in the DD should be terminated.

DD modules should have connections for the appropriate number of feeds and distribution outputs (outlets) as required by the RSCS being installed.

The following are basic considerations for locating and installing the DD:

- Adequate space must be adjacent to the DD for the surge suppression device for each conductive cable entering and leaving the building.

- Access to the building's electrical ground must be provided within 1.5 m (5 ft) of DD.

- The DD should accept listed cabling components.

- The DD should be housed in a cabinet with a removable or fully accessible cover, either surface-mounted on the wall or flush-mounted in the wall.

- The DD should have adequate cable entry holes and knockouts to accommodate the quantity of cables required for full capacity.

- The DD should be equipped with plastic grommets and plugs or standard-sized conduit connections to protect cabling and cover unused holes.

- The DD should have access to a dedicated, nonswitched 15 ampere (120 Vac) duplex electrical outlet, either adjacent to or within 1.5 m (5 ft) of the DD. The electrical outlet may be placed within the DD.

- The DD should be in a centralized location in the home to minimize length of cable runs.

- The DD should be easily accessible to the homeowner and the residential cabling installer.

- NID should be located in a space not subject to wide fluctuations in temperature and humidity.

The following are considerations for installing the DD in larger homes or using modular systems:

- Install two or more DDs in a large installation, segregating different types of media to each DD.

- Make sure multiple DDs can be readily connected via cabling to ensure future flexibility.

- Use a dedicated ER for a larger installation, with a backboard or a commercial grade 480 mm (19 in) equipment rack.

- Provide space for 25 to 50 percent growth in cable runs in the future.

- Use removable, clearly labeled modules in the DD.

- Provide space nearby for slack cable loops to facilitate future MACs.

Equipment rack.   The placement of equipment racks and cabinetry should be decided during the planning stage of the cabling system. The rack should be positioned so that access is available from both the front and back. The racks are generally used for home equipment (e.g., audio video receiver, satellite receiver, digital versatile disc [DVD], video cassette recorder, switches, and other home automation equipment). See Figures 2.11 and 2.12 for examples of wall-mounted and floor-mounted equipment racks.

Relay rack and cabinetry

The placement of relay racks and cabinetry should be decided on during the planning phase of RSCS projects. Location of this equipment is dependent on various factors. Entrance facility equipment (e.g., protection and grounding)

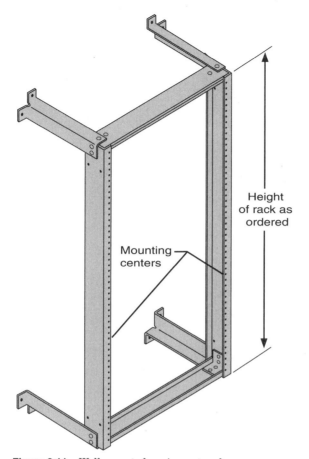

Figure 2.11   Wall-mounted equipment rack.

requires a reserved space. It may be necessary to plan for additional peripherals that require alternating current (ac) power. Many homes will designate one space to serve these facilities. By locating these facilities in one location, a neat, contained area will simplify the management and service of the installed systems.

Manufacturers have developed new self-contained products that will meet the requirements of the RSCS. These products house small patch panels that are specific to the installed media. VDV, security, and audio system connections can be made (see Figures 2.11 and 2.12).

Some of the devices that are mounted in these cabinets (e.g., minihubs and switches) may require ac power connections. The planning for ac power connections should be included, with close proximity to the location of these cabinets. Satellite and video electronics require space, which should be planned for when sizing the equipment cabinets. Shared video sources may require

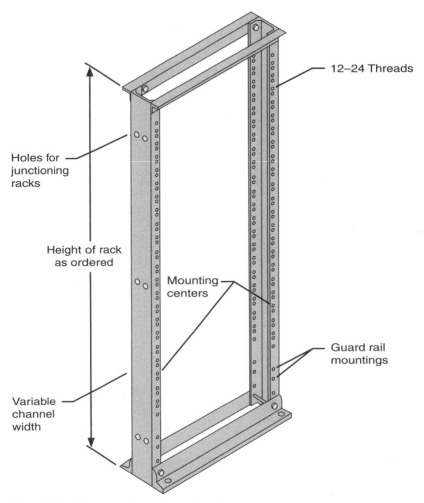

Figure 2.12   Floor-mounted equipment rack.

special consideration of space and power. Connections to the SPs should be considered in the planning phase. Most SPs specify a DP on the exterior of residential structures. Connections must be made in those demarcations and then fed to the area where the relay racks or cabinets are installed.

Outlet cable

Running from the central DD to individual rooms is the outlet cable, which provides the transmission path from the DD to the telecommunications outlet/connector. An outlet cable may be connected through a transition point (TP) or consolidation point (CP). The length of each outlet cable is not to exceed 90 m (295 ft).

> **IMPORTANT:** It is imperative that all outlet cables be clearly labeled at the DD.

Recognized cables. Recognized outlet cable includes:

- 4-pair twisted-pair Category 3, Category 5, or Category 5e cables. Category 5e is recommended.
- 50/125 or 62.5/125 μm multimode fiber.
- Singlemode fiber.
- Series 6 coaxial cable.

While Category 3 twisted-pair cable is recognized, it is highly recommended that a minimum of Category 5e twisted-pair cable be used for voice and data, and that Series 6 quad-shielded coaxial cable be used for video.

Multiple cables installed in a common sheath (e.g., hybrid, composite, or bundled cables) may be used provided they meet or exceed the hybrid and bundled cable requirements of ANSI/TIA/EIA-568-B.2, *Commercial Building Telecommunications Cabling Standard, Part 2: Balanced Twisted-Pair Cabling Components*.

Optical fiber cable may serve as a medium to accommodate new technologies. With new high-technology services being introduced in the marketplace (e.g., voice over Internet protocol and local area networking), it is best to coordinate distribution design with the technology SP.

Cabling topology for telecommunications outlet/connectors. Install outlet cable in a star topology, so that:

- One or more outlet cable runs extend from the DD to each telecommunications outlet/connector.
- A tenant, who may be located on more than one floor of an MDU, is cabled from a single DD serving that tenant.

> **NOTES:** It is not necessary that all outlets in a home be configured identically. Outlets in some rooms may contain more coaxial cable, and outlets in other rooms more twisted-pair.
>
> Consider placing two coaxial cable runs per outlet.
>
> Consider placing four 4-pair outlet cables in an area designed for home office use.
>
> Consider placing at least three coaxial cable runs and one twisted-pair cable run to outlets in a media/home theater room.
>
> Consider placing one multimode fiber cable containing either two 50/125 μm or two 62.5/125 μm multimode fiber strands per outlet.

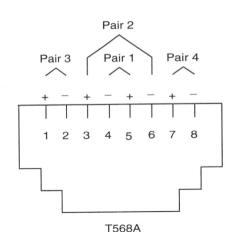

Figure 2.13   Multioccupant cabling layout.

Multioccupant cabling layout.   Typical diagrams of cabling up to the NID or ADO are shown. These diagrams are typically the designs used in distribution systems up to the ADO or NID that is colocated with the DD.

The following figure illustrates an example of multioccupant cabling layout. The NID may be located in the premises unit with the AP having responsibility for the backbone facility, depending on local jurisdiction.

> **NOTE:** The NID may be located at the protected terminal, the floor terminal, the ADO, or in another building or structure.

Telecommunications outlet/connectors

Types of outlet interfaces.   A telecommunications outlet/connector provides the means for the tenant to connect premises equipment. Telecommunications outlet/connectors should be designed for one or more of the following.

Suitable connecting hardware must be used to ensure proper signal transmission. The residential cabling installer must ensure that the connectors being installed are designed specifically for the media being connected. Various types of outlet interfaces are currently used in the industry, as shown in the following table.

TABLE 2.4 Outlet interfaces designed for media to be connected

Media	Outlet Interface
Twisted-pair	8-pin modular connector S-video (super video home system [SVHS]) Subminiature-D RCA
Screened twisted-pair	8-pin modular shielded connector
Coaxial	BNC series F series N series
Optical fiber	Various connectors meeting requirements of ANSI/TIA/EIA-568-B.3, *Optical Fiber Cabling Components* standard
Audio	RCA Binding posts Banana plugs Spring clips
Video	S-video (SVHS)

TABLE 2.5 Telecommunications outlets/connectors for residences

Grade 1	Grade 2
One T568A connector	Two T568A connectors
One F Connector (75 Ω coaxial cable)	Two F connectors (75 Ω coaxial cable)
	Two-fiber connector meeting ANSI/TIA/EIA-570-A (small form factor [SFF] connectors are allowable throughout homes)

Telecommunications outlet/connectors:

- May be either flush-mounted or surface-mounted.

- Must be weather-resistant when installed on building exteriors.

- Shall match the category of the cable being installed.

A sufficient number of telecommunications outlet/connector locations should be planned to prevent the need for extension cords. A telecommunications outlet/connector location should be provided in each living area, and additional telecommunications outlet/connector locations provided within unbroken wall spaces of 3.7 m (12 ft) or more. Additional telecommunications outlet/connector locations should also be provided so that no point along the floor line in any wall space is more than 7.6 m (25 ft), measured horizontally, from a telecommunications outlet/connector location in that space. Telecommunications out-

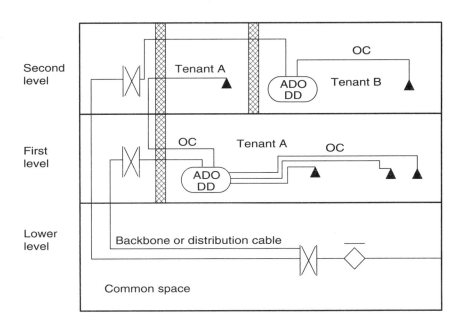

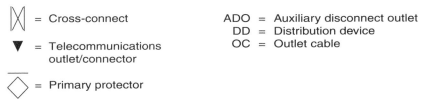

Figure 2.14   T568A pair and pin assignments.

let/connector mounting heights shall be in accordance with applicable codes or standards.

Some networks or services require application-specific electrical components (e.g., splitters, amplifiers, or impedance-matching devices) at the telecommunications outlet/connector. These application-specific electrical components are to be placed external to the telecommunications outlet/connector.

Outlets utilizing electrical boxes are recommended, at a minimum, to be dual-gang boxes with single-gang plaster rings (for the storage of cable slack or prewired cables). Optionally, low-voltage brackets may be installed to aid in maintaining proper cable bend radius. Consult with applicable local codes to verify compliance.

The following are considerations for locating and installing outlet cable and telecommunications outlet/connectors:

- Outlet cable may be 4-pair twisted-pair, singlemode or multimode fiber, or Series 6 coaxial cable.

- Outlet cable may be connected through a TP or CP.
- The length of each outlet cable is not to exceed 90 m (295 ft).
- Outlet cable runs should be star-wired from the DD to outlets.
- The quantity of outlet cables and telecommunications outlet/connectors may vary room to room, depending on the applications to be supported.
- Telecommunications outlet/connectors may be flush-mounted or surface-mounted.
- Telecommunications outlet/connectors must be weather-resistant when installed on building exteriors.
- Telecommunications outlet/connectors shall match the category of the twisted-pair cable being installed.
- It is recommended that the highest category of twisted-pair cable or the cabling medium with the highest bandwidth be installed to accommodate future growth.

> **NOTE:** Alternatively, design and install pathways (e.g., flexible conduit) with pull string to assist in additional cable pulls with easy access for future MACs.

Patch cords and equipment cords

Patch cords or equipment cords run from the telecommunications outlet/connector to specific CPE in each room. This equipment can include devices such as telephones, video equipment, and computers.

Twisted-pair equipment cords and patch cords, other than device-specific cords (i.e., telephone cords), shall be constructed of 4-pair, 100 Ω, 24 AWG [0.51 mm (0.020 in)] stranded copper cable that meets the category performance requirements of the installed system. These cords should be terminated with eight-position modular plugs (8P8C [RJ-45]).

Coaxial equipment cords and patch cords, other than device-specific cords, should be Series 6 coaxial cables, and should be connectorized with male F connectors that comply with SCTE IPS-SP-401, *"F" Port (Male Feed Thru) Physical Dimensions*.

When planning for residential-space cabling, keep the following items in mind:

- Patch cords are designed to provide easy routing changes.
- The maximum outlet cable length has been specified with the assumption that a maximum length of 3 m (10 ft) of patch cord is used in the residential space.
- Twisted-pair and coaxial patch cord with identical connectors on both ends are most commonly used. Patch cords should be factory manufactured.

- Twisted-pair patch cordage should be constructed with stranded cable. This allows for flexibility but also causes 20 percent more insertion loss in the circuit.

- When application-specific adapters (e.g., baluns or modular adapters) are needed in the residential space, ANSI/TIA/EIA-570-A requires that they be external to the telecommunications outlet/connector.

Cabling adapters in the residential space may have detrimental effects on the transmission performance of the cabling system.

Backbone cable

Recognized cables.   Recognized backbone cables for the home include:

- 100 Ω twisted-pair cable
- 50/125 μm multimode fiber
- 62.5/125 μm multimode fiber
- Singlemode fiber
- Trunk distribution and feeder coaxial cable
- Series 6 and Series 11 coaxial cable

Topology

A star topology shall be implemented for twisted-pair cabling. Coaxial backbone cable may be implemented using a star or bus topology. Optical fiber cabling shall be implemented in a star or ring topology.

Interbuilding cabling protection

When buildings are connected with interbuilding metallic cabling, the applicable fusing and overvoltage protection codes shall be followed. Fusing and overvoltage codes should be verified with the AHJ. Typically, these requirements can be found in the *National Electrical Code®* and the *National Electrical Safety Code®*.

Recommended Minimum Cabling Configuration

This section specifies the recommended minimum cabling quantities and other items related to the implementation of an RSCS. Additional cabling may be installed in accordance with ANSI/TIA/EIA-570-A. When cabling media (e.g., optical fiber or coaxial cable) are installed for these applications, coordinate distribution design with the technology SP.

Distribution Devices (DDs)

One DD with the following configuration is recommended:

- The DD should be of sufficient capacity and functionality to handle the minimum system requirements. The panel should include the following as a minimum:
  - Series 6 quad-shielded coaxial cable video distribution
  - Category 5e data network distribution
  - Category 5e telephone distribution
  - Appropriately sized conduit from the DD to the nearest accessible space allowing for MACs
- Cabling needed to meet the minimum is not required to be in conduit unless specified by local codes. However, the builder should install conduit for additional cable runs in the future.

Auxiliary Disconnect Outlet (ADO) cables

The following ADO cables are recommended:

- Voice/data—Two Category 5e cables
- CATV—Two Series 6 quad-shielded coaxial cable with F connectors
- Network connections—One optical fiber cable and conduit for optical fiber cable
- Video (DBS)—Two Series 6 quad-shielded coaxial cable modules, prewire only. Up to six quad-shielded coaxial cables may be required for some systems
- Off-air antenna—One Series 6 quad-shielded coaxial cable module

The MDU resident's needs will ultimately determine the minimum ADO cables required.

Outlets

Outlet locations are described within ANSI/TIA/EIA-570-A. The number and type of media that are brought to an outlet are not detailed. This is due to the fact that in a Grade 1 installation, it is possible that only a telephone and a television outlet are necessary in the home, thereby requiring only a wall telephone equipped with one twisted-pair cable run, and one television outlet equipped with one Series 6 cable run. For a Grade 2 installation, it is possible that an outlet will require two Category 5e cables, one two-fiber cable, and two Series 6 coaxial cables to feed the computer, network printer, telephone, and television. The judgment of the number of media to each outlet is at the discretion of the decision maker, whether it is the homeowner, builder, or telecommunications distribution designer.

For Grade 1 cabling, consideration should be given to wall telephones and other telecommunications outlet/connectors in the home. Wall telephones typically will have one outlet/connector (see Figure 2.15), thereby only one twisted-pair cable runs back to the DD. The other outlets should have multiple outlet connectors for a variety of services. For instance, an outlet by the television

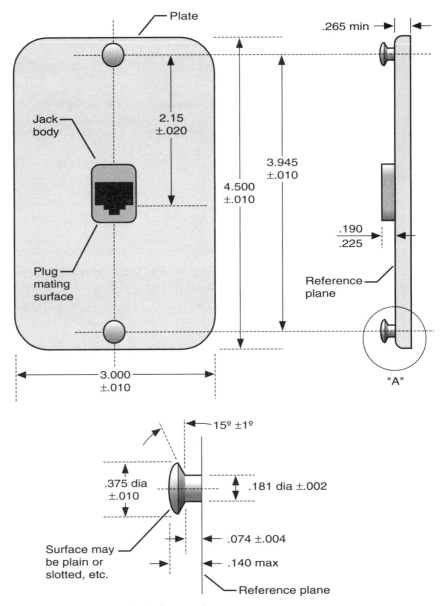

Figure 2.15   Typical wall telephone outlet.

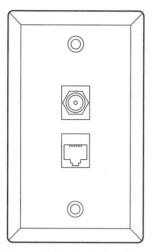

Figure 2.16   Grade 1 outlet.

should include a minimum of one twisted-pair connector and one Series 6 connector allowing the ability to use the coaxial cable for the television and the twisted-pair for dial-up to order a movie.

Homeowners like to rearrange their furniture, so installing outlets that are strategically placed to allow this aids in the value to the homeowner. See Figures 2.16 and 2.17 for examples of Grade 1 and Grade 2 outlet configurations.

Like Grade 1 cabling, consideration should be given to wall telephones and other telecommunications outlets in the home for Grade 2 cabling. Wall tele-

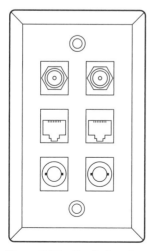

Figure 2.17   Grade 2 outlet.

phones cable runs are the same as for Grade 1 unless there is need for data at the phone, in which case another twisted-pair or optical fiber cable should be run. The other outlets in the home should have multiple outlet connectors for a variety of services and be located so that when moving furniture in the room, services can be extended without difficulty.

3

Codes, Standards, and Regulations

Introduction

Because residential telecommunications cabling installers work with other disciplines such as home automation and the building industry, they must follow a variety of rules in their everyday business. These rules are contained in codes, standards, and regulations.

While codes address minimum safety requirements that must be adhered to, standards are intended to ensure system performance by providing specifications and guidelines for proper installation. Regulations are considered mandatory rules issued by government agencies, but are becoming less numerous. More specifically:

- Codes are intended to ensure safety during the installation, use, and/or disposal of materials, components, fixtures, systems, premises, and related subjects. Codes are typically invoked and enforced through government regulation.

- Standards are a collection of requirements that encompass the properties of components and systems. Standards establish an accepted degree of functionality, interoperability, and longevity.

- Regulations are usually issued by government agencies and are considered mandatory rules. A typical government agency that issues regulations is the Federal Communications Commission (FCC). Regulations are becoming less numerous because governmental agencies are moving away from the regulation of telecommunications. The Telecommunications Act of 1996 removed most of the competitive regulations involving telecommunications; however, many state and some FCC rules still apply to telecommunications access providers (APs) and interexchange carriers (IXCs).

Construction in virtually all countries is regulated by building codes, standards, and regulations. In North America, codes and regulations are normally enforced by an authority having jurisdiction (AHJ) and encompass most, if not all, aspects of the construction industry. Installation methods, materials, and electrical products must conform to the requirements of the AHJ. Regulations are issued by governmental agencies, usually on state or federal government levels. Codes and standards contain two important words, "shall" and "should," which can have a major impact on how tasks are accomplished. These are defined as:

- Shall—A mandatory requirement.
- Should—An advisory recommendation.

Purpose of Codes, Standards, and Regulations

Building codes, standards, and regulations govern the minimum requirements for the installation and handling practices and type of materials used when constructing residential telecommunications facilities. Codes normally protect life, health, property, and the environment. Standards and regulations define construction quality.

Codes are not intended to cover all measures that may be required to protect telecommunications circuits and equipment from intrusion, induced noise, or events that disrupt the flow of information.

In general, standards are established as a basis to quantify, compare, measure, or judge:

- Capacity
- Quantity
- Value
- Quality
- Performance
- Limits
- Interoperability

Several independent organizations specialize in establishing and maintaining telecommunications standards for the residential cabling industry. A list of these organizations is provided at the end of this chapter.

Codes Affecting Residential Cabling

Codes address the safety of persons, property, and the environment associated with the residential cabling installation. They include electrical codes, building codes, fire codes, environmental codes, and all other safety codes. When adopted by AHJs, codes have the force of law.

The *National Electrical Code® (NEC®)* and the *Canadian Electrical Code® (CEC®)* are the most widely adopted sets of electrical safety requirements within North America. In addition, state, provincial, municipal, and local codes may add more restrictive provisions than national codes and, therefore, take precedence. There is a movement within the government of Mexico to adopt the *NEC* as its national code.

The National Fire Protection Association® (NFPA®) develops and produces codes relating to telecommunications that include the *NEC*. The Canadian Standards Association (CSA) publishes the *CEC*.

National Fire Protection Association® (NFPA®)

NFPA is an international organization with members from 70 countries. Established 100 years ago, the NFPA develops and produces fire safety codes used throughout the world. The following fire codes relate to telecommunications:

- American National Standards Institute (ANSI)/NFPA-70, *National Electrical Code® (NEC®)*

- ANSI/NFPA-71, *Installation, Maintenance and Use of Signaling Systems for Central Station*

- ANSI/NFPA-72, *National Fire Alarm Code*

- ANSI/NFPA-75, *Protection of Electronic Computer/Data Processing Equipment*

National Electrical Code® (NEC®), 2002 Edition

The NFPA sponsors, controls, and publishes the *NEC* within the United States' jurisdictional area. The *NEC* is intended to protect people and property from electrical hazards. The *NEC* specifies the minimum provisions necessary to safeguard persons and property from electrical hazards.

Most federal, state, and local municipalities have adopted the *NEC*, completely or in part, as their legal electrical code. Some states or localities adopt the *NEC* and add requirements that are more stringent.

The *NEC* is used by:

- Lawyers and insurance companies to determine liability.

- Fire marshals and electrical inspectors in loss prevention and safety enforcement.

- Telecommunications distribution designers and cabling installers, including residential cabling installers, to ensure a compliant installation.

The *NEC* is revised every three years, with the current edition published in 2002. Also known by its official alphanumeric designation of ANSI/NFPA-70, the *NEC* is arranged by chapter, article, and section (e.g., a citation may read Chapter 8, Article 800, Section 800.52).

Portions within the code having major importance to the residential cabling installer are:

TABLE 3.1 *NEC*, 2002, codes affecting telecommunications

NEC Reference	Title	Description
Section 90.2	Scope	The Scope provides information about what is covered in the *NEC*. The section references the *National Electrical Safety Code® (NESC®)* for industrial or multibuilding complexes.
Section 90.3	Code Arrangement	This section notes that Chapter 8 covers communications systems and is independent of the other chapters except where they are specifically referenced therein.
Article 100-Part I	Definitions	Definitions are those not commonly defined in English dictionaries. Some terms of interest include accessible, bonding, explosion-proof apparatus, ground, premises wiring, and signaling circuit.
Section 110.26	Spaces About Electrical Equipment (600 Volts, Nominal, or Less)	This section explains the space for working clearances around electrical equipment. This information is useful when placing a terminal in an electrical closet or electronic components on a communications rack.
Article 250	Grounding	This article is referenced from Article 800. It contains specific requirements for the communications grounding and bonding network.
Chapter 3	Wiring Methods and Materials	Chapter 3 is referenced in Article 800.48, "Raceways for Communications Wires and Cables" and covers wiring and cabling installation in raceways.
Article 640	Audio Signal Processing, Amplification, and Reproduction Equipment	This article covers equipment and wiring for audio signal generation, recording, processing, amplification, and reproduction; distribution of sounds; public address; speech input systems; temporary audio system installations; and electronic musical instruments.
Article 645	Information Technology Equipment	Article 645 contains information on equipment, power-supply wiring, equipment interconnecting wiring, and grounding of information technology equipment and systems, including terminal units.
Article 725	Class 1, Class 2, and Class 3 Remote-Control, Signaling, and Power-Limited Circuits	Article 725 specifies circuits other than those used specifically for electrical light and power. This Article contains requirements for security, control, and audio systems.
Article 760	Fire Alarm Systems	Article 760 contains requirements of wiring and equipment of fire alarm systems.

continued on next page

TABLE 3.1   *NEC*, 2002, codes affecting telecommunications (continued)

NEC Reference	Title	Description
Article 770	Optical Fiber Cables and Raceways	Article 770 contains the requirements for listing of optical fiber cable, marking, and installation.
Article 800	Communications Circuits	Article 800 contains the requirements for listing of copper communications cable, marking, and installation.
Article 810	Radio and Television Equipment	Article 810 contains requirements for radio and television receiving equipment.
Article 820	Community Antenna Television and Radio Distribution Systems	Article 820 contains requirements for community antenna television, including satellite, and radio distribution systems.
Article 830	Network-Powered Broadband Communications Systems	This section provides information on network-powered broadband communications systems that provide combinations of voice, audio, video, data, and interactive services through a network interface unit.

Product ratings

There are three categories used within the United States to rate products. These are:

- Listed
- Classified
- Recognized

A product is Listed after it successfully completes a series of mechanical, electrical, and thermal characteristics tests that simulate all reasonable, foreseeable hazards. The Listed classification is exclusive to the product for the specific applications for which it was tested and is not valid for other applications.

A product is Classified after it is evaluated and passes tests for one or more of the following:

- Specific hazards only
- Performance under specified conditions
- Regulatory codes
- Other standards, including international standards

The Classified rating is generally restricted to industrial or commercial products.

A product is Recognized after it is tested for use as a component in a Listed package and passes. These component products are tested for electrical, mechanical, and thermal characteristics.

The Recognized classification is a more general-purpose approval than Listed because it allows a product to be certified for a category of equipment uses. An example is insulated wire, which is Recognized as appliance wiring material, a category of uses that includes:

- Data communications
- Telecommunications
- Instrumentation

Canadian Standards Association (CSA)

The CSA produces several documents and standards that affect telecommunications in Canada. The *CEC* is updated every four years. Among the most important are:

- CSA-C22.1-1998, *Canadian Electrical Code® (CEC®)*, Part 1, 1998 (to be revised in 2002).

- CSA-T525, *Residential Wiring for Telecommunications*, 1994. Reaffirmed in 1999.

- CSA-T527, *Grounding and Bonding for Telecommunications in Commercial Buildings*, 1994 (harmonized with ANSI/TIA/EIA-607). Reaffirmed in 1999.

- CSA-T528, *Design Guidelines for Administration of Telecommunications Infrastructure in Commercial Buildings*, 1993 (harmonized with ANSI/TIA/EIA-606). Reaffirmed in 1997.

- CSA-T529, *Design Guidelines for Telecommunications Wiring Systems in Commercial Buildings*, 1995 (harmonized with ANSI/TIA/EIA-568-A).

- CSA-T530, *Commercial Building Standard for Telecommunications Pathways and Spaces*, 1999 (harmonized with ANSI/TIA/EIA-569-A).

***Canadian Electrical Code®*, Part 1 (*CEC®*, Part 1).**   The CSA sponsors, controls, and publishes the *CEC*, Part 1. The intent of this code is to establish safety standards for the installation and maintenance of electrical equipment, including telecommunications. As with the *NEC*, the *CEC*, Part 1, is a voluntary code that may be adopted and enforced by provincial and territorial regulatory authorities.

Residential cabling installers shall be familiar with all sections of the *CEC*, Part 1. The following chart lists sections of the *CEC*, Part 1, that are of primary interest to the residential cabling installer. The *CEC*, Part 1, is the equivalent of the *NEC* and *NESC* in the United States.

TABLE 3.2 *CEC* sections

CEC Reference	Title	Description
2	General Rules	Provides information on: • Permit • Marking of cables • Flame spread requirements for electrical wiring and cables
10	Grounding and Bonding	Contains detailed grounding and bonding information and requirements for using and identifying grounding and bonding conductors.
12	Wiring Methods	Involves the requirements for installing wiring systems. It outlines: • Raceway systems • Boxes • Other system elements
54	Community Antenna Distribution and Radio and Television Installation	Applies to circuits employed to distribute video and other information frequency signals.
56	Optical Fiber Cables	Contains the requirements for installing optical fiber cables.
60	Electrical Communication Systems	Contains the requirements for installing copper communications circuits.
82	Closed-Loop Power Distribution	Contains the requirements for controlling the signal between the energy controlling equipment and the utilization equipment.

Local building codes

Building codes establish strict requirements and regulations designed to safeguard life and property from fire and other hazards. In the United States, residential cabling installers and contractors must adhere to federal, state, and local codes for new construction and renovation projects. These codes are based on accepted safety and fire protection principles.

Three private organizations in the United States have developed "model" codes:

- The Building Officials and Code Administrators International, Inc. (BOCA), which publishes the *National Building Code*

- The International Conference of Building Officials (ICBO), which publishes the *Uniform Building Code*

- The Southern Building Code Congress International, Inc. (SBCCI), which publishes the *Standard Building Code*

Most government agencies adopt part or all of one of these model codes.

Identification of the municipal, country, state, and national codes that are applicable within a specific geographic locality is beyond the scope of this manual. The best way to identify codes applicable to any given construction project is to consult the local authorities. Such authorities can usually be found in city halls or municipal buildings, and are almost always found in local governmental listings of the local telephone book. Appropriate job titles to look for include the:

- Local fire marshal or inspector
- Electrical inspector
- Building inspector

Another source of information, if subcontracting on a project, is the general contractor, who may have already pulled all of the necessary permits for the job. If there is no evident local source of information on codes and permitting, consider consulting the International Association of Electrical Inspectors (IAEI) for the address of the electrical inspector.

Rules of the Federal Communications Commission (FCC)

The FCC is the Washington, D.C.-based federal agency responsible for monitoring and regulating telecommunications in the United States. The FCC is a large and complex administrative and regulatory body responsible for the U.S. activities of such telecommunications carriers as AT&T and MCI Worldcom.

However, the FCC is also the regulatory body overseeing the manufacture, performance, installation, and operation of the telecommunications equipment that is installed in individual residences. This equipment, given the label customer premises equipment (CPE), includes:

- Telephone headsets
- Modems
- Wireless devices

At the urging of telecommunications trade associations (e.g., TIA) and private professional organizations (e.g., BICSI), the FCC adopted in early 2000 an order (CC Docket No. 88-57) that set a new minimum quality standard for telephone inside wiring using Category 3 cable. The enforcement of installing this quality of cabling was due to inferior cabling being installed that was causing crosstalk between telephone lines and affecting the service provider. Defining telephone inside wiring as wiring located on the customer premises side of the telephone network, the order is aimed at promoting consumer access to existing and advanced telecommunications services.

The category system of identifying cabling performance is explained in Chapter 4: Cabling Media. Manufacturers verifying that their cable is Category 3 or higher and Part 68-compliant is adequate for installation in single-family residences and multi-dwelling units (MDUs).

Standards Affecting Residential Cabling

The purpose of a standard is to ensure a minimum level of performance. As defined in the TIA *Engineering Manual*, a standard is "a document that establishes engineering and technical requirements for processes, procedures, practices, and methods that have been decreed by authority or adopted by consensus."

Standards may also be established for selection, application, and design criteria for material. Standards are established as a basis to quantify, compare, measure, or judge capacity, quantity, value, quality, performance, limits, and interoperability. A significant benefit of standards in the telecommunications industry is to aid the improved interoperability of components and systems of multiple manufacturers.

Test standards provide uniform rules for what is to be tested, how it is to be tested, and which results are acceptable.

American National Standards Institute/Telecommunications Industry Association/Electronic Industries Alliance (ANSI/TIA/EIA)

TIA and EIA are organizations that develop and submit standards to ANSI for approval. The standards are then published and made available to the industry. TIA/EIA publishes standards for the manufacturing, installation, and performance of electronic and telecommunications equipment and systems. Five of these TIA/EIA standards, adopted by ANSI, govern telecommunications cabling in buildings. Each standard covers a specific part of intrabuilding cabling. The standards address the required cable, hardware, equipment, design, testing, and installation practices. In addition, each ANSI/TIA/EIA standard lists related standards and other reference materials that address the same topics.

Most of the standards also include sections that define important terms, acronyms, and symbols. The ANSI/TIA/EIA standards that govern telecommunications cabling in buildings are:

- ANSI/TIA/EIA-568-B, *Commercial Building Telecommunications Cabling Standard*, 2001.

- ANSI/TIA/EIA-569-A, *Commercial Building Standard for Telecommunications Pathways and Spaces*, 1998.

- ANSI/TIA/EIA-570-A, *Residential Telecommunications Cabling Standard*, 1999.

- ANSI/TIA/EIA-606, *Administration Standard for the Telecommunications Infrastructure of Commercial Buildings*, 1993.

- ANSI/TIA/EIA-607, *Commercial Building Grounding and Bonding Requirements for Telecommunications*, 1994.

> **NOTE:** Only ANSI/TIA/EIA-570-A focuses on residential cabling infrastructure.

The TIA/EIA Residential Cabling Standard.   The primary cabling standard covering residential cabling practices, including small offices and home offices (SOHO), is ANSI/TIA/EIA-570-A, *Residential Telecommunications Cabling Standard*.

The standard gives an overview of premises cabling systems and defines important concepts and components. It also provides example illustrations of a typical:

- Premises cabling system

- Residential cabling layout

- Multi-dwelling cabling layout

Section 4 of ANSI/TIA/EIA-570-A establishes a cable grading system based on the type of service to be provided.

Section 5 of ANSI/TIA/EIA-570-A describes single residential unit cabling systems. Included is information about installing:

- An auxiliary disconnect outlet (ADO).

- The distribution device (DD), including its input and output connections.

- Outlet cable, including recognized cables and the requirements for the maximum length of 90 m (295 ft). The operational length is 100 m (328 ft), including patch cords or jumper wire and equipment cords.

- Telecommunications outlet/connectors.

- Equipment cords, patch cords, and jumpers.

Section 6 of ANSI/TIA/EIA-570-A provides information on multi-dwelling and campus pathways and spaces from the demarcation point to the MDU-telecommunications room.

The specifications for premises cabling components are included in Section 7 of ANSI/TIA/EIA-570-A. The standard includes requirements for:

- Unshielded twisted-pair (UTP), optical fiber, and coaxial cable

- Equipment and patch cords

- Connecting hardware

- Telecommunications outlet/connector termination

The section on installation requirements (Section 8 of ANSI/TIA/EIA-570-A) provides warnings related to cabling operations. The section also specifies the minimum bend radius and maximum pulling tension for both UTP copper cable and optical fiber cabling. Other subsections cover cable placement and electromagnetic compatibility (EMC).

The standard also contains four annexes:

- Annex A—Optical Fiber Connector Performance Specifications. This annex is normative and is part of the standard.

- Annex B—Field Test Requirements. This annex is normative and is part of the standard.

- Annex C—Cabling Residential Buildings. This annex is informative and is not considered part of the standard.

- Annex D—Bibliography and References. This annex is informative and is not considered part of the standard.

Many of the terms used in the description above may be unfamiliar to first-time residential cabling installers, residential property owners, builders, architects and engineers, or those entering the residential cabling market. More complete descriptions of the components of a residential cabling system are covered in Chapter 2: Residential Structured Cabling Systems, while other terms and concepts are defined and explained throughout this manual.

Related standards

Other organizations provide standards related to residential cabling installation that may impact installation work.

Institute of Electrical and Electronics Engineers, Inc.® (IEEE®). As the world's largest professional engineering society, IEEE provides:

- Standards for rating the performance of electrical and electronic equipment and materials.

- Standards for installation and maintenance of such equipment.

- Courses to allow engineers to keep abreast of developments in the electrical and electronic engineering fields.

Residential cabling installers should be familiar with IEEE 802.3, Ethernet.

IEEE 1394, *Standard for a High Performance Serial Bus*—FireWire™. The IEEE 1394 specification, also known as FireWire™, is now an open-architecture standard supported by the IEEE and managed by its IEEE 1394 Working Group. The standard is designed to support high-bandwidth requirements of devices such as digital video equipment and high-capacity mass storage. For further information in this manual, refer to Chapter 10: Alternative Infrastructure Technologies.

Society of Cable Telecommunications Engineers Inc. (SCTE).   Recently changing the word "television" to "telecommunications" in its name, the SCTE is a nonprofit professional association dedicated to advancing the careers and serving the industry of telecommunications professionals by providing technical training, professional certification, and industry-wide standards.

Because SCTE is now and has in the past mostly been involved with video delivery protocols and equipment, its standards-making activities, adopted by ANSI, are aimed at the broadband industry—primarily providers of cable television services.

Standards relating to the delivery of information over the coaxial cabling systems taken as its province by the SCTE include:

- SCTE IPS-SP-001 for Series 6 and Series 11 coaxial cable in nonbackbone applications, 1996
- SCTE IPS-SP-100 for trunk, feeder, and distribution coaxial cable, 1997

Insulated Cable Engineers Association (ICEA).   ICEA is a wire and cable manufacturers' organization that writes telecommunications specifications for the communications and electrical cable industries. ICEA publications are adopted in the public interest and are designed to:

- Eliminate misunderstanding between manufacturer and user.
- Assist users in selecting and obtaining proper products for their particular needs.

ICEA specifications of interest are:

- ANSI/ICEA S-80-576, *Communications Wire and Cable for Premises Wiring*, 1994
- ANSI/ICEA S-83-596, *Fiber Optic Premises Distribution Cable*, 1994

National Research Council of Canada, Institute for Research in Construction (NRC-IRC).   NRC-IRC produces several documents and standards that affect telecommunications. Among the most important are:

- NRCC 38726, *National Building Code of Canada*, 1995
- NRCC 38727, *National Fire Code of Canada*, 1995
- NRC/AT&T 555-400-021, *A Guide to Premises Distribution*, April 1988

International Organization for Standardization/International Electrotechnical Commission (ISO/IEC).   In an international community, there exists the requirement for the interoperability of electronic networks connecting all parts of this community. One of the functions of the ISO/IEC is to define the international cabling standards that make this interoperability possible. Of primary importance today is the ISO/IEC Standard 11801, *Information Technology–Generic Cabling for Customer Premises*. This standard accepts both 100 and 120 Ω twisted-pair cabling and also allows for shielding of both of these types as an option.

Other residential cabling standards groups and associations. Because residential and SOHO cabling involve so many different applications, the standards picture is much more complicated than is the case with commercial cabling, which traditionally provides infrastructure for only voice, data, and video (VDV) services.

This manual covers many of the standards and standards bodies that might have an impact on residential and SOHO cabling. This section lists some of the applications involved and lists the major organizations governing these applications. A more complete description of some of these organizations can be found in the bibliography, which lists additional informational resources:

- Home automation—The Continental Automated Buildings Association (CABA), the Home Automation and Networking Association (HANA, formerly known as HAA, or the Home Automation Association), and the LonMark Interoperability Association

- Consumer electronics—Custom Electronic Design and Installation Association (CEDIA) and Video Electronics Standards Association (VESA)

- Wireless technology—The Wireless Data Forum (WDF) and the International Wireless Telecommunications Association (IWTA)

- Installer groups—The Electronics Technicians Association (ETA) and National Systems Contractors Association (NSCA)

- Related services—The American Society of Heating, Refrigerating, and Air Conditioning Engineers (ASHRAE)

Other Areas of Regulation in the United States

Occupational Safety and Health Act

Passed by Congress in 1970, the Occupational Safety and Health Act attempts to ensure a safe and healthful environment for every working person in the United States. Under this statute, the Occupational Safety and Health Administration (OSHA) was created within the U.S. Department of Labor. The provision and requirements of OSHA are set forth in the *Code of Federal Regulations* (CFR).

While OSHA is responsible for the administrative work relating to the statute, most field work has been passed down to each state's Department of Labor. As a result, each state is responsible for field inspections and enforcement.

OSHA is responsible for job site inspections and has the authority to shut down a job site and levy fines against individuals and companies for noncompliance with OSHA regulations. Additionally, OSHA is responsible for the development, publication, and enforcement of safety standards.

More information on commonly accepted safety practices can be found in Chapter 9: Safety.

Licensing

At this writing, only a few states require that telecommunications cabling contractors doing low-voltage installations be licensed. However, there are nationwide trends, both in the United States and Canada, toward this type of licensing.

The best procedure for determining if licensing is a requirement in a state or region is to contact the appropriate arm of the state government, which is often listed in the local telephone book. In the states and provinces where licensing is required at this time, the most likely governmental arm responsible for this function is the board of electrical inspectors. In any case, an inquiry directed to that board should clarify whether low-voltage licensing applies in a state or province, or if it is pending in the legislature.

Another source of information is the database compiled by BICSI's Governmental Relations Committee, and posted on the organization's Web site, www.bicsi.org. This database includes state-level licensing requirements for low-voltage cabling contractors in the United States, as well as province-level requirements for Canada.

Recognized Regulatory and Reference Bodies

Various organizations publish codes, standards, and methods for materials and testing. Much of this is adopted by manufacturers to ensure standardization. Some local enforcement agencies adopt or adhere to such standards as evidence of quality in installation. Some of the major recognized agencies and organizations are:

Alliance for Telecommunications Industry Solutions (ATIS)
1200 G St. NW, Ste. 500
Washington, DC 20005 USA
202-434-8837; fax: 202-393-5453
e-mail: atispr@atis.org
Web site: www.atis.org

American Institute of Architects (AIA)
1735 New York Ave. NW
Washington, DC 20006 USA

202-676-7300
Web site: www.aiaonline.com

American Insurance Association (AIA)

1130 Connecticut Ave. NW, Ste. 1000
Washington, DC 20036 USA
202-828-7100; fax: 202-293-1219
e-mail: webmaster@aiadc.org
Web site: www.aiadc.org

American Insurance Service Group (AISG) (National Building Code)

Customer Service Division
545 Washington Blvd.
Jersey City, NJ 07310 USA
800-888-4476; fax: 201-748-1472
e-mail: info@iso.com
Web site: www.iso.com/aisg/index.html

American National Standards Institute (ANSI)

11 W 42nd St., 13th Flr.
New York, NY 10036 USA
212-642-4900; fax: 212-398-0023
e-mail: info@ansi.org
Web site: www.ansi.org

American Society for Testing and Materials (ASTM)

100 Barr Harbor Dr.
W. Conshohocken, PA 19428-2959 USA
610-832-9585; fax: 610-832-9555
e-mail: service@astm.org
　　　　infoctr@local.astm.org
Web site: www.astm.org

Bellcore (See Telcordia™ Technologies)

Building Officials and Code Administrators International, Inc. (BOCA)

4051 W Flossmoor Rd.
Country Club Hills, IL 60478-5795 USA
708-799-2300; fax: 708-799-4981
e-mail: info@bocai.org
　　　　boca@aecnet.com
Web site: www.bocai.org

Cable Television Laboratories (CableLabs)

400 Centennial Pkwy.
Louisville, CO 80027
303-661-9100; fax: 303-661-9199
Web site: www.cablelabs.com

Ceilings and Interior Systems Construction Association (CISCA)
1500 Lincoln Hwy., Ste. 202
St. Charles, IL 60174 USA
630-584-1919; fax: 630-584-2003
e-mail: 75031.2577@compuserve.com
Web site: www.cisca.org

Committee T1 (See ATIS)

Construction Specifications Institute (CSI)
99 Canal Center Plaza, Ste. 300
Alexandria, VA 22314 USA
800-689-2900 or 703-684-0300
fax: 703-684-0465
Web site: www.csinet.org

Consumer Electronics Association (CEA)
(formerly known as Consumer Electronics Manufacturers
Association—CEMA)
2500 Wilson Blvd.
Arlington, VA 22201
703-907-7600; fax: 703-907-7675
Web site: www.ce.org

CSA International
Canadian Standards Association (CSA)
178 Rexdale Blvd.
Toronto, ON M9W 1R3, Canada
800-463-6727 or 416-474-4058
fax: 416-747-4149
e-mail: info@csa.ca
Web site: www.csa-international.org

Electronic Industries Alliance (EIA)
2500 Wilson Blvd.
Arlington, VA 22201-3834 USA
703-907-7500; fax: 703-907-7501
e-mail: publicaffairs@eia.org
Web site: www.eia.org

ETL SEMKO
70 Codman Hill Road
Boxborough, MA 01719 USA
800-967-5352; fax: 800-813-9442
e-mail: info@etlsemko
Web site: www.etlsemko.com

ETSI InfoCentre
European Telecommunications Standards Institute
650 Route des Lucioles
F-06921 Sophia Antipolis Cedex, France
+33-4-92-94-42-22; fax: +33-4-93-65-43-33
e-mail: infocentre@etsi.fr
Web site: www.etsi.org

European Committee for Electrotechnical Standardization (CENELEC)
Rue de Stassart, 35
B-1050 Brussels, Belgium
+32-2-519-68-71; fax: +32-2-519-69-19
e-mail: general@cenelec.org
          cenelec@cenelecbel.org
Web site: www.cenelec.org

Federal Communications Commission (FCC)
445 12th St. SW
Washington, DC 20554 USA
888-225-5322; fax: 202-418-0232
e-mail: fccinfo@fcc.gov
Web site: www.fcc.gov

Federal Information Processing Standards (FIPS)
Information Technology Laboratory Publications
100 Bureau Dr., Stop 8900
Gaithersburg, MD 20899-8900 USA
301-975-2832; fax: 301-840-1357
Web site: www.itl.nist.gov/fipspubs

General Services Administration (GSA)
1800 F St. NW
Washington, DC 20405 USA
202-501-0705; fax: 202-501-1300
e-mail: public.affairs@gsa.gov
Web site: www.gsa.gov

Institute of Electrical and Electronics Engineers, Inc.® (IEEE®)
445 Hoes Ln.
P.O. Box 1331
Piscataway, NJ 08855-1331 USA
732-981-0060; fax: 732-981-9667
e-mail: customer.service@ieee.org
Web site: www.ieee.org

Insulated Cable Engineers Association (ICEA)
P.O. Box 440
S Yarmouth, MA 02664 USA
508-394-4424; fax: 508-394-1194
e-mail: icea@capecod.net
Web site: www.icea.net

International Association of Electrical Inspectors (IAEI)
P.O. Box 830848
Richardson, TX 75083
901 Waterfall Way, Ste. 602
Richardson, TX 75080
972-235-1455; fax: 972-235-3855
Web site: www.iaei.com

International Conference of Building Officials (ICBO)
5360 Workman Mill Rd.
Whittier, CA 90601-2298 USA
800-284-4406 or 562-699-0541
fax: 888-329-4226
e-mail: order@icbo.org
Web site: www.icbo.org

International Electrotechnical Commission (IEC)
3, Rue de Varembé
P.O. Box 131
CH-1211 Geneva 20, Switzerland
+41-22-919-02-11; fax: +41-22-919-03-00
e-mail: pubinforl@iec.ch
Web site: www.iec.ch

International Organization for Standardization (ISO)
1, Rue de Varembé
Case Postale 56
CH-1211 Geneva 20, Switzerland
+41-22-749-01-11; fax: +41-22-733-34-30
e-mail: central@iso.ch
Web site: www.iso.ch

International Telecommunication Union (ITU—formerly CCITT)
Place des Nations
CH-1211 Geneva 20, Switzerland
+41-22-730-51-11; fax: +41-22-733-72-56
e-mail: itumail@itu.int
Web site: www.itu.int

Intertek Testing Services (ITS)
NA Inc.
3933 U.S. Rte. 11
Cortland, NY 13045 USA
800-967-5352; fax: 800-813-9442
e-mail: www.info@ETLSEMKO

National Building Code—Canada (NBCC)
Publication Sales, M-20
Institute for Research in Construction
National Research Council of Canada
Ottawa, ON K1A 0R6 Canada
800-672-7990 or 613-993-2607
fax: 613-952-7673
e-mail: irc.client-services@nrc.ca
Web sites: www.nrc.ca/irc
            www.ccbfc.org

National Electrical Contractors Association (NECA)
3 Bethesda Metro Ctr., Ste. 1100
Bethesda, MD 20814 USA
301-215-4521 or 301-657-3110
fax: 301-215-4500
e-mail: brooke@necanet.org or abv@necanet.org
Web site: www.necanet.org

National Electrical Manufacturers Association® (NEMA®)
1300 N 17th St., Ste. 1847
Rosslyn, VA 22209 USA
703-841-3200; fax: 703-841-3300
e-mail: webmaster@nema.org
Web site: www.nema.org

National Fire Protection Association® (NFPA®)
1 Batterymarch Park
P.O. Box 9101
Quincy, MA 02269-9101 USA
617-770-3000; fax: 617-770-0700
e-mail: custserv@nfpa.org
        library@nfpa.org
Web site: www.nfpa.org

National Institute of Standards and Technology (NIST)
100 Bureau Dr., Stop 3460
Gaithersburg, MD 20899-3460 USA
301-975-6478; fax: 301-926-1630
e-mail: inquiry@nist.gov
Web site: www.nist.gov

National Research Council of Canada (NRC-IRC)
Institute for Research in Construction
Bldg. M-20, Montreal Rd., Campus
Ottawa, ON K1A 0R6, Canada
613-993-2607; fax: 613-952-7673
e-mail: irc.client-services@nrc.ca
Web site: www.nrc.ca/irc

National Technical Information Services (NTIS)
U.S. Department of Commerce
5285 Port Royal Rd.
Springfield, VA 22161 USA
800-533-6847 or 703-605-6000
fax: 703-321-8547
e-mail: info@ntis.fedworld.gov
Web site: www.ntis.gov

Occupational Safety and Health Administration (OSHA)
200 Constitution Ave. NW
Washington, DC 20210 USA
800-321-6742 or 202-693-1999
fax: 202-219-5986
Web site: www.osha.gov

Rural Utilities Services (RUS—formerly REA)
AG-Box 1522
14th and Independence Ave. SW
Washington, DC 20250 USA
202-720-8674; fax: 202-250-3654
Web site: www.usda.gov/rus

Society of Cable Telecommunications Engineers (SCTE)
140 Philips Rd.
Exton, PA 19341
610-363-6888; fax: 610-363-5898
Web site: www.scte.org

Southern Building Code Congress International, Inc. (SBCCI)
900 Montclair Rd.
Birmingham, AL 35213-1206 USA
205-591-1853; fax: 205-591-0775
e-mail: info@sbcci.org
Web site: www.sbcci.org

Telcordia™ Technologies (formerly Bellcore)
8 Corporate Pl. #3A 184
Piscataway, NJ 08854-4120 USA
800-521-2673 or 732-699-5800

fax: 732-336-2559
e-mail: m-webleads@notes.cc.telcordia.com
Web site: www.telcordia.com

Telecommunications Industry Association (TIA)

500 Wilson Blvd., Ste. 300
Arlington, VA 22201-3836 USA
703-907-7700; fax: 703-907-7727
e-mail: tia@tia.eia.org
Web site: www.tiaonline.org

Underwriters Laboratories Inc.® (UL®)

33 Pfingsten Rd.
Northbrook, IL 60062 USA
847-272-8800; fax: 847-272-8129
Web site: www.ul.com

Uniform Building Code (UBC)

See International Conference of Building Officials (ICBO)

To order copies of the standards, contact:

Global Engineering Documents

15 Inverness Way
Englewood, CO 80112 USA
800-854-7179; fax: 314-726-6418
e-mail: global@ihs.com
Web site: www.global.ihs.com

4

Cabling Media

Overview of Residential Cabling Media

The ANSI/TIA/EIA-570-A, *Residential Telecommunications Cabling Standard*, specifies a variety of different cabling media for use in residences. The standard defines a cabling media as wire, cable, or conductors used for telecommunications.

This chapter lists and describes the various cabling media used in residential applications, including brief discussions of their physical and electrical properties. The chapter concludes with information on some of the cable support accessories used to hold and route cabling media.

The basic types of cabling media in use in residences are twisted-pair copper cable, coaxial cable, and optical fiber cable. Wireless technology, plastic optical fiber between audio electronic components, and other transmission media are also more likely to be found in residences, where a wider range of applications (e.g., security, home automation, and entertainment systems) exist.

The cabling media used in residential outlet cabling applications include:

- 4-pair, 100 Ω, twisted-pair Category 3, Category 5, or Category 5e cables (ANSI/TIA/EIA-568-B.2, *Commercial Building Telecommunications Cabling Standard; Part 2: Balanced Twisted-Pair Cabling Components*).

- 50/125 or 62.5/125 μm multimode fiber (ANSI/EIA/TIA-492AAAB, *Detail Specification for 50 μm Core Diameter/125 μm Cladding Diameter Class Ia Multimode, Graded-Index Optical Waveguide Fibers* and ANSI/TIA/EIA-492AAAA, *Detail Specification for 62.5 μm Core Diameter/125 μm Cladding Diameter Class Ia Graded-Index Multimode Optical Fibers*).

- Singlemode fiber (ANSI/TIA/EIA-492CAAA, *Detail Specifications for Class IVa Dispersion—Unshifted Single-Mode Optical Fibers*).

- 75 Ω Series 6 coaxial cable (SCTE IPS-SP-001, *Flexible R.F. Coaxial Drop Cable*).

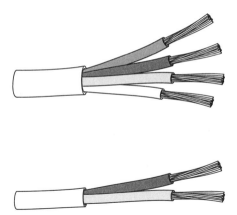

Figure 4.1   Typical stranded high-quality audio cable.

The cabling media used for alternative systems include:

- 14 AWG (1.6 mm [0.063 in]) or 16 AWG (1.3 mm [0.051 in]) two- or four-conductor stranded.
- 22 AWG (0.64 mm [0.025 in]) four- or six-conductor stranded shielded.
- 18 AWG (1.0 mm [0.039 in]) or 22 AWG (0.64 mm [0.025 in]) two- or four-conductor solid or stranded.
- Other vendor-specific cabling as required.

TABLE 4.1   Cable distance based on current draw of conductors

Wire Size	Total Current Drawn by All Devices Connected to a Single Zone							
AWG	50 mA m (ft)	100 mA m (ft)	300 mA m (ft)	500 mA m (ft)	600 mA m (ft)	1 Amp m (ft)	2 Amp m (ft)	3 Amp m (ft)
22	152 (500)	76 (250)	25 (83)	15 (50)	13 (41)	8 (25)	4 (12)	3 (8)
20	213 (700)	107 (350)	36 (116)	21 (70)	18 (58)	11 (35)	5 (17)	4 (11)
18	396 (1300)	198 (650)	66 (216)	40 (130)	33 (108)	20 (65)	10 (32)	7 (21)
16	457 (1500)	229 (750)	76 (250)	46 (150)	38 (125)	23 (75)	11 (37)	8 (25)
14	975 (3200)	488 (1600)	163 (533)	98 (320)	81 (266)	49 (160)	24 (80)	16 (53)
12	1554 (5100)	777 (2550)	259 (850)	155 (510)	130 (425)	78 (255)	39 (127)	26 (85)

NOTE: 1- to 3-Amp ratings are intended for sounders.

The cabling media used in residential backbone applications include:

- 100 Ω, twisted-pair Category 3, Category 5, or Category 5e cables (ANSI/TIA/EIA-758, *Customer-Owned Outside Plant Telecommunications Cabling Standard* and ANSI/TIA/EIA-568-B.2).
- 50/125 μm multimode fiber (ANSI/TIA/EIA-492AAAB).
- 62.5/125 μm multimode fiber (ANSI/TIA/EIA-492AAAA).
- Singlemode fiber (ANSI/TIA/EIA-492CAAA).
- Trunk distribution and feeder (hard-line) coaxial cable (SCTE IPS-SP-100, *Specification for Trunk, Feeder, and Distribution Coax Cable*).
- 75 Ω Series 6 and Series 11 coaxial cable (SCTE IPS-SP-001).

Residential cabling installers must be aware of the various types of cabling media they might be called upon to install. Each type and configuration has specific uses and defined methods of installation that must be employed as defined in ANSI/TIA/EIA-570-A.

Grades of Residential Cabling

The U.S. residential cabling standard, ANSI/TIA/EIA-570-A, defines two grades of residential service, a basic service called Grade 1 and a more advanced service labeled Grade 2. These grades are based on the services expected at each telecommunications outlet/connector. For details, see ANSI/TIA/EIA-570-A and Chapter 2: Residential Structured Cabling Systems.

Figure 4.2 shows a typical Grade 1 cable and Figure 4.3 shows a typical Grade 2 cable.

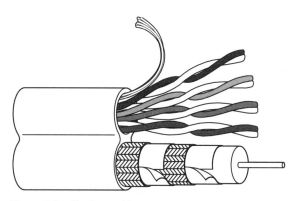

Figure 4.2   Grade 1 cable (consisting of one Category 5e unshielded twisted-pair cable and one quad-shielded Series 6 coaxial cable).

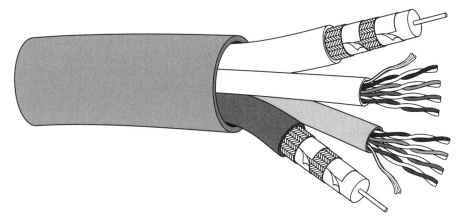

Figure 4.3 Grade 2 cable (consisting of two Category 5e unshielded twisted-pair cables and two quad-shielded RG-6 coaxial cables).

NOTE: Grade 2 cables are available with two optical fibers.

Category Classifications

Telecommunications Industry Association (TIA) has applied a category-based classification system to twisted-pair copper cabling, both unshielded and screened.

In ANSI/TIA/EIA-568-B.2, specifications for performance levels of twisted-pair cable and associated connecting hardware are established, as follows:

- Category 3—Twisted-pair cables and associated connecting hardware whose transmission characteristics are specified up to 16 megahertz (MHz).

- Category 5e—Twisted-pair cables and associated connecting hardware whose transmission characteristics are specified up to 100 MHz, and which also provide additional improvements in overall transmission characteristics.

Category 3 twisted-pair cabling is the minimum grade recommended in ANSI/TIA/EIA-570-A for new installations, and is now the required minimum specified by the Federal Communications Commission (FCC) for new home construction.

Twisted-Pair Copper Cable

The basic categories of twisted-pair copper cable currently in use are Categories 3, 5, and 5e. Category 6 is being proposed.

Twisted-pair cable is used for both voice and data cabling (see Figure 4.4). It:

- Is composed of pairs of wires twisted together.

- Is commonly available in various pair counts (2–4200 pairs).

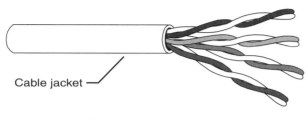

Unshielded twisted-pair

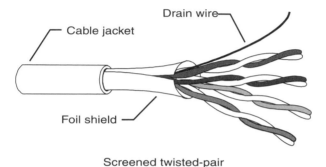

Screened twisted-pair

Figure 4.4   Typical twisted-pair cable—unshielded and screened.

- Is normally not shielded below 200 pairs and has an overall aluminum-steel shield up to 2400 pairs.
- Reduces electrical interference by twisting of conductors.
- Has a characteristic impedance of 100 Ω at 100 MHz.
- Has recommended conductor sizes of 22 AWG (0.64 mm [0.025 in]) to 24 AWG (051 mm [0.020 in]).

  Category 5e twisted-pair copper cable:

- Is rapidly replacing Category 5 cable within the marketplace.
- Provides a minimum 3 decibel (dB) improvement in near-end crosstalk (NEXT) loss compared with Category 5.

Twisted-Pair Patch Cords

Twisted-pair patch cords:

- Should have stranded conductors for flexibility.
- Are allowed to exhibit up to 50 percent more attenuation than solid conductors.
- Should meet the same category as the outlet cabling in use.
- Usually have 8-position, 8-contact (8P8C [RJ-45]) connectors on the ends.

If using stranded conductor cable, make sure connectors are designed for these conductors. The wrong kind of connector may damage the conductor and make a faulty connection.

TABLE 4.2   Color codes for 100 Ω twisted-pair patch cables

Pair	Identification	Conductor Color Code—Option 1	Conductor Color Code—Option 2
Pair 1	Tip	White-Blue (W-BL)	Green (G)
	Ring	Blue-White (BL-W)	Red (R)
Pair 2	Tip	White-Orange (W-O)	Black (BK)
	Ring	Orange-White (O-W)	Yellow (Y)
Pair 3	Tip	White-Green (W-G)	Blue (BL)
	Ring	Green-White (G-W)	Orange (O)
Pair 4	Tip	White-Brown (W-BR)	Brown (BR)
	Ring	Brown-White (BR-W)	Slate (S)

Screened Twisted-Pair (ScTP) Cable

ScTP copper cable makes use of the same category system as twisted-pair cable: Categories 3, 4, 5, and 5e.

ScTP cable is widely used in Europe, while in the United States its use is slowly growing. Although more expensive than UTP cable, ScTP cable offers a higher immunity to electromagnetic interference (EMI), a common problem in the electrically noisy residential environment (e.g., electrical wiring, microwave ovens, fans, generators).

ScTP cables feature a foil and/or braided shield with a nonconductive material on one side. Screened cabling blocks the reception and transmission of signals through the screened cable jacket. The drain wire runs adjacent to the conductive side of the overall screen. Proper installation practices vary by manufacturer. ScTP cable can increase noise levels and EMI if not properly bonded to a ground.

ScTP cable has:

- A characteristic impedance of 100 Ω or 120 Ω.
- Four pairs of 22 AWG (0.64 mm [0.025 in]) to 24 AWG (0.51 mm [0.020 in]) solid conductors.
- Aluminum foil and/or braided shield around all conductor pairs.
- A drain wire that must be properly bonded to a ground.
- Similar electrical performance characteristics to the UTP equivalent category.
- May be used where extra protection from EMI is desired.

> **NOTE:** When installing ScTP cabling systems, all components of the system must be screened.

Coaxial Cable

Coaxial cable provides a much higher bandwidth and degree of protection against EMI than twisted-pair cable. Coaxial cable consists of a centered inner conductor insulated with a dielectric material. This is surrounded by a foil and/or braid shield, which acts as a second conductor and is covered with an overall jacket. Coaxial cable (commonly referred to as coax) is used for computer networks, community antenna television (CATV), and video systems. Historically, coaxial cable was designated as "RG" (radio grade) cable. Coaxial cables are now referred to as Series-X cables. The "X" designates the construction of the cable with such factors as:

- Center conductor diameter.
- Center conductor being solid or stranded.
- Dielectric composition.
- Outer braid percent of coverage.
- Impedance.

A dual-shielded coaxial cable typically uses an aluminum foil for the high-frequency shielding and an aluminum braid for the low-frequency shielding. For additional shielding, a tri-shielded cable employs two foils and a braid. Quad-shielded cables typically have a foil, 60 percent braid, foil, and 40 percent braid. This construction is recommended since it provides the best shielding characteristics. See Figure 4.5.

Series 6 coaxial cable has:

- A characteristic impedance of 75 Ω.
- More usable bandwidth, which is needed for the transmission of digital video signals.
- A coated/aluminum foil shield over the dielectric to shield against high frequencies.

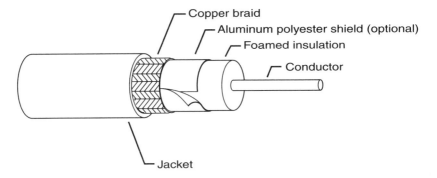

Dual-shielded

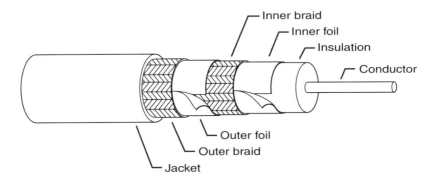

Quad-shielded

Figure 4.5   Series 6 coaxial cable—dual-shielded and quad-shielded.

- A braided shield over the coated/aluminum foil to shield against low frequencies.
- A solid-center conductor.

Series 11 coaxial cable is used in video backbone distribution. It has less signal attenuation than Series 6 coaxial cable, making it the preferred choice for longer runs. Series 11 coaxial cable should not be used for outlet cables, due to large minimum bend radius.

Series 11 coaxial cable has a:

- Characteristic impedance of 75 Ω.
- Coated/aluminum foil shield over the dielectric to shield against high frequencies.

- Braided bare copper shield over the coated/aluminum foil to shield against low frequencies.
- 14 AWG (1.6 mm [0.063 in]) solid center conductor.

Closed-circuit television (CCTV [surveillance video]) coaxial cables use a bare copper inner conductor and a 95 percent bare copper braid shield. The bare copper inner conductor facilitates remote powering of cameras, while the bare copper braid provides low-frequency shielding for the baseband signal.

CATV (also satellite) coaxial cables use a copper clad steel center conductor for reliable connector performance, since the center conductor functions as the connector center pin. Since CATV signals exceed 1 gigahertz (GHz) and satellite signals exceed 2 GHz, both low- and high-frequency shielding is required for the broadband signal. Coaxial cables of each series are available in a variety of configurations, jacket types, shield configurations, bandwidths, and attenuations. A coaxial cable cannot be selected by simply identifying the physical size (series); a full understanding of the application is necessary. Many supply houses and most manufacturers offer consultation services to assist in the selection of the cable best suited to the needs of the job.

Coaxial equipment and patch cords

Equipment and outlets serviced by coaxial cable require coaxial equipment and patch cords. The cords typically provided with consumer audio/video (A/V) equipment are only suitable for use as equipment cords.

- Equipment cords are used to connect equipment to the outlets and may be Series 59 or Series 6 coaxial cabling and should be factory terminated with an F connector. Threaded and nonthreaded (push-on) connectors are acceptable.
- Patch cords are used to connect outlet cabling to the distribution system and must be the same cable that is used in the outlet cabling. These cords can be either factory or field terminated with F connectors.

Optical Fiber Cable

The optical fiber cables most commonly used today are 50/125 μm and 62.5/125 μm multimode fiber and 8–9/125 μm singlemode fiber. (The abbreviation "μm" stands for "micrometer," and utilizes the lowercase Greek letter mu, along with the English letter, m. A more commonly used term is micron.)

Optical fiber allows the transmission of light impulses instead of electrical signals. The two key elements of an optical fiber are its core and cladding. The core is the innermost part of the fiber through which light pulses are guided. The cladding surrounds the core to keep the light in the center of the fiber. Light traveling along the core that strikes the cladding at a glancing angle is confined in the core by internal reflection.

Optical fiber cable may be:

- Multimode.
- Singlemode.
- Hybrid or composite (a combination of two different types of fiber cable).

Multimode optical fiber cable:

- Is the most common fiber medium for backbone and horizontal runs within buildings and campus environments.
- Has a 50 μm or 62.5 μm core and a 125 μm cladding diameter.
- Has a distance limitation of 2000 m (6560 ft) for structured cabling systems.
- Generally utilizes a light-emitting diode (LED) light source.
- Supports wavelengths of 850 nanometer (nm) and 1300 nm. See Figure 4.6.

Singlemode optical fiber cable:

- Has an 8–9 μm core, depending on the manufacturer.
- May be used for distances up to 3000 m (9840 ft) for residential structured cabling systems. Longer distances are common for service providers.
- Normally uses a laser light source.
- Supports wavelengths of 1310 nm and 1550 nm. See Figure 4.7.

All of the above types of optical fiber cable are available in the following versions:

- Simplex (1-fiber) cable—Commonly used for single-direction transmission of signals (e.g., video cameras).
- Duplex (2-fiber) cable—Commonly used for bidirectional transmission of data.

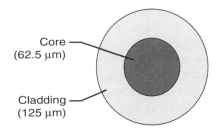

Structure of a
multimode optical fiber

Figure 4.6   62.5/125 μm multimode fiber.

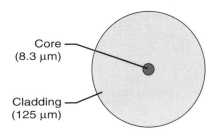

Core
(8.3 µm)

Cladding
(125 µm)

Structure of a
singlemode optical fiber

Figure 4.7   Singlemode fiber.

- Multiple fibers—Commonly used for backbone cabling. Typical constructions include six-, twelve-, and twenty-four strands.

  Optical fiber cables may be:

- Tight-buffered. See Figure 4.8.
- Loose tube.

  Tight-buffered optical fiber cable:

- Is primarily used inside buildings.
- Is available with various jacket types to meet building codes.

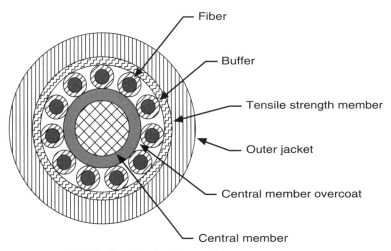

Fiber

Buffer

Tensile strength member

Outer jacket

Central member overcoat

Central member

Figure 4.8   Tight-buffered optical fiber cable.

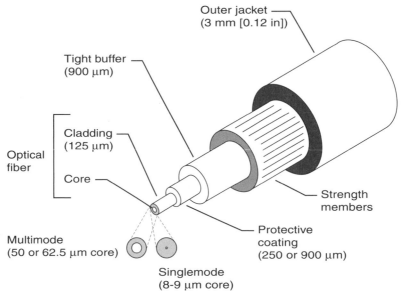

Figure 4.9   Single-stranded, tight-buffered optical fiber cable.

- Protects the fiber by supporting each strand of glass in a tight-buffered coating, increasing the strand's diameter to 900 μm.
- Is easily connectorized.
- May be singlemode or multimode. See Figures 4.9, 4.10, and 4.11.

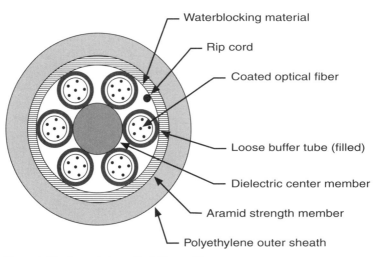

Figure 4.10   Loose-tube optical fiber cable.

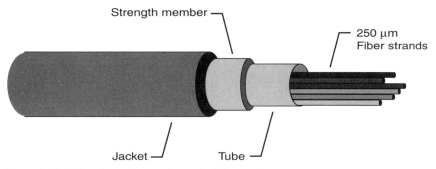

Figure 4.11   Side view of loose-tube optical fiber cable.

Loose-tube optical fiber cable:

- Is used primarily outdoors.
- When specified for outdoors, contains a water block to prevent water from penetrating into a fiber cable, freezing and expanding.
- May be singlemode or multimode.
- May require a furcation (break-out) kit to allow connectorization. See Figure 4.12.

900 µm Outer diameter furcation (breakout) kit

Figure 4.12   Loose-tube furcation (breakout) kit.

Multimode and singlemode optical fiber cables should perform as shown in the following table.

TABLE 4.3   Optical fiber cable transmission performance parameters

Optical Fiber Cable Type	Wavelength (nm)	Maximum Attenuation (dB/km)	Minimum Information Transmission Capacity for Overfilled Launch (MHz•km)
50/125 µm	850	3.5	500
Multimode	1300	1.5	500
62.5/125 µm	850	3.5	160

continued on next page

TABLE 4.3   Optical fiber cable transmission performance parameters (continued)

Optical Fiber Cable Type	Wavelength (nm)	Maximum Attenuation (dB/km)	Minimum Information Transmission Capacity for Overfilled Launch (MHz•km)
Multimode	1300	1.5	500
Singlemode	1310	1.0	N/A
Inside plant cable	1550	1.0	N/A
Singlemode	1310	0.5	N/A
Outside plant cable	1550	0.5	N/A

Reprinted with permission of Telecommunications Industry Association.

NOTE: The information transmission capacity of the fiber, as measured by the fiber manufacturer, can be used by the cable manufacturer to demonstrate compliance with ANSI/ICEA S-83-596, *Fiber Optic Premises Distribution Cable*.

High Pair-Count Twisted-Pair Cable

High pair-count twisted-pair cable (of more than 4 pairs) is used for backbone service in multi-dwelling unit environments and is often referred to as multi-pair cable. This cable is generally Category 3 rated. The characteristics of this type of cable include that it is:

- Commonly available in various pair counts up to 4200 pairs.

- Not normally shielded in cables up to 200 pairs, and has an overall aluminum-steel shield in cables of 600 to 2400 pairs.

Color Coding

When working with twisted-pair cables, the residential cabling installer should be able to identify individual pairs and conductors within the cable.

Standard color codes for copper and optical fiber cables have been developed. These color codes enable the residential cabling installer to quickly identify any pair within a copper cable or strand within an optical fiber cable.

Residential cabling installers must also be able to identify individual pairs within the cable and the individual conductors within each pair. In telephony, the individual conductors are referred to as tip and ring conductors. Each pair has a tip and a ring conductor, or a positive and negative conductor respectively.

The terms tip and ring originated from the earliest types of telephone systems, where the operator had to physically use switchboard cords to route the calls. The operator's switchboard plug had three conductors: tip, ring, and sleeve.

In twisted-pair cabling, it takes two conductors to make a pair. Each pair has a tip conductor and a ring conductor. The colors used to identify tip conductors are different from the colors used to identify ring conductors. There are five colors associated with tip conductors, and five different colors associated with ring conductors.

When the two conductors are paired, the two different colors identify the pair number. There are 25 possible color combinations when the five tip and five ring color codes are mixed, as can be seen in the color-coding Table 4.4. Two tips or two rings are never used to make a pair.

TABLE 4.4 High pair-count cable, color-coded pairs

	Ring Conductors				
Tip Conductors	Blue 1st Pair	Orange 2nd Pair	Green 3rd Pair	Brown 4th Pair	Slate 5th Pair
White (1–5)	1	2	3	4	5
Red (6–10)	6	7	8	9	10
Black (11–15)	11	12	13	14	15
Yellow (16–20)	16	17	18	19	20
Violet (21–25)	21	22	23	24	25

The tip colors are white, red, black, yellow, and violet. There are five pairs that use each tip color in a 25-pair cable. Tip colors indicate in which group a pair is located. For example, white indicates the pair group is 1, 2, 3, 4, or 5 and black indicates that it is pair group 11, 12, 13, 14, or 15.

NOTE: The tip color does not identify a specific pair number until the combination of tip and ring colors are matched.

The ring colors are blue, orange, green, brown, and slate. The ring color identifies the position of the pair within the group of five tip possibilities (e.g., blue is the first pair and green is the third pair within the tip group).

To identify the pair number, the tip and ring colors must be viewed together (e.g., the white/blue pair is actually a white tip and a blue ring). White indicates the pair is between pairs 1 through 5, while blue indicates it is the first pair. The first pair of 1 through 5 is the "#1" pair.

NOTE: Another example is the black/brown pair. Black indicates pairs 11 through 15, while brown indicates the fourth pair. The fourth pair in the third 5-pair group is the fourteenth pair in the binder group.

The standard color code can be used to identify up to 25 pairs without duplicating any pair color combinations.

For cables of over 25 pairs, the first group of 25 pairs is formed into a bundle and has a colored binder (tape or thread) placed around it. Additional bundles of 25 pairs have their own uniquely colored binders. The colored binder wraps follow the same color code as the individual pair color code. They are referred to as binder groups or binders.

The white/blue binder wrap is the first one to be used and contains cable pairs 1–25, the white/orange binder wrap is second and contains cable pairs 26–50, and the red/blue binder wrap is the sixth and contains cable pairs 126–150. This system of 25-pair binder groups works for a cable containing up to 625 pairs of cable (25 binders × 25 pairs per binder = 625 total pairs); however, the system stops at 600 pairs. The violet/slate binder wrap is not used. On some smaller cables (100 pairs or less), the manufacturer may indicate binders with only the ring color, since it can be assumed that its tip color is white.

Once cable exceeds 600 pairs, each group of 600 pairs (24 binders) is wrapped in a colored super binder. The colored super-group binder denotes that 600 pairs are present. The color code changes from the normal tip/ring identifiers to a simple tip color. For example, white indicates 1–600 pairs, red indicates 601–1200, and black indicates 1201–1800 pairs.

TABLE 4.5   High pair-count cable, color-coded pairs, and binder groups

	Pair		Binder Group	
Number	Tip	Ring	Color	Pair Count
1	White	Blue	White-Blue	001–025
2	White	Orange	White-Orange	026–050
3	White	Green	White-Green	051–075
4	White	Brown	White-Brown	076–100
5	White	Slate	White-Slate	101–125
6	Red	Blue	Red-Blue	126–150
7	Red	Orange	Red-Orange	151–175
8	Red	Green	Red-Green	176–200
9	Red	Brown	Red-Brown	201–225
10	Red	Slate	Red-Slate	226–250
11	Black	Blue	Black-Blue	251–275
12	Black	Orange	Black-Orange	276–300
13	Black	Green	Black-Green	301–325
14	Black	Brown	Black-Brown	326–350
15	Black	Slate	Black-Slate	351–375
16	Yellow	Blue	Yellow-Blue	376–400
17	Yellow	Orange	Yellow-Orange	401–425

continued on next page

TABLE 4.5   High pair-count cable, color-coded pairs, and binder groups (continued)

	Pair		Binder Group	
Number	Tip	Ring	Color	Pair Count
18	Yellow	Green	Yellow-Green	426–450
19	Yellow	Brown	Yellow-Brown	451–475
20	Yellow	Slate	Yellow-Slate	476–500
21	Violet	Blue	Violet-Blue	501–525
22	Violet	Orange	Violet-Orange	526–550
23	Violet	Green	Violet-Green	551–575
24	Violet	Brown	Violet-Brown	576–600
25	Violet	Slate	25th binder is not used.	

ANSI/TIA/EIA-598-A, *Optical Fiber Cable Color Coding,* covers the color coding of optical fiber cables. The standard states:

- Strands 1–12 shall be uniquely color-coded as shown in the following table.
- Strands 13–24 shall repeat the same color code as 1–12, with the addition of a black tracer.
- The black tracer may be a dashed line or solid line.
- The black strand has a yellow tracer.

TABLE 4.6   Optical fiber color code chart

Fiber Number	Color	Fiber Number	Color
1	Blue	13	Blue/Black Tracer
2	Orange	14	Orange/Black Tracer
3	Green	15	Green/Black Tracer
4	Brown	16	Brown/Black Tracer
5	Slate	17	Slate/Black Tracer
6	White	18	White/Black Tracer
7	Red	19	Red/Black Tracer
8	Black	20	Black/Yellow Tracer
9	Yellow	21	Yellow/Black Tracer
10	Violet	22	Violet/Black Tracer
11	Rose	23	Rose/Black Tracer
12	Aqua	24	Aqua/Black Tracer

Optical fiber cables larger that 24 strands shall be color coded by means of binder tapes, ribbons, or threads consisting of two binders: one to match the base color and one to match the tracer color. Refer to the manufacturer's specific color scheme for details.

Connectors

Overview

A connector is a mechanical device, such as a plug or jack, attached to the end of a cable or to a piece of equipment. When two connectors are mated, electrical signals (in copper systems) or light impulses (in fiber systems) can be transferred from one cable or piece of equipment, through the connectors, to the other cable or piece of equipment.

Cable connectors are as essential to the integrity of a residential telecommunications network as is the cable itself. Together the cable, connector, and device form the communications link by which information moves. The connector provides the physical attachment from the media to:

- A transmitter.
- A receiver.
- Another media of the same or similar type.
- An active telecommunications device.
- A passive telecommunications device.

Along a media path, connections contribute to the loss of signal from transmitter to receiver. The role of the connector is to provide a coupling mechanism that keeps loss to a minimum. This enables the outlet cable to be terminated at the outlet and distribution device (DD), allowing for patch cords to be used to better manage the cabling system.

Some of the key factors in establishing a good connection and minimizing loss of signal include:

- Fit—The attachment must be sufficiently secure to prevent the connectors from separating unintentionally.

- Alignment—The device must provide the physical alignment necessary to complete the link.

- Functionality—The connection must provide efficient transfer of light or electricity from one connector to the other.

- Configuration—Connectors must be designed specifically for the media and the devices being connected.

Durability is demonstrated by a connectors' ability to withstand hundreds of insertion and withdrawal cycles without failing. This is calculated as mean time between failures.

Connectors are classified as "male" and "female." In general, outlets or jacks are female connectors, and plugs are male connectors.

ANSI/TIA/EIA-570-A, *Residential Telecommunications Cabling Standard*, provides recommendations concerning the choice of proper connectors for use in residential structured cabling systems (RSCS).

Copper Insulation Displacement Connector (IDC) Termination

Proper cable termination practices are vital for the complete and accurate transfer of both analog and digital information signals. IDC termination is the recommended method of copper termination recognized by ANSI/TIA/EIA-568-B.1, *Commercial Building Telecommunications Cabling Standard, Part 1: General Requirements*, for twisted-pair cable terminations. This method removes or displaces the conductor's insulation as it is seated in the connection.

Commonly used IDC termination blocks and panels include:

- 66-type.

- 110-style.

- BIX™.

- Krone LSA.

For any IDC-type modular termination, the T568A wiring scheme is required. For more information, refer to Chapter 15: Trim-Out Finish.

66-Block termination

A 66M1-50 block provides the means of terminating two 25-pair or twelve 4-pair cables per block. These blocks have two rows of contacts that are mechanically connected to provide cross-connection capability. The 66M1-25 blocks have four rows of connected contacts. Voice applications may use bridging clips to make a connection between the left and right set of contacts on a 66M1-50 block. By lifting the bridging clips, which opens the circuit, it is easy to test the voice circuit in both directions when troubleshooting. The 66-type termination block is normally mounted on backboards with an 89-style bracket.

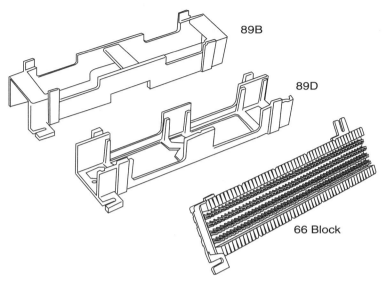

Figure 5.1 66-Type block and 89-style brackets.

> **NOTE:** Bridging clips are not Category 5e compliant. This description is provided for information purposes only.

110-Style hardware

The 110-style IDC termination block is used in both voice and data cabling applications.

> **NOTE:** 110-style blocks are smaller than 66-type blocks and are typically mounted horizontally.

BIX™ hardware

BIX-type termination hardware is similar to the 110-style hardware previously described. Unlike 110-style hardware, however, which places clips on top of the wiring block, BIX equipment uses a one-piece, pass-through unit that is reversed in its mount after termination of the cable to expose the opposite side to provide a cross-connect field.

Krone LSA hardware

Krone LSA-type termination hardware provides silver-plated IDC contacts at a 45-degree angle, with the conductor being held in place by tension in the contacts. This hardware is available in patch panels, telecommunications out-

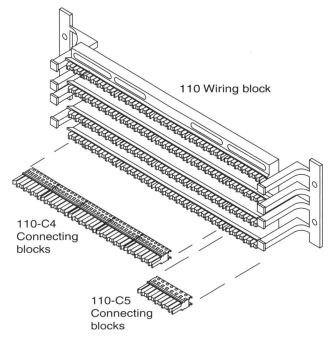

Figure 5.2   110-Style hardware.

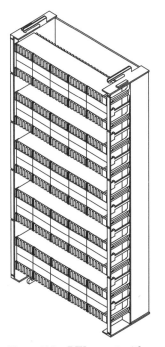

Figure 5.3   BIX mount with connector.

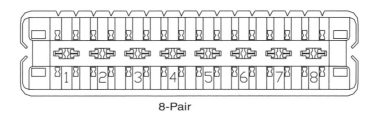

8-Pair

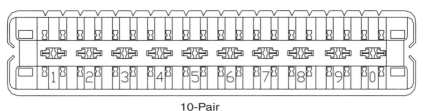

10-Pair

Figure 5.4   Krone LSA 8-pair and 10-pair blocks.

let/connectors, and termination blocks. Krone LSA hardware also provides disconnect modules, connect modules, switching modules, and feed-through modules.

Twisted-Pair Connectors

Twisted-pair cable requires IDCs. An IDC connection permits termination of an insulated conductor without stripping insulation from the conductor. As the insulated conductor is inserted between two or more sharp edges of the contact, the insulation is displaced, allowing contact to be made between the conductor and the connector terminal.

Many key and private branch exchange (PBX) telephone systems, most likely to be found in multi-dwelling units, use 25-pair cables to interface the entrance facility and cross-connect fields. Telephone system manufacturers often specify the demarcation point to be an RJ-21X.

These connectors can be ordered as complete cable assemblies but are usually assembled (or connectorized) by the telecommunications cabling installer. Field installation is generally less expensive and allows for custom cable lengths that can enhance the installation's appearance.

> **NOTE:** It is common to buy double-ended cables twice as long as needed and cut them in half. This provides a factory connector and minimizes the field time and labor expense to terminate.

Telco 50-pin connectors are available in several versions and require a special tool, called a butterfly tool, for termination. During assembly, the tool's two

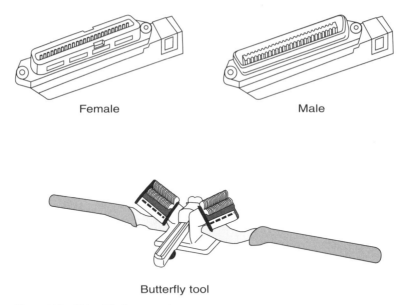

Female   Male

Butterfly tool

Figure 5.5   Telco 50-pin connector.

actuating arms and conductor fanning strips extend outward and are then raised to terminate and cut the conductors.

Connections between data equipment often utilize D-subminiature connectors, so called because of the D-shaped metal skirt surrounding the connector's pins. These connectors are also referred to as DB-## connectors. The "##" indicates the number of pins in the connector. There are both male and female connectors, but the name does not indicate gender and must be specified when ordering. The more common DB-type connectors are:

- DB-9—Standardized 9-pin serial port used on laptop computers.

- DB-15—Standardized 15-pin connector used for Ethernet transceivers, video graphics array (VGA) monitors, and joy sticks (ANSI/TIA/EIA-422-B, *Electrical Characteristics of Balanced Voltage Digital Interface Circuits,* and ANSI/TIA/EIA-485-A, *Electrical Characteristics of Generators and Receivers for Use in Balanced Digital Multipoint Systems*).

- DB-25—Standardized 25-pin connector used for serial data communications (ANSI/TIA/EIA-232-F, *Interface Between Data Terminal Equipment and Data Circuit-Terminating Equipment Employing Serial Binary Data Interchange*).

Twisted-pair modular connectors

Modular plug connectors have IDC contacts designed for either stranded or solid conductors, as well as connectors having universal contacts that accept

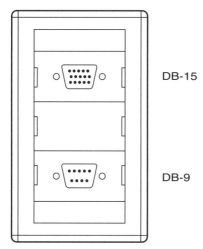

Figure 5.6   DB-9 and DB-15 connectors shown within an outlet faceplate.

either stranded or solid conductors. Modular plug contacts for stranded conductors are designed as a single blade, intended to penetrate the conductor's
insulation and slide between the individual conductor strands. If this type of
contact is used with a solid conductor that has no strands to slide between, the
solid conductor is usually nicked or broken. Nicked conductors soon break
from fatigue and can become nonfunctional or suffer from intermittent opens
(no electrical pathway).

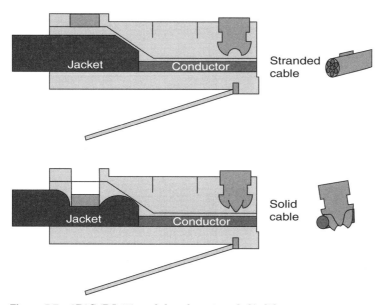

Figure 5.7   8P8C (RJ-45) modular plug, stranded/solid.

NOTE: It is rare to achieve a reliable connection using a connector designed for stranded cable on a solid conductor.

Modular plug contacts for solid conductors are designed with three fingers that penetrate the insulation on both sides of the conductor. They provide an electrical connection by trapping the conductor between the fingers.

Some manufacturers market solid-conductor, modular-plug contacts as multipurpose connectors capable of terminating both stranded and solid conductors. These connectors work well for solid conductors.

Modular plugs and connectors are 8-position, 8-contact (8P8C [RJ-45]) and are available in various sizes and shapes (keyed/nonkeyed). The number of positions (8P) indicates the connector's width, while the number of contacts (8C) installed into the available positions indicates the maximum number of conductors the connector can terminate.

A connector may be sized for eight positions, but only have four contacts installed. This saves on manufacturing costs (e.g., connectors are available as 8P2C, 8P4C, and 8P6C). They are all the same physical size but have different numbers of contacts to terminate conductors.

In the 1970s, the universal service order code (USOC) system was developed for the identification of parts and services. The Federal Communications Commission (FCC) adopted the USOC codes that related to some of the interface jacks used to connect various devices to the public telephone network. Designations such as RJ-21X were used to define not only the physical connector, but also how it is to be wired. The specific codes as adopted in part by the FCC are:

- RJ—Registered jack.
- C—Flush- or surface-mounted jack.
- W—Wall-mounted jack.
- X—Complex or multiline connector.

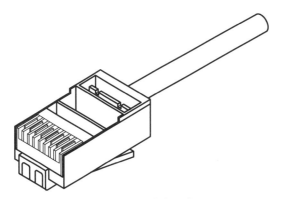

Figure 5.8   8P8C (RJ-45) modular plug.

Today's use of the term RJ-45 has come to represent the physical appearance more than any specific wiring configurations. The cabling industry has adopted the term RJ-45 to describe all the 8-position connectors. However, in the United States, FCC Part 68 has adopted some of the USOC characters. Refer to FCC Part 68 and the current edition of the BICSI *Telecommunications Distribution Methods Manual (TDMM)* for additional information.

To reduce costs, telephone and modem manufacturers often supply line cords composed of 6P2C plugs with a two-conductor cable. Per ANSI/TIA/EIA-570-A, to prevent damage of the eight-position outlet/connector when mating to a six-position plug, the tab width for a six-position plug shall be 6 mm (0.238 in) to 6.2 mm (0.243 in).

CAUTION: All 6-position plugs do not meet the requirements of ANSI/TIA/EIA-570-A, especially imported plugs and old cords. Inserting a noncompliant plug into an 8-position modular jack can spread the outer pins of the jack, causing it to become ineffective for some systems.

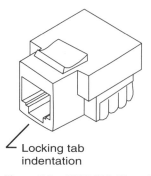

Locking tab
indentation

Figure 5.9   8P8C (RJ-45) modular jack.

8P8C (RJ-45) modular jacks are designed to interface twisted-pair cables with equipment lines or patch cords. The cables are terminated on the rear of the jack, while the connector's front provides access for the modular 8P8C (RJ-45) plug.

The jack should be installed with the contact pins on top to reduce the effects of dust and the locking tab indentation on the bottom. When viewed from the front, the pin numbers run from left to right consecutively, 1 through 8. Twisted-pair outlet connectors are used in residential as well as office environments. Figure 5.10 shows twisted-pair outlet cables mated to twisted-pair outlet connectors on a faceplate, as they would typically look in the room of a home.

Modular jacks are capable of accepting pin assignments compatible with all known data applications operating over 100 Ω twisted-pair cable. The jack and

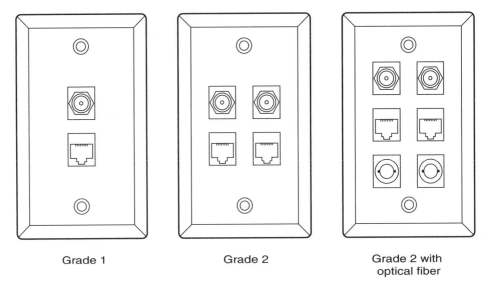

Grade 1   Grade 2   Grade 2 with
                                                        optical fiber

Figure 5.10   Telecommunications outlet/connectors.

pin/pair assignments for the T568A modular jack as approved in ANSI/TIA/
EIA-570-A are shown in Figure 5.11.

Terminations for twisted-pair modular connectors should be 8P8C (RJ-45)
modular jacks or plugs and are designed to:

- Terminate 100 Ω, 4-pair, 22–24 AWG (0.64 mm [0.025 in]–0.51 mm [0.020 in]).
- Support up to 100 MHz, Category 5e standards.
- Use IDC terminations.

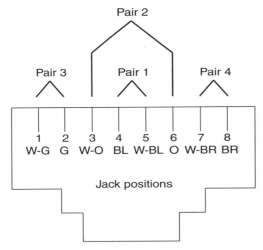

Figure 5.11   Eight-position jack pin/pair assignments, T568A.

> **NOTE:** Harsh environments and outdoor applications may require the use of weather-resistant outlet connectors.

Alternative twisted-pair connectors

Other types of twisted-pair connectors that may be used in residential applications include:

- S-video—These connections allow higher picture quality in video displays than most other video standards.
- RCA jacks—For line level audio and/or composite or component video routing.

> **NOTE:** The above connections may also require additional support equipment or interfaces (such as baluns) to eliminate group looping or impedance mismatching issues. See manufacturer documentation regarding the use of these connectors.

Screened twisted-pair (ScTP) connectors

ScTP cabling installations require a shield around every component of the cabling system. Shield continuity is maintained from the first connector in the DD through the outlet cables to the telecommunications outlet/connector through the user's equipment cord to the equipment's chassis ground. A metal shield surrounds each modular jack. A drain wire in contact with the cable screen must be attached to the connector's shield.

ScTP patch cords and equipment cords must be used to extend the shield from the shielded modular jacks. Each cord is comprised of a stranded ScTP cable with a shielded modular plug on each end. Each plug has a metal shield surrounding it. A drain wire in contact with the cable screen must be attached to each plug's shield.

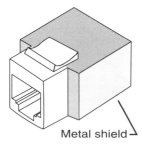

Metal shield

Figure 5.12 Screened twisted-pair modular jack.

ScTP connectors are shielded 8P8C (RJ-45) modular jacks or plugs that are designed to:

- Terminate 100 Ω, 4-pair, 22–24 AWG (0.64 mm [0.025 in]–0.51 mm [0.020 in]) cable.
- Support up to 100 MHz, Category 5e standards.
- Use IDC terminations.
- Be covered by a metallic shield.
- Be connected to the ScTP shield, using the drain wire.

Coaxial Connectors

Most coaxial cable should be terminated with male F connectors. At outlet locations, a feed-through bulkhead connector allows connections from male-terminated cables. Coaxial patch cables may be purchased with connectors already attached.

F connectors

Use the proper male F coaxial connector for the media type that is being installed.

F connectors are:

- Economical.
- Used to connect Series 59, Series 6, and Series 11 coaxial cable to video, cable television, and security.
- The most widely used connector for residential cable service.
- Available for indoor and outdoor applications.

> **NOTE:** Coaxial cable is sometimes identified as radio grade (RG) rather than series.

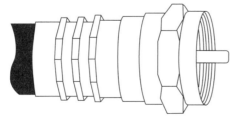

Figure 5.13   Terminated F connector.

N connectors

N coaxial connectors are used in data, radio, and sometimes video applications. The connector consists of a threaded coupling nut similar in design to that of an F connector, except that it is larger.

N connectors have a center pin that must be installed over the cable's center conductor. The N male connector uses a threaded outer collar to mate with the female connector.

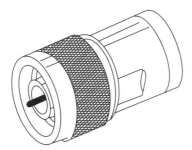

Figure 5.14   N connector.

Bayonet Neil-Concelman (BNC) connectors

Named for its designers, the BNC coaxial connector has been in use since the 1940s. BNC connectors have a center pin that must be installed over the cable's stranded center conductor. BNC connectors are available for 75 Ω products as well as Series 59, 6, and 11. In some cases, video equipment uses a BNC rather than an F connector.

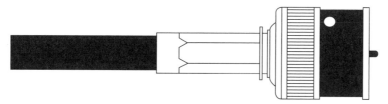

Figure 5.15   Terminated Bayonet Neil-Concelman connector.

Optical Fiber Connectors

Optical fiber connectors perform the same function as copper connectors but use a different medium (light impulses through optical fiber as opposed to electrical signals through a metallic wire conductor). Optical fiber connectors must be precisely aligned to allow maximum signal transfer between the connectors. Most optical fiber connectors require an adapter to precisely align two connectors tip to tip. The adapters are sometimes referred to as couplers or sleeves.

Connectors and adapters provide low-loss coupling by ensuring alignment of the two elements being joined. Any misalignment of optical connectors increases the loss.

The most commonly used connectors in optical fiber systems are the straight tip (ST) and subscriber connectors (SCs), but other connector types such as small form factor (SFF) connectors that are in conformance with ANSI/TIA/EIA-570-A may be used. SFF connectors require less space in the outlet and the DD.

Figure 5.16   Terminated ST connector.

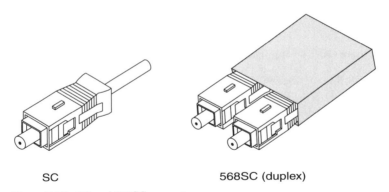

SC   568SC (duplex)

Figure 5.17   SC and 568SC connectors.

6

Consumer Component System Integration

Introduction

In this chapter, basic implementation issues affecting the consumer electronics that will be connected to residential structured cabling systems (RSCSs) are described. This chapter addresses only the most basic and common design and residential cabling installation issues.

While the RSCSs described earlier are by definition generic (i.e., designed without regard to proprietary hardware requirements), the requirements for working with various consumer components are often highly specific to the brand and model of equipment being used.

There are often differences among products of vendors that affect their interoperability. The residential cabling installer must be aware of these unique features, which are beyond the scope of this manual. Vendor-specific training and detailed product manuals should be consulted prior to and during installation of residential electronics.

Infrastructure for Integration of Multiple System Components

The benefits of the residential telecommunications technologies covered in this manual are greatly enhanced by applying an integrated approach to design and installation.

In an integrated approach, all of the residential technologies are connected into a single functional network or system. A large multipurpose distribution device (DD) or separate individual DDs may be used.

Functional examples of an integrated system include:

- Flashing the exterior lights if the home security system is activated to alert police and fire services more quickly that there is a problem in the home.

- Linking the lighting control to the home theater to automatically dim or extinguish lights when the video screen is lowered.

- Lowering the volume of the whole-home audio system when the doorbell or intercom is in use.

- Controlling and managing various home control and security functions remotely via the Internet or telephone connections.

Networking Electronics

There are two main building blocks that are required to provide high-speed data communications within the home. First, broadband Internet access is needed to bring two-way communications into the home. Then, the various computers, printers, and other networked devices in the home must be integrated.

Broadband Internet access

The primary differences between broadband Internet access and the traditional dial-up Internet access available to most individuals over the past decade are:

- "Always-on" functionality.
- Increased bandwidth.

By being always-on, broadband technologies are much more useful to the family. There is no need to wait for the dial-up modem to connect to the Internet service. There is also no need to sacrifice telephone or facsimile (fax) connections in order to access the Internet.

The increased bandwidth to the home made available through a broadband service:

- Makes the retrieval of information off the Internet much faster.
- Supplies the capability for true multimedia, video-rich applications.

File transfers take a fraction of the time required via dial-up technologies, and real-time audio/video services can be implemented.

While broadband data rates vary widely, a general definition suggests that a broadband service should provide always-on, two-way functionality, with data throughput rates significantly exceeding that of dial-up connections.

There are five common technologies used today for receiving and transmitting broadband Internet signals in the home:

- Cable modem
- Satellite

- Digital subscriber line (DSL)
- Wireless
- Fiber-to-the-home (FTTH)

Supported technologies.   All of the broadband technologies are not necessarily available to every residence (e.g., many older communities may not be supported by fiber optic cabling). Wireless broadband access is limited geographically to within a relatively short distance of a transmitting tower. These towers have not been widely deployed throughout the United States. DSL service is widely available, but the technology is available in many different forms, not all of which may be available in all locations.

Data rates.   In general, the higher the data rate, the better the system performance and the quicker the user can "surf the Web." However, there are two factors that make data-rate comparisons between different technologies difficult at best.

The information most often quoted by an access provider (AP) is the maximum data rate allowed by the particular technology. However, the true usable data rate is dependent on many factors. Cable modem technology, for example, uses a coaxial cable infrastructure that is shared by many customers. If multiple customers use large amounts of bandwidth at the same time, the throughput for each customer will be lower than the potential maximum.

Satellite broadband services are also constrained by the number of users on a system who share the limited bandwidth capabilities of a fixed number of satellites. With both community antenna television (CATV) and satellite systems, the total available bandwidth is also affected by the requirements of the video services being transmitted simultaneously with the data stream.

Equally important to the homeowner is not only the data rate bringing information into the home, but also the upstream data rate available for the homeowner to transmit information. It has long been assumed that homeowners require much higher data rates into the home than out of it. This makes sense as long as the majority of the bandwidth-consuming information is generated outside the home. However, as more homes become offices, the upstream bandwidth requirements out of the home will continue to grow. A healthy upstream data rate capability will become a more important determining factor when selecting broadband Internet service.

Multiple locations.   The true power of broadband is its ability to change the lifestyle of each member of the family. Parents can more easily work from home, and children can more readily study from home. The Internet can bring a higher level of entertainment to everyone.

To accomplish this, however, requires access to high-bandwidth broadband services throughout the home. Access can no longer be limited to a single computer or device. While all broadband technologies can technically be set up for use in multiple locations, the homeowner should review the monthly charges that are accessed for additional connections before choosing an AP.

Typically, the AP provides one broadband connection to the residence. Multiple connections can be established by adding electronic equipment (e.g., a router) or obtaining additional service from the AP.

Internetworking technologies

Cable modems.   Cable modems use the CATV cabling system present in most homes to transmit high-bandwidth, broadband Internet communications. The system provides very high data rate downstream (i.e., from the CATV's central office [CO], often referred to as the "headend" to the home). There is also a slower, upstream data path from the home to the headend.

The broadband Internet provider supporting cable modems brings service into the home via the coaxial auxiliary disconnect outlet (ADO) cable, and into the DD. Once in the DD, the cable modem service requires a data interface to turn the incoming broadband data stream into a format that is usable by the home's networked computers and peripherals, commonly known as the cable modem.

Figure 6.1 shows the traditional set-up of a cable modem within the home, linking a single computer to the broadband Internet connection.

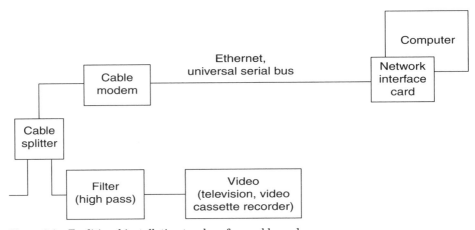

Figure 6.1   Traditional installation topology for a cable modem.

Figure 6.2 illustrates the added benefits gained by the homeowner who installs an RSCS within the home. In this case, multiple personal computers (PCs) and other data devices can share the bandwidth of the system.

In Figure 6.2, a residential gateway (RG) incorporates a router and switch for delivery of broadband Internet services throughout the home. The RG takes the output of the cable modem as its input. The RG then distributes data

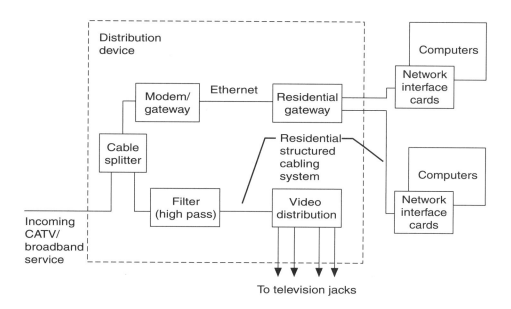

CATV = Community antenna television

Figure 6.2 Installation topology for a cable modem through a residential structured cabling system.

over multiple paths, either as coaxial cable outputs or twisted-pair outputs. Each twisted-pair output then becomes the input to a network interface card (NIC) in a PC, providing a shared portion of the system's total bandwidth to multiple users.

> **NOTE:** The router functionality is also important because most units provide firewall protection against potential external network intruders.

The RG may be mounted in the DD. Short patch cables are then used to couple the outputs from the switch portion of the gateway to the point where the outlet cables are terminated. There will likely be additional hardware, such as NICs, needed in each PC. Software and addressing issues may have to be considered, depending on each manufacturer.

Satellite. A satellite system uses an additional, or modified, dish to draw broadband Internet signals being transmitted via an orbiting communications satellite. The main benefit of such a system is its ability to access the satellite network from virtually anywhere in the country, even in locations that are not served by CATV systems (e.g., new construction).

The installation of a satellite-based broadband system is similar to the installation of a satellite system designed for video programming. A satellite dish is mounted on the home and aimed at the satellite.

Satellite grade coaxial cable (quad-shielded RG-6) is extended between the dish and the DD. Once the data-carrying cable reaches the DD, the remainder of the network connections are similar to those described in a cable modem network.

Digital Subscriber Line (DSL).   DSL technology has as its main goal the provision of broadband Internet speeds and services over the existing base of installed telephone cable. DSL is available in many forms, including high-bit-rate DSL (HDSL), asymmetric DSL (ADSL), and very-high-bit-rate DSL (VDSL). DSL is typically asymmetrical and provides a higher data rate for the downstream path into the home than for the upstream path out of the home.

If it is determined by the telephone company or another AP that DSL service can be brought to a particular home, the residential cabling technician should only be concerned about how to properly install the DSL service within the home.

One possible implementation is shown in Figure 6.3. This design is efficient when the home has a DD and an RSCS adhering to the criteria outlined in ANSI/TIA/EIA-570-A, *Residential Telecommunications Cabling Standard* and this manual.

DSL can also be run over the home's existing telephone cable. However, experience has shown that a high percentage of existing cabling may not adequately support DSL throughput. If new cable is needed, Category 5e twisted-pair cable should be installed, per ANSI/TIA/EIA-570-A. It is also possible to run both DSL and plain old telephone service (POTS) signals over the same twisted-pair, but noise can be coupled from the DSL portion of the service to the POTS portion of the service. Separate cables are recommended; however, filters may also be placed to mitigate the noise problem.

Wireless.   Fixed wireless systems use a transmitting tower mounted in a residential community. The tower is used to transmit and receive high-bandwidth

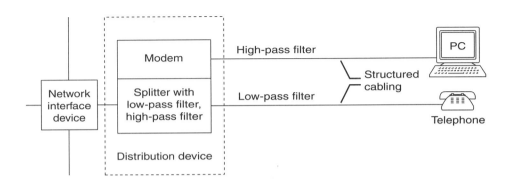

PC = Personal computer

Figure 6.3   Installation of a digital subscriber line modem.

Internet services to individual homes. Each home is provided with a separate wireless transmitter/receiver to communicate with the main system.

Some wireless solutions use an antenna mounted to the exterior of the home to receive the broadband signals from the central antenna, in much the same way as satellite systems work. Cable then connects the antenna to the DD.

Other wireless solutions use a desktop mounted receiver, much like the wireless telephone cradle found in many homes, to communicate to the base station. Twisted-pair cable then runs from this cradle to the DD for further distribution of the signals throughout the home.

Fiber-to-the-Home (FTTH).   Optical fiber cable can transmit enormous bandwidth over very long distances. Optical fiber cabling systems are also immune to electromagnetic interference (EMI) and are highly reliable, thus lowering long-term maintenance costs.

The main benefit of optical fiber in the cabling system, however, is the high data rate. Newly installed optical fiber cabling from a headend or CO to the home can provide the highest possible data rates and support intensive applications (e.g., high definition television [HDTV]). The symmetric (i.e., high bandwidth in both the upstream and downstream directions) nature of the optical fiber cabling systems can support home offices that in the future will transmit much more data than at present. In short, optical fiber cabling may prove to be the most popular medium for delivering information to the home of the future.

Residential Gateways (RGs)

RGs perform a number of functions. Used in conjunction with structured cabling, they provide a powerful and straightforward solution for distributing content and providing access to various communications media. There are less obvious functions that they perform as well, including to create and interconnect subnetworks. For example, a residential gateway can be used to link a home automation subnetwork to local PCs and the Internet.

As one of its functions, an RG may be designed to provide a firewall between the data kept within the home and the data being transmitted through the broadband Internet connection. Most agree that firewall protection is a common requirement of these products.

Another common feature of RGs is the means to distribute a single broadband Internet connection to multiple devices within a home.

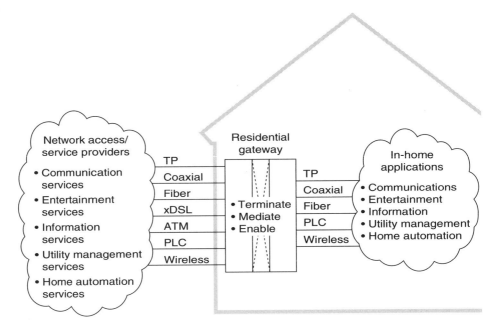

ATM   = Asynchronous transfer mode
Fiber = Optical fiber
PLC   = Power line carrier
TP   = Twisted-pair
xDSL   = xDigital subscriber line

Figure 6.4   Example of a residential gateway.

7

Transmission Fundamentals and Cabling Topologies

Transmission Fundamentals

This chapter describes the terms, methods, and issues related to the transmission of information over various media. It also identifies the impact of telecommunications cabling installation methods. The residential structured cabling system (RSCS) is described, and the various arrangements, or topologies, in which it can be configured are outlined.

Transmission, as described in this chapter, is the movement of information as electrical or optical signals from one point to another via a medium. The medium can be air, copper cable, optical fiber cable, or whatever else might be used to carry the signal. The most commonly used media today are copper conductors and glass optical fiber strands that pass transmission signals.

The choice of a specific medium is influenced by economics and technical considerations, such as the:

- Bandwidth requirements.
- Type of services to be provided (e.g., voice, data, and video [VDV]).
- Size of the network.
- Transmission path.
- Distance.

All media have certain characteristics that limit their performance. Media selection should be based on current and future transmission and bandwidth requirements.

To ensure compliance with cabling standards, manufacturers of copper and optical fiber cable have instituted design changes in the manufacture of their products as the standards have continued to require higher transmission performance. On the copper side, the assignment of a category performance level rating (Category 3, Category 5, and Category 5e) provides a simple means to select a cable construction suitable for the modest requirements of a voice circuit or one meeting the strict demands of high-speed data transmission. Optical fiber is available in multimode and singlemode in which multimode is more than sufficient for bandwidth requirements in a home. Coaxial cable is specified with a series number and is typically used for delivery of community antenna television (CATV) and satellite signals.

Terminology

This chapter defines the major terms associated with electronic transmission. The objective is to show the impact that improper residential cabling installation can have on media transmission characteristics.

Power

Power (P) in watts (W) is a term that applies to the energy that is consumed to operate an electrical device such as a motor, amplifier, or telephone transmitter. Its main components are voltage (V) in volts and current (I) in amperes—there is no power unless both are present. Voltage may be thought of as being similar to water pressure, while current is similar to the quantity of water delivered. Millions of volts with no current will not provide power, and thus will not energize a device to perform the desired function.

Power is equal to the voltage multiplied by the current, or $P = VI$.

Alternating Current (AC) and Direct Current (DC)

The standard commercial power frequency in the United States is 60 hertz (Hz [cycles]). Thus, ac voltage completes 60 sine wave cycles per second.

For each sine wave in ac, the voltage begins at zero level and increases to its maximum positive value before dropping back to cross the zero level, where the voltage becomes negative or changes direction. The voltage continues in the negative direction to reach a maximum negative value; it then becomes less negative until it again reaches the zero level and completes the cycle (see Figure 7.1). Although the power has gone to zero twice during each cycle, the eyes are not sensitive enough to perceive the rapid increase and decrease of the power and the accompanying flicker of reading lights or lamps.

Direct current (dc) refers to a steady value that does not change the direction of voltage or current flow. Direct current rises to its maximum value when switched on and remains there until the circuit is interrupted. A battery is an example of a dc source.

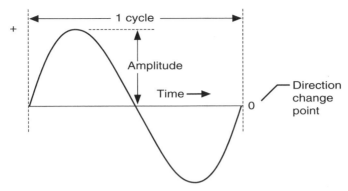

Figure 7.1   Analog sine wave.

Frequency

Frequency is defined as the number of cycles or sine waves occurring in a given time period. If the unit of time is equal to one second, the frequency is stated in hertz (Hz). One Hz is equal to one cycle per second.

- 1 cycle in 1 second = 1 Hz
- 60 cycles in 1 second = 60 Hz
- 1000 cycles in 1 second = 1 kHz (kilohertz)
- 1 million cycles in 1 second = 1 MHz (megahertz)
- 1 billion cycles in 1 second = 1 GHz (gigahertz)

Although humans can hear frequencies that range from 20 Hz to 20,000 Hz, typical voice-grade copper transmission lines are generally limited to the range of 300 to 3400 Hz, which covers the majority of the spoken voice range.

Analog

Analog transmission uses continuously varying electrical signals that correspond to the changes in loudness (amplitude) and frequency (tone) of the input signal (the human voice). An example of analog transmission is where the transmitter of a telephone converts sound waves to an analog electrical signal that varies in amplitude (signal strength) directly with the loudness of the talker and varies in frequency directly with the tone of the talker's voice. See Figure 7.2.

Digital

A digital signal is a discontinuous signal that changes from one state to another in a limited number of discrete (single) steps. In its simplest form, there are two states or signal levels—on and off, where the "on" translates to a digit "1"

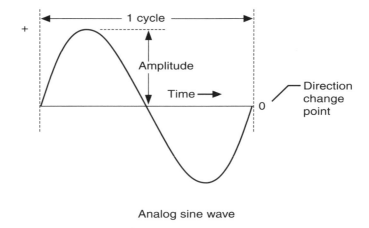

Analog sine wave

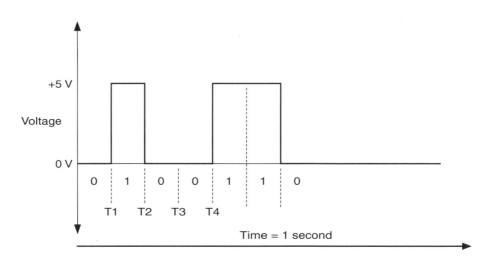

Digital pulse train

Figure 7.2   Analog sine wave vs. digital pulse train.

and "off" is seen as a zero level corresponding to the digit "0." Other means can also be used to represent the two states (e.g., a positive and negative voltage).

The digital message is made up of a sequence of these digital pulses (or bits) transmitted at regular and defined time intervals if the signal is synchronous. These bits are usually square in shape. Thus, digitally transmitted information is an encoded representation of input, unlike analog transmission, which reproduces an analog input in both frequency and amplitude.

If the input is analog, the transmission of a digital signal generally involves a sampling process in which the analog voltage is sampled often enough that, when replicated at the receiver, there is an acceptable facsimile of the original signal.

Bits and bytes

The term bit, a shortening of the expression binary digit, is a single on or off state, representing a "1" or a "0." Because only two states are possible, such a transmission scheme is described as binary. One bit cannot convey much information, so bits are commonly grouped together and transmitted as strings or sequences of 1s and 0s. A sequence of eight bits is called a byte, and the byte is accepted universally as the standard package size for data transmission.

Eight binary or two-state bits can be arranged in 256 different combinations, which is more than adequate for conveying alphanumeric character sets used in the Western world. The capital letter "A," for instance, might be conveyed by the sequence: 00000001; the lowercase letter "a" might then be encoded as: 00000010.

Digital transmission speed

The correct measurement term for the rate of flow (speed) of binary digit (bit) information over a transmission line is bits per second (bps or b/s). Some incorrectly use the term "baud" when describing this information flow rate. The baud technically is the signal-change rate on the line, and is the rate at which the signal changes from one state to the other. While in a simple binary digital system the baud and bit rates are equal, the terms are not interchangeable. For example, a system that uses one timing bit for every four bits of data may be set to operate at 125 megabaud while only transmitting 100 Mb/s of information.

The transmission of digital data may be accomplished on a direct digital basis or with modulators/demodulators (modems). Modems first convert the digital 1s and 0s into two separate frequencies, one representing a 1 and the other a 0, for serial transmission over telephone lines; the modem then reverses the process at the destination. Totally digital lines do not need any conversion and transport the information bits in the native binary digital mode.

Synchronous and asynchronous transmission

Another fundamental distinction between methods of conveying information is that between synchronous and asynchronous transmission.

Synchronous transmission occurs when each bit of data is transmitted during the same time period, regulated by a timing pulse from a transmitter. A steady flow of bits can be the most efficient way of transmitting certain types of information (e.g., large files of data).

Asynchronous transmission takes place when the information to be conveyed is "bursty" or erratic in its timing. A telephone conversation is a good example of an asynchronous situation—long periods of silence (in terms of the speeds at which computers operate) are sandwiched between the spoken words of the two speakers.

Megahertz (MHz) and Megabit (Mb)

It is important to highlight the relationship between Hz and bits, or MHz and Mb, since the two are sometimes confused and may erroneously be used interchangeably.

MHz, one million hertz, as in 100 MHz Category 5 cable or 500 MHz optical fiber cable, refers to the bandwidth of the cable. Coaxial cable is usually measured in the GHz range. This relates to the information-handling capability of the media. Another way to view it is as the size of the pipe. Mb, or one million bits of information, refers to the number of bits of information that can flow through the pipe.

Decibel (dB)

A measure of analog signal strength is the bel, named in honor of telephone pioneer Alexander Graham Bell. The term is inappropriately large for telecommunications work, so dB (1/10 of a bel) is used. It is not an absolute value. It is the ratio of two signals and may be used to compare power, current, or voltage.

The dB is a logarithmic function that allows large variations to be shown in very small increments. With respect to residential cabling installation efforts, it is most important to understand that a 1 dB change under ideal conditions can be heard.

Within the telephony industry, a standard-level ratio of measurement is 0 dBm. This represents a zero (0) dB reading from one milliwatt of power at a frequency of 1004 Hz across an impedance of 600 ohms.

Increases or decreases of 3 dB will result in a doubling or halving of the power, respectively. Increases or decreases of 6 dB will cause a doubling or halving of the voltage.

Increases or decreases of 10 dB for power and 20 dB for voltage or current will cause a change of ten times (x10). This moves the decimal point one place.

Decibels are commonly used to measure the following cabling characteristics:

- Insertion loss (formerly known as attenuation for copper systems while optical fiber systems use the term attenuation)—The smaller the dB number, the better.

- Connector insertion loss—The lower the dB number, the better.

- Crosstalk loss (copper systems)—The larger the dB number, the better (as a negative number).

- Structural return loss (copper systems)—The higher the dB number, the better.

Copper Cable Media

With any copper cable, the proper signal level must be coupled through the media from transmit to the receive end and properly drive the receiver. The most effective and efficient transfer of energy occurs at interfaces and connection points where all connections have the same characteristic impedance.

This chapter describes the electrical properties that make up characteristic impedance and how characteristic impedance can be impacted by hardware, design, and cabling installation methods or procedures.

Copper wire size

For metallic cables, the electrical energy of the signals is transmitted over copper conductors. Copper is preferred since it is a better conductor of electricity than most other metals and is relatively economical. Copper is readily converted into a round wire through a drawing process whereby it is drawn, or pulled, through a series of successively smaller round holes (dies) until the desired diameter is reached.

In the United States, copper conductors are categorized through an accepted numbering system called American wire gauge (AWG). The AWG number assigned to a particular wire size approximates the number of steps required to draw it to that diameter. For example, a 14 AWG (1.6 mm [0.063 in]) wire is drawn through 14 different-sized die openings during the drawing process.

The smaller the wire diameter, the larger the gauge number. Conductors larger than 1 AWG (7.4 m [0.29 in]) are labeled 0, 00, 000, and 0000, and are called "one ought," "two ought," "three ought," and "four ought." Conductors larger than 0000 are measured by their diameter referenced in circular mils; the customary abbreviation is kcmil (thousand circular mils).

Bandwidth

Bandwidth is defined as the range between the highest and lowest frequencies used for a particular function over a given distance. Although humans can hear portions of the frequencies from 20 to 20,000 Hz, speech transmitted by analog telephone systems carries voice signals between 300 and 3400 Hz, which reflects the majority of what one actually hears and recognizes as speech.

V = Voltage in volts
 I = Current in amps
R = Resistance in ohms

Figure 7.3   Ohm's law.

Ohm's law

The resistance to the flow of electrical energy is measured in ohms (Ω). By definition, one Ω is the resistance that allows one ampere, or amp, (electrical current unit) to flow when one volt is applied. This is known as Ohm's law and can be stated mathematically as:

$$V = IR$$

Where:

$$V = \text{Voltage in volts}$$
$$I = \text{Current in amps}$$
$$R = \text{Resistance in ohms}$$

This formula can be restated to calculate any one value when the other two are known.

- Ohm's law is usually shown as in Figure 7.3. Values are easily solved for reference. Place your finger over the I and you see V is over R. Place your finger over the R and you see V is over I. Place your finger over the V and you see that I is next to R.

- Higher temperatures increase the conductor resistance by approximately two percent for each 5.5 °C (10 °F) rise.

- Resistance is increased in proportion to any increase in length (e.g., doubling the length of a cable doubles its resistance).

- Higher frequencies cause the resistance to increase (e.g., increasing the frequency by four times results in a doubling of the resistance).

Inductance

Inductance, measured in henrys (H), is the property of an electrical force field built around a conductor when current flows through it. When dc current flows through a conductor, the field is steady. An ac current causes the lines of force to constantly build and collapse.

Inductance is a resistance to a change in current. Inductive coupling is the transfer of energy from one circuit to another (e.g., power lines on a utility pole can inductively couple a power surge onto telephone cables).

Capacitance

Capacitance, measured in farads (F), is the tendency of an electronic component to store electrical energy. Capacitance is the resistance to a change in voltage. Pairs of wires in a cable tend to act as a capacitor, which has two conductors, or plates, that are separated by a dielectric.

Cable normally exhibits some level of capacitance. Typically, twisted-pair cabling will have a value of 15 to 20 picofarad per foot (pF/ft).

Capacitance in cable is considered undesirable and must be minimized. If cable is improperly installed, capacitance can be affected. Because it is an important element of the cable's characteristic impedance, changes in capacitance and inductance can degrade the quality of transmission through the cable.

Characteristic impedance

Characteristic impedance is made up of three components:

- Capacitance
- Inductance
- Resistance

Every cable or transmission line has characteristic impedance. The value of the characteristic impedance is determined by the diameter of the conductors and the insulating value (or dielectric constant) of the materials separating them.

Cables are designed to achieve constant characteristic impedance independent of cable length. For example, if a 30 m (100 ft) length of cable has characteristic impedance of 100 Ω, it will still have a value of 100 Ω if the length is doubled or tripled for a given frequency. (The Ω symbol is the Greek capital letter omega; it is traditionally used to represent the electrical unit ohm.)

In general, the most important point to remember about characteristic impedance or impedance, is that all components of a system must have the same value of input or output impedance if maximum signal (energy) transfer at the interface point is to occur. Impedance matching, as this procedure is called, becomes even more critical at higher frequencies.

Twisted-pair and screened twisted-pair (ScTP) cables used for residential cabling exhibit a characteristic impedance of approximately 100 Ω at 1 MHz and above.

Insertion loss

Because insertion loss causes loss of usable signal or power to the load or receiver, the lower the dB number the better. Higher insertion loss means less available signal.

There are times when an attenuator is added so that the signal is not too high at the receiver end. This is called padding down the circuit.

Return loss

Return loss results when impedance discontinuities exist along the cable and at the termination points. Such discontinuities result from connectors and from variations within and between cables. Such variations are expressed in terms of return loss. Minimizing impedance variations along the transmission line reduces the power reflected back to the transmitter.

Insulation

Copper conductors must be physically separated from each other. In a single pair of conductors, contact of the two conductors (short circuit) prevents the signal from traveling down the transmission line. The choice of material to cover the copper is one that involves economics as well as a trade-off in the characteristics desired for the application and installation environment.

An electrically efficient insulation is nearly always desired, but a trade-off may be required (e.g., to obtain insulation capable of meeting plenum cable requirements).

> **NOTES:** Efficient insulation is defined as a material where any losses of the transmitted signal due to losses associated with the insulation are minimal.
>
> Insulation common in residential cabling consists of such materials as polyethylene, polyvinyl chloride, and fluorinated ethylene propylene. The type selected can affect the physical size of the completed cable and determine mutual capacitance of a pair.

Shielding

The specific shielding material used, its thickness, and relative coverage (number of holes) influences the degree of effectiveness. Types of shields include copper braids, metallic foils, and solid tubing. The most common use of shielded cabling in a residential application is for audio signal transmission.

The shield is employed to:

- Reduce the level of radiated signal from the cable.
- Minimize the effect of external electromagnetic interference (EMI) on the cable pairs.

> **IMPORTANT:** Shields are effective only when continuity is assured and they are properly grounded.

Pair twist

In order to comply with the transmission quality level demanded of telecommunications cables installed in residential applications, the two conductors of an insulated copper conductor pair are uniformly twisted together. This twist (or lay) varies and is determined by the cable manufacturer.

Both the effects of capacitance unbalance and electromagnetic induction are improved by twisting of the pairs.

Balanced and unbalanced cables and baluns

There are two fundamental methods of sending signals over a copper transmission media: one is described as balanced and the other as unbalanced. A balanced transmission path is one in which the same signal is carried over two conductors of opposite polarity. At the receiver, a differential amplifier subtracts the negative signal value from the positive, doubling the signal strength.

In the United States, the primary balanced media in use in the telecommunications industry is the unshielded twisted-pair. Balance provides a level of noise immunity in lieu of the shielding used in more elaborate and expensive cable constructions.

An unbalanced transmission path relies on a single conductor to carry the signal. This is the case in coaxial cable, where the central conductor conveys the information.

It is often necessary in a network to convey both balanced and unbalanced signals. Video signals, for instance, are often carried by unbalanced coaxial cable, while data is usually transmitted over balanced twisted-pair cable. At junctions where one form of transmission must be converted to the other, a special device is required. This device is called a balun, a shortened form of the terms balanced and unbalanced, because it converts balanced to unbalanced and unbalanced to balanced signaling formats.

The majority of residential consumer audio/video equipment uses unbalanced signaling for audio and video, including super video home system (S-video). For long interconnecting cabling runs, baluns must be used.

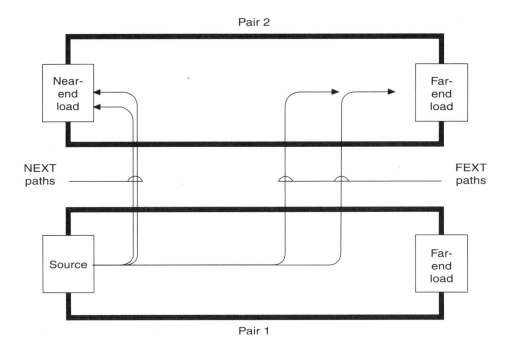

FEXT = Far-end crosstalk
NEXT = Near-end crosstalk

Figure 7.4 Crosstalk loss paths.

Crosstalk loss

The unwanted transfer of signal from one or more circuits to other circuits is called crosstalk loss. This transfer may be between conductor pairs in close proximity, or between adjacent cables. Crosstalk loss is decreased by pair twists, cable lay, shielding, and physical separation during the cable manufacturing process.

Two types of crosstalk loss are of interest for twisted-pair copper cable systems—near-end crosstalk (NEXT) loss and far-end crosstalk (FEXT) loss. See Figure 7.4.

Electromagnetic compatibility and interference

Not all interference affecting electrical signals being carried by copper conductors comes from other, nearby electrical supply cables. Noise can also be coupled onto conductors from the environment (e.g., electrical motors, transformers, or the ballasts of fluorescent lighting). Radio frequency interference (RFI) is generated by transmitters operating in the radio spectrum.

A particularly powerful but short-lived source of interference, as well as potential damage to electronic equipment and circuits, is the electromagnetic pulse (EMP). The most common source of such pulses is lightning strikes to power-distribution infrastructure (e.g., aerial power lines). These powerful electrical surges may then be coupled onto telecommunications lines and transmitted into buildings.

The Federal Communications Commission (FCC) has established guidelines for the manufacture of electronic and communications equipment that are designed to shield such equipment from this kind of environmentally coupled interference. These products are sometimes described as complying with the FCC's electromagnetic compatibility (EMC) directive. EMC establishes acceptable limits for electromagnetic emissions from an electronic device, as well as specifying the amount of EMI such a device should be able to tolerate.

EMI and RFI can be minimized by the selection of appropriate cabling components and by proper installation techniques. Shielded cable, for instance, may be more resistant to chronic, high-intensity interference than unshielded cable. As to residential cabling installation procedures, proper grounding is essential in reducing noise and interference. In particularly noisy environments, optical fiber may be a better choice; since fiber conveys light pulses rather than electrical signals, it is immune to EMI and RFI.

Common sources of RFI are:

- Radio transmitters.
- Television stations.
- Cordless telephones.
- Radar television stations.
- Portable two-way radios or any radio transmitting device.

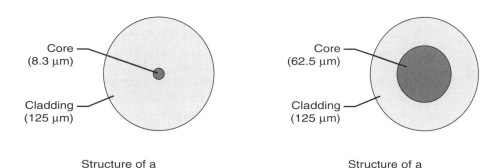

Structure of a
singlemode optical fiber

Structure of a
multimode optical fiber

Figure 7.5   Structure of optical fiber—singlemode and multimode.

Optical Fiber Cabling Media

Transmission of information through optical fiber cables is free from some of the detrimental effects inherent with copper cables (e.g., crosstalk loss, noise, interference from lightning, and EMI problems). However, attenuation (loss) of the transmitted signal along the cable route and environmental considerations are of concern for both copper and fiber systems. Optical fiber attenuation is significantly less than copper.

The primary difference between copper and fiber as transmission media is that pulses of light composed of photons are injected into the fiber instead of electrical energy such as is used in a copper-based cable. "On" and "off" digital light pulses are generated by either lasers or light-emitting diodes (LEDs).

The inner portion of the glass fiber used to carry the light pulse is called the core, and the surrounding glass layer is called the cladding. The cladding confines the light within the core by creating a reflective zone.

Singlemode fiber systems are frequently used for long-haul transmission lines, extremely high-speed data systems, and fiber-based video distribution systems. The lasers used to drive the signal over singlemode fiber are generally more expensive than LED- and vertical-cavity surface-emitting laser-based multimode systems, providing a major economic incentive for the continued use of this medium.

Multimode cable is described in this manual since it is the type normally selected for short distances (< 2 km [1.2 mi]).

There are two characteristics of particular importance in the transport of information over fiber media—bandwidth and attenuation.

Bandwidth and dispersion

The bandwidth of an optical fiber provides a measure of the amount of information a fiber is capable of transporting at a given error-free bit rate. Bandwidth for multimode fiber is described in terms of MHz•km. Increasing either the length of the cable or the modulation frequency of the light source decreases the bandwidth and lessens the information-carrying capacity of the transmission path. The bandwidth measurement is normally the maximum carrying capability of the fiber. One would not increase the modulation (frequency) unless one first decreased the distance. The bandwidth measurement is based on 1 km (3280 ft) of glass.

Dispersion is the widening or spreading out of the modes in a light pulse as it progresses along the fiber. Should the pulse widen too much, it can overlap with other pulses at the receiver and make it impossible to distinguish one pulse from another. As errors occur in reading the pulses (which represent data bits), there is an increase in the bit error rate (BER). Bandwidth is limited by total dispersion, which is the sum of modal dispersion and chromatic dispersion.

Modal dispersion in multimode fibers is the major factor contributing to total dispersion. It is the result of the different lengths of the light paths taken by the many modes (multimode) as they travel down the fiber from source to receiver. Since portions of the signal arrive ahead of or following other portions of the signal, an individual light pulse may spread out (take too much time) in transmission, making its identification at the far end questionable.

Chromatic dispersion begins at the light source. The source creating the light pulses is either a laser or an LED. Neither furnishes perfectly monochromatic (single wavelength) light, so the light injected into the fiber medium contains a number of slightly differing wavelengths. Since the index of refraction of the glass fiber is not the same for different wavelengths, each wavelength travels through the fiber at a slightly different speed. Modal dispersion tends to further broaden the pulses, and together these dispersion factors increase the BER and lower the effective bandwidth. Bandwidth and dispersion are inversely proportional.

Modal dispersion affects multimode and chromatic dispersion affects single-mode.

Attenuation

Light pulses in optical fiber cables are also subjected to a loss in power as they travel along the fiber. This attenuation occurs:

- As a result of absorption of power, and
- Because of scattering of the light by impurities within the glass itself.

Loss due to scattering is caused by such factors as:

- The glass material.
- Macrobends and microbends in the strands.

Fiber attenuation is measured in decibels per kilometer (dB/km). The measurement is proportional to cable length and is affected by the wavelength of the light used for the pulse.

The improper installation of optical fiber cable is the cause of most unexpected attenuation. Maintaining adequate bend radius, using the proper cable for the application, and conducting the work as defined by standards and manufacturer guidelines are musts in order to provide a fully functional fiber transmission system.

Electrical Protection Systems

Overview

This chapter provides an overview of electrical protection, grounding, and bonding for both the single-family and multi-dwelling unit (MDU). Protection, grounding, and bonding requirements are somewhat different for the single-family residence than they are for the MDU. Initially, single-family protection was typically provided by the access provider, but with the advent of residential structured cabling systems (RSCSs) and networking, secondary protection and a single point ground are necessary to protect sensitive home computing, home automation, and consumer electronics. This protects against sneak currents, power surges, and floating ground potential.

The MDU is considered to be a commercial building, and falls under the jurisdiction of the *National Electrical Code® (NEC®)* and should also be installed to meet the requirements of ANSI/ TIA/EIA-607, *Commercial Building Grounding and Bonding Requirements for Telecommunications*, the grounding and bonding standard for commercial buildings. Because of their complexity, this chapter will focus on the electrical protection, grounding and bonding concepts, equipment, and procedures as they apply to MDU buildings (e.g., apartment complexes and high-rise condominiums).

A primary responsibility of the residential cabling installer is safeguarding people, property, and equipment from "foreign" electrical voltages and currents. Foreign refers to electrical voltages or currents that are not normally carried by, or expected in, the telecommunications distribution system.

The results of such disturbances could be:

- Death or injury.
- Destruction of electronic equipment.
- Downtime of equipment.
- Degradation of telecommunications signals.

Grounding, bonding, and earthing are terms used to define the practice of connecting all components of a system together to a main building ground electrode, for the purposes of reducing or eliminating the differences of potential between all of the utilities inside a structure.

A grounding conductor is defined by the *NEC* as:

"A conductor used to connect equipment or the grounded circuit of a wiring system to a grounding electrode or electrodes."

Bonding is defined by the *NEC* as:

"The permanent joining of metallic parts to form an electrically conductive path that ensures electrical continuity and the capacity to conduct safely any current likely to be imposed."

The combination of grounding and bonding culminates in a system that equalizes the difference of electrical potentials among all components of a telecommunications network to as close to zero volts as possible.

Grounding and bonding provide additional safety factors where equipment and people are involved. They protect people from being shocked by voltage potentials and provide a point of discharge for static electricity used by technicians prior to working on electronic equipment. In telecommunications installations, grounding and bonding also reduce or eliminate stray voltage and current that might interfere with signals traveling through telecommunications cables.

Grounding and bonding also reduce the effects caused by lightning, static electricity, and ground faults in electrical equipment. Properly grounding the shields of cables can help reduce noise and alien crosstalk loss from adjacent cables.

Electrical grounding and bonding are covered throughout the *NEC*. For telecommunications grounding and bonding requirements, refer to Article 800, which references other articles in the *NEC* that should be reviewed as needed.

Keep in mind, however, that the *NEC* is written primarily for personal safety and the protection of equipment. Manufacturers may also require additional bonding and grounding. Always review the manufacturer's specifications for equipment grounding and bonding instructions. When a conflict exists between a code and a manufacturer's specification, request an interpretation from the local authority having jurisdiction (AHJ) or ask for a variance.

Safety in Grounding and Bonding

Avoiding electrical shock

The most common electrical shock occurs from inadvertent, accidental contact with energized devices or circuits. Other common conditions that create potential shock hazards are:

- Touching a faulty or improperly grounded electrical component.
- Standing on a damp floor.
- Poor clearance.
- Using or being near conducting material during a lightning storm.

Preventing electrical shock

Until verified, always assume equipment is energized.

Residential cabling installers must be especially watchful for irregular or abnormal conditions during the construction phase of a project. Use prudent electrical safety measures, such as insulated rubber gloves for personal protection, and use appropriate test equipment to verify the presence or absence of dangerous voltages on all exposed:

- Cables.
- Conductors.
- Metal.

Properly installed residential cabling is almost never dangerous. During installation verify that exposed conductors, cable shields, and metal equipment are grounded or free of fault potentials, and otherwise generally safe.

Overall electrical protection must take into account:

- Direct lightning strikes.
- Ground potential rise.
- Contact with power circuits.
- Induction.

Examples of some conditions causing these kinds of disturbances are shown in Figure 8.1.

Additional safety practices and information relating to electrical shock can be found in Chapter 9: Safety.

Standards and Local Code Requirements

Always review the local code requirements before proceeding with an installation. This review includes what edition of the code is adopted and what, if any, exceptions to the code are adopted by the AHJ. Most of the code requirements

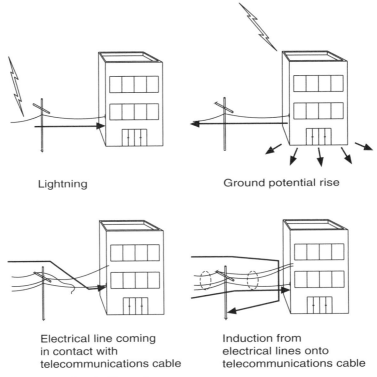

Lightning

Ground potential rise

Electrical line coming
in contact with
telecommunications cable

Induction from
electrical lines onto
telecommunications cable

Figure 8.1   Electrical hazards.

for the job should be included in the telecommunications distribution designer's documents. The residential cabling installer should never take this information for granted, since the contractor is fully responsible for all work on the project.

If no code has been adopted locally, consult with the state fire marshal's office to determine which AHJ is responsible for that geographic area and what codes are in effect. Do not depend on other residential cabling installers, contractors, or even company personnel in making these determinations.

Except when local codes are in conflict, follow the national codes and standards. Familiarize yourself with the *NEC* and ANSI/TIA/EIA-607. Determine whether manufacturers have requirements that exceed the *NEC*, ANSI/TIA/EIA-607, and local code requirements. If a conflict exists, obtain an interpretation from the code AHJ.

Local code information can be obtained from the local AHJ charged with the responsibility for code enforcement. Some local government agencies may not have adopted a code, in which case state government agencies may be charged with implementing code restrictions or enforcing the *NEC*. The AHJ is the only entity that can interpret the code and make exceptions of the code. Do not rely on code information from other contractors or people who claim to know the local restrictions. They may not be totally familiar with the local requirements.

Electrical Exposure

Exposure in telecommunications

In addition to personal protection from electrical hazards, there is the question of the telecommunications system's susceptibility to such hazards. This susceptibility is usually minimized by the installation of electrical protection systems on both private residences and MDUs. The main components of these systems are twofold: primary and secondary protectors. Protectors located on the exteriors of buildings are, however, only the first step in full electrical protection. The telecommunications system within the building must also be properly grounded and bonded.

Section 800.30(A) of the *NEC* covers safety code requirements for protectors. Additionally, it defines "exposed" as the situation that occurs when a circuit is in such a position that, in case of failure of supports or insulation, contact with another conductor may result.

The *NEC* requires a listed primary protector (at both ends of a cabling path) whenever outside plant (OSP) cable may be exposed to lightning or accidental contact with power conductors operating at over 300 volts (V).

Exposure refers to an outdoor telecommunications cable's susceptibility to electrical power system faults, lightning, or other transient voltages. A cable is also considered exposed if any of its branches or individual circuits is exposed.

Lightning exposure

Lightning strikes can cause severe damage to telecommunications systems that have not been properly installed. Even with a properly installed grounding infrastructure, there is no guarantee that a direct lightning strike will not damage a system. However, it will have a better chance of sustaining uninterrupted and undamaged operation than will an improperly installed system.

A Lightning Exposure Guideline is included as a fine print note (FPN) in the *NEC*'s Section 800.30(A), FPN No. 2. It states, "Interbuilding circuits are considered to have a lightning exposure unless one or more of the following conditions exist:

1. Circuits in large metropolitan areas where buildings are close together and sufficiently high to intercept lightning.

2. Interbuilding cable runs of 42 m (140 ft) or less, directly buried or in underground conduit, where a continuous metallic cable shield or a continuous metallic conduit containing the cable is bonded to each building grounding electrode system.

3. Areas having an average of five or fewer thunderstorm days per year and earth resistivity of less than 100 ohm-meters. Such areas are found along the Pacific coast."

If cable exposure is in question, consider it exposed and protect it accordingly.

Additional exposure factors are:

- Aerial cable usually has power cables routed above it that will intercept and divert direct lightning strikes. This can help but does not negate the need for protectors.

- Buried cable collects ground strikes within a distance determined by soil resistance (typically 1.8–6 m [6–20 ft]). High soil resistance intensifies this problem. Without the proper protection, a system could be receiving repeated ground-strike surges without the evidence of damaged cable associated with an aerial strike.

- Rooftop antennas, such as:

  - Aerial television (TV) antennas.

  - Direct broadcast satellite (DBS) dish antennas.

  - Radio frequency (RF) antennas.

Lightning is so powerful and unpredictable that the best insurance against damage is a properly grounded and bonded system. Lightning may strike at any time with the potential of:

- A direct current charge, pulsating between 100 kHz and 2 MHz.

- Greater than 10 million V.

- An average of 40,000 amps and can peak as high as 270,000 amps.

- Temperatures in excess of 27 760 °C (50,000 °F).

Grounding and earthing systems must be designed and installed that can safely carry these unwanted voltages to the earth.

There are several types of electrical protection systems within every commercial building. Although the residential cabling installer is not usually responsible for systems other than the telecommunications grounding and bonding system, it may not be safe to work in a building without recognizing and understanding the purpose of each system. Although electrical protection systems for commercial buildings are beyond the scope of this manual, they are covered in detail in BICSI's *Telecommunications Distribution Methods Manual (TDMM)* and *Telecommunications Cabling Installation Manual (TCIM)*. Refer to these resources for more information if regularly involved in commercial building cabling installation.

Exposure in building structures

NEC Article 100, *Definitions*, defines a ground in much the same way as does ANSI/TIA/EIA-607. Likewise, *NEC* Article 100 and Section 250.70 define bonding. Specific bonding requirements are found in Part V of Article 250 and in other sections of the Code as referenced in Section 250.4."

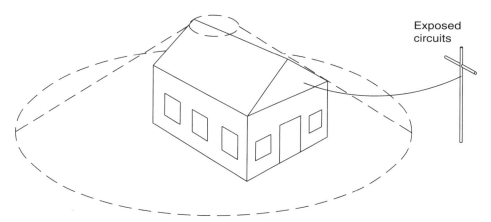

Figure 8.2   Cone of protection.

Bonding conductors are not intended to carry electrical load currents under normal conditions, but must carry fault currents so that electrical protection, (e.g., circuit breakers) will operate properly.

Another important safety application for bonding is to limit hazardous potential differences between different systems during lightning or power faults. In this case bonding protects against arcing between different system (or equipment) grounds and safeguards people who may be exposed to both systems at the same time. For example, an individual talking on a telephone handset (other than a wireless handset) and operating a personal computer at the same time could be exposed to a potential difference between the telephone and electrical power lines.

Protector Technology

Primary protector

There are many different components used by manufacturers to implement protection functions. The following components are typical of primary protectors.

Carbon block protector.   The carbon block is the original protector. An air gap between carbon elements is set to arc at about 300 to 1000 V and conducts surge current to a grounding conductor. When the surge current finally drops low enough, the arc stops and the protector resumes its normal isolation of ground.

A fail-safe function causes carbon blocks to short permanently to ground when an extended hot surge or permanent fault current overheats them. They are typically installed as pairs and are the lowest-cost option; however, carbon blocks tend to wear out quickly under extreme conditions and can cause leakage and noise in voice circuits.

Gas tube protector. The gas tube is an improved arrester that basically operates in the same way as carbon blocks, arcing over a gap to a grounding conductor. It has a wider gap because it contains a special gas and is of higher reliability than the carbon block. It also has a tighter tolerance on arc breakdown voltage and is typically set to arc at a lower voltage, providing better protection than carbon blocks.

Another type, the dual-gap gas tube, provides a common arc chamber that grounds both wires of a pair together and minimizes metallic surges that would otherwise occur from the operation of individual arresters.

Solid-state protector. The solid-state protector is the newest type of arrester. It relies on high-power semiconductor technology. Though more expensive than either the carbon block or gas tube, the cost is recovered over the extended life of the protector. These devices are fast-acting and well-balanced, and they do not deteriorate with age below a rated maximum surge current.

Fuses and fuse links. The *NEC* identifies the two types of primary protectors as fused and fuseless. In a case where overcurrent is of extended duration, the exposed side of the protector must fuse open without damaging the ground conductor or indoor circuit. The fused type of device accomplishes this with an integral line fuse. The fuseless type must be installed with fine-gauge fuse wire (a fuse link) on the exposed line side (provided by the manufacturer). Both types, when installed, will operate in the same way.

Secondary protectors

Secondary protectors are required to coordinate with the lightning-transient and power-fault capabilities of primary protectors. In some cases, secondary protectors may include one of the previously described arrester components. For this reason, cost-effective secondary protection is typically available as an option on primary protectors (qualified to both Underwriters Laboratories Inc.® [UL®] UL 497, *Protectors for Paired Conductor Communications Circuits*, and UL 497A, *Secondary Protectors for Communications Circuits*).

Secondary protectors must handle sneak currents. The components of sneak current are different from those to which primary protectors are exposed.

Heat coil. The heat coil detects sustained low-level current by the heat it generates. The heat melts a spring-loaded shorting contact that permanently shorts the line to ground, requiring manual inspection and replacement.

Sneak current fuses. This is a fuse that opens the station-circuit wiring under sustained low-level current, requiring manual inspection and replacement. This fusing is on the station side of the arrester and should not be confused with the primary protector line fusing.

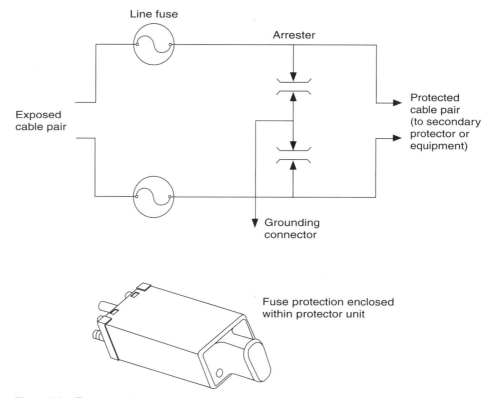

Figure 8.3   Fuse protection.

Positive Temperature Coefficient (PTC) resistors.   The PTC resistor is used in place of a sneak current fuse and will limit sustained current as it heats. It does not require replacement after the sneak-current fault is cleared.

Enhanced protection

Enhanced protection uses additional components to provide protection that is typically suited for specific low-voltage data circuits. In most cases, equipment manufacturers have designed such protection into their equipment, but others prefer to rely on available external protectors.

In some cases, these protectors are UL 497-qualified and can substitute in a primary-protector panel for special applications of individually exposed circuits.

In other cases, protectors may be qualified according to UL 497A (or even UL 1863, *Communications Circuit Accessories*, as telecommunications circuit accessories) for use in office locations where the protected station or terminal is powered, or behind, primary protectors.

Primary protector installation practices

Suggested cabling installation practices for primary protectors include:

- Primary protectors must be installed immediately adjacent to the exposed cable's point of entry into a building. The associated grounding conductor must be routed as straight as possible directly to the closest approved ground.
- A noncorrosive atmosphere is required for long-term reliability.
- Adequate lighting is important for personnel safety at protector locations.
- When a protector is installed in a metal box, bond the box with an approved grounding conductor directly to the protector ground.
- Make certain there are no obstructions around or in front of protectors and ensure that protector locations will not be used for temporary storage.

> **WARNING:** Do not locate primary protectors near any hazardous or easily ignitable material.

Grounding of Telecommunications Equipment

A building may be grounded at different points depending on its size, age, and location. Smaller buildings are typically grounded at the alternating current (ac) service meter base. A ground conductor is installed from the power neutral bus in the meter base to a man-made electrode (typically a driven ground rod). Larger buildings are grounded at the main distribution panel (MDP) in the building. A grounding electrode conductor is installed from the equipment grounding bus in the MDP to a man-made electrode called the grounding electrode system. The different electrode systems in use are described in detail in the *NEC*.

Multistory buildings also should have a telecommunications bonding backbone (TBB) that appears on each floor of the building. The TBB is connected to the telecommunications main grounding busbar (TMGB), which is connected in turn to the grounding electrode conductor at the main electrical distribution panel (EDP) via the bonding conductor for telecommunications (BCT). The TBB is then connected to a telecommunications grounding busbar (TGB) in each telecommunications room (TR) via a pigtail that has been spliced onto the TBB. This splice must be accomplished using an irreversible mechanical connector or an exothermic welded connection. An example of an irreversible mechanical connector is called a C or H connector.

The specific grounding and bonding needs of a particular installation will vary due to:

- Building size.
- Equipment designs.

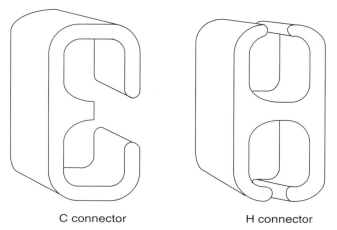

C connector   H connector

Figure 8.4   Example of C and H connectors.

- Special manufacturer's requirements.
- Local code requirements.

Other things to consider when planning for bonding and grounding are:

- Multiple TRs on each floor of a large building.
- Multiple vendors' products to be installed.
- Special electronic equipment.

There can be as few as one and a maximum of four telecommunications grounding and bonding points within a building structure. They are the:

- Main building ground electrode (earth electrode).
- Main electrical distribution panel.
- Telecommunications main grounding busbar (TMGB).
- Telecommunications grounding busbar (TGB).

TABLE 8.1   Sizing grounding conductors American wire gauge (AWG)

Conductor AWG Size	Conductor Resistance (ohm/ft)	Maximum Length of Bonding Conductors
6 (4.1 mm [0.16 in])	.0051	1433 m (4700 ft)
4 (5.2 mm [0.20 in])	.0032	2286 m (7500 ft)
3 (5.8 mm [(0.23 in])	.0025	2926 m (9600 ft)
2 (6.5 mm [0.26 in])	.0020	3657 m (12,000 ft)

NOTES: This table was derived from information contained in Table 8, *NEC* 2002.

The 2002 *NEC* allows a maximum of 25 Ω resistance from the farthest point in the grounding and bonding infrastructure to the earth grounding electrode.

1.0 Ω maximum resistance is allowed for the various connections in the bonding infrastructure.

Bonding conductor size must be engineered to compensate for heavy load factors such as active equipment (private branch exchange [PBX]), which may be using the same infrastructure.

Refer to Figure 8.5 for an overview of building grounding infrastructure. Connection of the BCT to the bus inside the main EDP requires the services of a licensed electrician. Only a licensed electrician should install ground conductors to the safety ground busbar inside the main electrical distribution panel. This is the only point inside a building where the neutral (current-carrying conductor) and the equipment grounding conductor (green-wire ground) are connected together. If they are connected at any other place in the building, a ground loop will exist, resulting in stray current.

The size of the BCT provided by an electrician from the main electrical distribution panel to the TMGB varies depending on the service requirements of the building and the complexity of the telecommunications distribution system. For the BCT, a minimum size of 6 AWG (4.1 mm [0.16 in]), insulated (solid or stranded) wire with green insulation should be installed for this purpose, with consideration being given to a BCT of 3/0 AWG (10 mm [0.39 in]). If this wire is installed in metallic conduit, the wire should be bonded to the conduit where it exits, using conduit grounding bushings and a conductor that equals its size. Any grounding busbar should contain enough multiple points of connection to facilitate the connection of ground and bond wires from the different components inside the TR to the TMGB or TGB.

The TBBs are connected to the TMGB using a two-hole connector that is attached to the TBB via an irreversible mechanical connection or an exothermic weld.

To determine the effectiveness of a ground wire, measure the resistance between the ground source and a reference ground that is isolated from the other ground system. The closer the residential cabling installer can get to one Ω (preferably less), the better the system.

Building entrance protectors must be connected to the ground system to protect circuits from lightning and power faults. Only building entrance protectors installed on customer-owned physical plant should be grounded and bonded by residential cabling technicians employed by the customer. Physical plant installed by the local regulated telephone company, cable TV company, or other regulated entity can only be maintained, modified, or otherwise changed by that company's employees. Work by technicians other

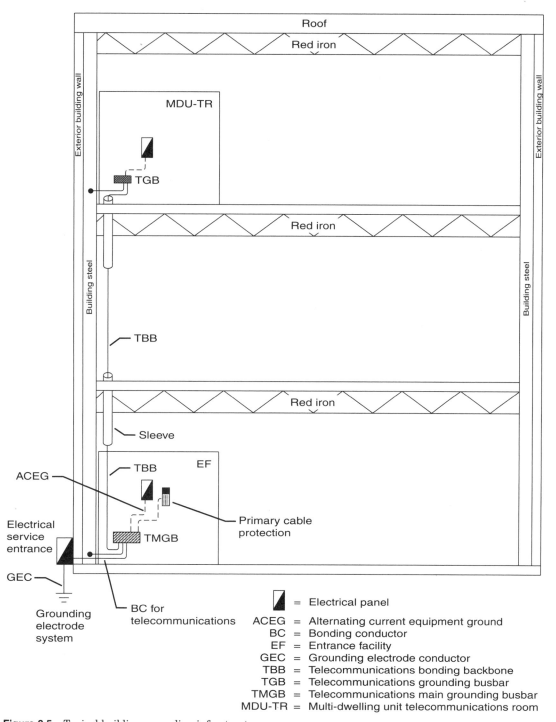

Figure 8.5 Typical building grounding infrastructure.

than regulated company employees may result in service outages and billing rendered to the customer for unauthorized tampering with regulated company facilities.

Avoid splicing a ground or bond conductor. Ground and bonding conductors should be installed along the shortest, straightest route between the equipment being grounded or bonded and the points being connected. If the residential cabling installer must splice a ground or bond conductor, use an irreversible mechanical connector or an exothermic weld.

Ground source

If architectural blueprints that contain electrical drawings are available, refer to them to determine what has been provided and installed by the electrical contractor. Refer to the electrical riser diagram and the electrical engineer's notes that accompany it to determine what grounding electrode system has been designed and installed. Consult with the electrical contractor on the job to determine whether the original design has been followed or whether changes have been made since the drawings were issued.

If the electrical contractor is still on the job, consult with the contractor to determine where the ground electrode is and whether a BCT to the main TR has been provided for use by telecommunications cabling technicians. If the electrical contractor is not on the job, consult with the general contractor.

The size of the BCT to the main TR will vary, but should be not less than 6 AWG (4.1 mm [0.16 in]). ANSI/TIA/EIA-607 recommends consideration of a 3/0 AWG (10 mm [0.39 in]). The BCT must originate at the main distribution panel for the building electrical system. This is sometimes referred to as the main ac switch. This location can be readily identified because it is the only location in a building where the neutral bus and the equipment ground bus are tied together. The BCT will then be routed along the shortest, straightest path to the telecommunications entrance facility (TEF). This is the location that houses the building entrance protectors. If the TEF is equipped with a BCT, determine whether a TMGB is provided.

Types of grounding

Single-point equipment grounding.   Some PBX manufacturers and other large equipment suppliers require single-point equipment grounding. When equipment is grounded through only one point, surges that are conducted through the building ground will not pass through the equipment. Manufacturer's documentation should provide details for acceptable implementations.

Receptacle outlet grounding.   As described in *NEC* Sections 250.114 and 250.138, receptacle outlet grounding is the sole protection of some equipment. Such a system relies on a receptacle (safety) grounding conductor, electrical cord, and plug. This is typical of small equipment and may provide for suitable performance and reliability if the receptacle ground contact is adequately bonded and free of electrical disturbances.

Additionally, the receptacle outlet ground should be bonded to other metallic systems if the receptacle is not located close to the grounded electrical service equipment. Equipment grounding through a receptacle outlet is not suitable for large equipment or equipment manufactured with bonding terminals.

Telecommunications circuit protectors.   *NEC* Article 800 covers telecommunications circuit protection, a primary responsibility of the residential network designer. The basic functions of protectors are:

- Arresting surges or overvoltages that come from exposed circuit pairs by diverting them to ground.
- Protecting against sustained hazardous currents that may be imposed.

Based on Underwriters Laboratories (UL) standards, there are three types of telecommunications circuit protectors:

- Primary protectors as qualified by UL 497—These are intended for application on exposed circuits according to *NEC* requirements. They must be installed as near as possible to the point at which the exposed cables enter the building, and the grounding conductor must safely carry lightning and power-fault currents.
- Secondary protectors as qualified by UL 497A—These devices are not required by the *NEC* but are typically used for additional protection behind primary protectors. In addition to voltage protection, this type of device must also protect against sneak current. Sneak current can be caused by:
  - Power faults that are too low in voltage to engage primary protectors.
  - Station equipment that draws excess current and overheats station wire.
  - Induction from power lines.

The importance of secondary protection becomes important in the residential area because of the need to protect the homeowner's computer equipment, home automation, and consumer electronics. In some cases, homeowners may require uninterruptable power supplies or possibly a power generator to allow sensitive residential systems to work during temporary power outages.

- Data and fire-alarm protectors as qualified by UL 497B, *Protectors for Data Communication and Fire Alarm Circuits*—These modules are more sensitive and have a lower fault threshold because they do not have to allow for the 90 V associated with a telephone's ring voltage. Though not required by the *NEC*, these modules must perform primary protection against lightning transients. They do not have the ability to protect against power faults. These devices should be used according to manufacturer's guidelines.

There can be some overlap in function with available protector products. The manufacturers should be consulted for specific applications; however, the following rules generally apply:

– A primary protector is required where a circuit is exposed to electrical power faults and lightning. Other protector types are not qualified to protect under these conditions.

– Where a circuit is exposed to lightning surges, a primary protector or a data/alarm protector is required as dictated by equipment manufacturers.

– Where sneak currents are hazardous, a secondary protector or primary protector with secondary protection is required (as directed by equipment manufacturers). The basic secondary protector function (sneak current protection) can be included by manufacturers in some primary protectors.

– Manufacturers may include additional protection functions, sometimes called enhanced protection, for specific applications.

Bonding and grounding principles

Telecommunications bonding and grounding are additional bonding and grounding installed specifically for telecommunications systems. From a safety code standpoint, the *NEC* and ANSI/NFPA-780, *Standard for the Installation of Lightning Protection Systems*, already cover such bonding and grounding.

There are many situations, however, where these minimal safety codes can be interpreted and implemented in different ways. Some methods may not be as suitable as others to ensure the reliability and performance of sensitive electronic equipment. Therefore, equipment manufacturers and network designers use a wide range of differing solutions to adapt to:

▪ Site variations.

▪ Prewired buildings.

▪ Differing manufacturer interpretations.

▪ Multi-vendor applications.

▪ Equipment design variations.

▪ Sensitive high-speed electronics.

Most such situations cannot rely solely on safety grounding methods for adequate protection. Instead, telecommunications systems require direct and dedicated grounding and bonding.

Telecommunications bonding and grounding is used to:

▪ Minimize electrical surge effects and hazards.

▪ Augment electrical bonding.

▪ Lower the system ground reference impedance.

This type of bonding and grounding does not replace the requirement for electrical power grounding but supplements it with additional bonding that generally follows communications pathways between telecommunications entrance facilities, equipment rooms (ERs), and multi-dwelling unit telecommunications rooms (MDU-TRs).

Telecommunications grounding practices

Establishing a suitable telecommunications ground is critical in properly grounding telecommunications equipment. A telecommunications ground is always required, and is typically found in the following:

- Telecommunications entrance facilities for sites with exposed metallic cable
- ERs
- MDU-TRs

Grounding choices.   Direct attachment to the closest point in the building's electrical service grounding electrode system is preferred because telecommunications cabling and electrical power cabling must be effectively equalized.
   Select the nearest accessible location on one of the following:

- The building ground electrode system
- An accessible electrical service ground

   If no electrical service exists, use one of the following:

- Driven ground rod
- Another grounding electrode system installed for the purpose

Using the electrical service ground.   A direct electrical service ground is one of the best points for grounding telecommunications systems. In new construction, an electrical contractor must provide accessible means. *NEC* Section 250.94 requires an intersystem bonding connection accessible at the electrical service equipment, such as:

- nonflexible metallic raceway (using an approved bonding connector).
- Exposed grounding electrode conductor.

   WARNING: This conductor is critical to the safety of the electrical power system. Do not move, modify, or disconnect it without the direct participation of personnel responsible for that system.

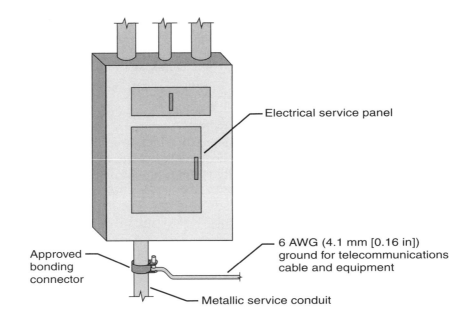

AWG = American wire gauge

Figure 8.6   Typical telecommunications ground.

- Approved external connection on the electrical power service panel

> **NOTE:** Ensure the grounding electrode system is properly installed. Verify this with the electrical contractor or have the system tested by a licensed electrician.

Installing a grounding electrode

The resistance of the grounding electrode system should be as low as possible (25 Ω or less) and measured annually with an earth megaohm-meter (megger). Many equipment manufacturers require less than 10 Ω resistance between the grounding electrode system and earth. The resistance of the grounding electrode system is not as important as the bonding between the different grounding electrode systems that may exist on site.

The *NEC Handbook*, Section 250.52, provides a few examples for the installation of a grounding electrode or electrode system. The installation of a telecommunications grounding electrode is allowed if:

- There is no electrical service ground, or

- Additional grounding is needed (per *NEC* Section 250.54, 62, 64, and 118). If so, the installed electrode must be bonded to the existing ground electrode system.

As a last resort, *NEC* Section 800.40(B)(2) specifies a minimum 12.7 mm (1/2 in) diameter, 1.5 m (5 ft) ground rod driven completely into the ground.

Another alternative is a ground ring electrode system. This system consists of noninsulated conductors that are buried in the shape of a ring. According to *NEC* 250.52(A)(4), conductors are buried at a minimum depth of 762 mm (30 in) for a minimum of 6 m (20 ft) and should not be smaller than 2 AWG [6.5 mm (0.26 in)].

The following conditions must be strictly observed:

- Any installed grounding electrode must be at least 1.83 m (6 ft) away from other existing electrodes.

- Electrodes or down conductors that are part of a lightning protection system are not allowed for use as an electrode for this purpose.

- Regardless of what alternative is selected for installing a ground rod, all other electrical system grounds, structural building steel, and metallic piping systems must be bonded together. This is required of the electrical service and is usually already accomplished, but it should be verified.

- Gas pipes, steam pipes, or hot water pipes are not allowed as a grounding electrode.

Three components determine the resistance of an electrode to earth:

- Resistance of the earth itself
- Contact resistance of the electrode to the surrounding earth
- Resistance of the electrode itself and its connections

Grounding hardware

The two types of grounding that apply to buildings are the:

- Grounding electrode system (also known as the earthing system).
- Equipment grounding system (also known as the safety ground).

The grounding electrode system consists of a grounding:

- Field (earth).
- Electrode conductor.
- Electrode.

The equipment grounding system consists of:

- Equipment grounding conductors.
- A main bonding jumper.

In addition to the building grounding systems, a separate grounding system for telecommunications is defined in ANSI/TIA/EIA-607. The electrical service entrance is outside the scope of this standard and is grounded and bonded in accordance with all applicable electrical codes. The telecommunications grounding and bonding infrastructure originates with a connection to the service equipment (electrical power) ground and extends throughout the building. The telecommunications grounding system is composed of the following five components:

- Bonding conductor for telecommunications.
- Telecommunication main grounding busbar (TMGB).
- Telecommunications bonding backbone (TBB).
- Telecommunications grounding busbar (TGB).
- Telecommunications bonding backbone interconnecting bonding conductor (TBBIBC).

Additional components that may also be included are:

- Lightning protector grounding system connections.
- A grounding electrode conductor (GEC).

If a TMGB is not provided, determine the maximum number of ground/bond wires to be installed in the TR and install a TMGB that will handle them. The minimum number of positions on a TMGB should be seven. It is necessary to be familiar with the manufacturer's requirements for each piece of equipment requiring grounding and bonding. Always follow local codes and ordinances, as well as national codes and standards.

The recommended connector is made of a copper alloy with a crimp connection for the ground/bond wire. It is equipped with a flat plate, featuring a hole(s), so that it can be screwed or bolted to the ground/busbar. A second type of connector is the H or C connector used to join two sections of bond wire together to form a connection. This type of connector can only be used on a bond wire. Ground wires should be continuous and without splices or connectors from the point of origination to the point of termination. Use a two-hole connector when the ground/bond wire will be connecting more than one device to the busbar. A single-hole connector can be used when connecting only one device to the busbar. When dissimilar metals are bonded to a busbar, a conductive paste or grease should be used to minimize electrolysis.

When grounding or bonding specific telecommunications hardware, always review the manufacturer's specifications.

Telecommunications Main Grounding Busbar (TMGB). Determine whether the electrical contractor has provided and installed a BCT from the main electrical distribution panel to the TEF. If the BCT is provided and installed, proceed with installing the TMGB. If there is no BCT installed, determine the location and makeup of the building's grounding electrode system.

The building's grounding electrode system may consist of:

- Building structural steel, where effectively grounded.

- Exposed metal service entrance raceway.

- Ground rods, pipes, conduits, and plates that have at least 2.4 m (8 ft) of contact with surrounding moist soil.

- Concrete encased electrodes.

- Metallic cold water pipe—This system must be supplemented by one of the four types listed above. (It is necessary for the metal piping to be in contact with moist soil for a minimum of 3 m [10 ft].) The interior cold water piping of a facility is acceptable, providing it is at the building's street side of the water meter. Connections to metal drain pipes are not an acceptable means of providing a grounding electrode connection, due to the high probability of polyvinyl chloride (PVC) piping used in these systems.

If the number of ground/bond wires exceeds the number of positions on the busbar employed at the site, a TGB must be installed to allow connection of the additional ground/bond wires. The TGB should be as close to the TMGB as feasible.

Use a minimum of 6 AWG (4.1 mm [0.16 in]), green-insulated, solid/stranded wire to connect the TMGB to the TGB. Mount the TGB at the same height as the TMGB. (The recommended height is 150 mm [6 in] below the top of the plywood backboard.)

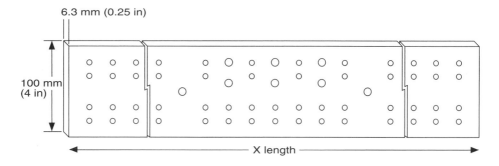

X = Variable

Figure 8.7 Telecommunications main grounding busbar.

Installing the TMGB

Step	Installing TMGB
1	Select the length of busbar to be installed. The TMGB should be 100 mm (4 in) high by 6.3 mm (0.25 in) thick by the length necessary to terminate all of the conductors to it.
2	Determine the exact position of the busbar on the backboard.
3	Prelocate the mounting hardware that comes with the TMGB. Mark the location where the two mounting holes for each bracket will be attached to the backboard.
4	Predrill holes for the screws using a drill and a 3.2 mm (1/8 in) bit. Install mounting brackets using the lag screws contained in the kit.
5	Install the insulators on the mounting brackets using the bolts contained in the kit.
6	Install the busbar on the insulators using the bolts contained in the kit.
7	Always install the BCT to the TMGB before proceeding with the additional grounding.
8	Identify the BCT installed from the main electrical distribution panel. Determine how much wire is required to position it onto the busbar. Cut off the excess wire using cable cutters.
9	Using an insulation-removal tool, remove the correct amount of jacket from the wire. **NOTE:** The correct amount is determined by the size of the connector and the gauge of the wire.
10	Insert the bare end of the copper wire into the end of the connector. Obtain a compression tool equipped with a die that matches the connector body.
11	Place the round connector body into the die of the compression tool. Close the jaws of the tool fully. This action may have to be repeated on another section of the connector to fully ensure contact with the wire. Leave approximately 6.3 mm (0.25 in) of copper showing between the end of the insulation on the wire and the end of the connector that the wire is installed into.
12	Position the ground wire and connector by routing them to the busbar. Ensure that a minimum 75 mm (3 in) radius bend is used where the wire changes direction. Create smooth bends in the wire. Attach the ground wire to the backboard using a staple gun or wire clamps. Install staples or wire clamps at intervals of not more than 457 mm (18 in).
13	Label the ground wire as the main ground source for the room and where it originates. The label should contain where it came from, where it is going to, and its unique identifier as described in ANSI/TIA/EIA-606, *Administration Standard for the Telecommunications Infrastructure of Commercial Buildings*. If this is the BCT, the residential cabling installer must also install a warning label on each end of the wire. Refer to ANSI/TIA/EIA-607, Chapter 5.
14	Upon completion of the installation of the TMGB, update all installation documents with the information that will allow a residential cabling installer to identify any ground source and the methods employed in providing the ground/bond.

The procedure outlined above is also used for connections between the TMGB and the other components of the telecommunications grounding system within the building. Refer to ANSI/TIA/EIA-607.

Other components that will have to be installed in a building with multiple TRs will be:

- Telecommunications bonding backbones (TBBs).
- Telecommunications grounding busbars (TGBs).
- Telecommunications equipment bonding conductors (TEBCs).
- Telecommunications bonding backbone interconnecting bonding conductors (TBBIBCs).
- Alternating current equipment ground (ACEG).

Telecommunications Bonding Backbone (TBB).   A TBB must be installed from the TMGB through each set of TRs in a multistory building or in a series of TRs in a low-wide building. The TBB must be a minimum of 6 AWG (4.1 mm [0.16 in]), green-insulated conductor. Consideration should be given for sizing the TBB at 3/0 AWG (10 mm [0.39 in]).

TBBs should always be terminated at the TMGB using a two-hole connector. In each of the other TRs, a short length of conductor, sized at the same size as the TBB, must be tapped onto the TBB using an H or C connector or an exothermic weld. This conductor will be extended to the TGB and terminated using a two-hole connector.

Telecommunications Grounding Busbar (TGB).   In each additional TR, a TGB must be installed. The TGB must be 50 mm (2 in) high by 6.3 mm (0.25 in) thick and as long as necessary to terminate all of the conductors onto it. It must also be insulated from its mounting (e.g., wall, plywood) by a 50 mm (2 in), or greater, insulator. The TGB must be installed in a manner similar to the TMGB.

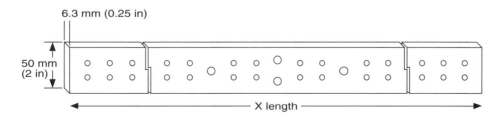

X = Variable

Figure 8.8   Telecommunications grounding busbar.

Telecommunications Equipment Bonding Conductor (TEBC). Each piece of telecommunications equipment should be bonded to the TGB to equalize potential within the grounding mesh. To do so, install a TEBC from each piece of equipment to the TGB. When equipment is mounted on a relay rack, a rack grounding kit can be installed on the rack and the various pieces of equipment bonded to it. Then a TEBC can be installed from the relay rack to the TGB. The minimum size conductors must be a 6 AWG (4.1 mm [0.16 in]), green-insulated wire. Normally, a TEBC can be terminated using a single-hole connector. However, if it connects multiple pieces of equipment (i.e., from a relay rack), it should be terminated using a two-hole connector.

Telecommunications Bonding Backbone Interconnecting Bonding Conductor (TBBIBC). Whenever two or more TBBs are run vertically in a building, on each third floor and the top floor of the building, a TBBIBC must be installed between the TGBs in the TRs. The TBBIBC must be a minimum of a 6 AWG (4.1 mm [0.16 in]), green-insulated copper wire. Like all grounding/bonding conductors, it should be routed in the shortest, straightest manner between its two points of termination. It should be terminated using a two-hole connector.

Alternating Current Equipment Ground (ACEG). Many times a TR will be equipped with an electrical distribution panel (EDP). This panel will serve all of the electrical service contained within the walls of the TR. It is usually on a dedicated feed from the MDP or from some other EDP. If such an electrical panel is located within a TR, an ACEG must be installed from the equipment grounding bus inside the panel to the TGB. This conductor must be a minimum of 6 AWG (4.1 mm [0.16 in]), green-insulated copper wire. It should be terminated at the TGB using a two-hole connector.

Steps in a ground test. Measure the resistance between the TMGB and the remaining ground points of the building grounding system with a megaohm-meter (megger). The maximum system resistance allowed, per *NEC* 250.56, is 25 Ω.

Step	Ground Test
1	A short jumper wire should be installed between the P and C terminals of the earth ground resistance field test instrument. This allows the meter to be used in the two-point configuration.
2	Terminate one cable at the E terminal of the field test instrument.
3	Terminate the other cable at the P or the C terminal of the field test instrument.
4	To compensate for the resistance of the instrument cables, connect the open ends of the instrument cables together. Depress the measurement button of the test instrument and record the indicated value. This is the resistance of the meter leads and should be subtracted from each reading during the bonding resistance measurements.

Step	Ground Test
5	Once the lead resistance has been obtained, measurements should then be made between selected points in the grounding system. Resistances that can be checked are between the following: ■ Metallic water pipe main and structural steel ■ Metallic water pipe main and any driven electrode ■ Service entrance conduit and structural steel ■ Electrical system grounding electrode and the building lightning rod grounds ■ Power transformer feeding the electronic equipment and the structural steel ■ Telecommunications main grounding busbar (TMGB) and the ac grounding electrode system ■ Telecommunications main grounding busbar (TMGB) and the telecommunications grounding busbar (TGB). ■ Telecommunications grounding busbars (TGBs) located in separate TRs Other combinations may exist depending on the size of the facility or the type of activity that takes place in the building. **NOTE:** To make this bonding verification complete, the residential cabling installer should verify the tightness of connections to the different grounding electrode systems. Corroded or loose connections cause the majority of deficiencies in the resistance between grounding electrode systems. At a minimum, the system should be checked annually.

Physical protection

If a chance of damage to the grounding conductor exists, some form of physical protection is required. If the grounding conductor is metallic conduit or raceway, both ends must be bonded to the grounding conductor or the same terminal or electrode to which the grounding conductor is connected [*NEC* Section 800.40(A)(6)].

Community Antenna Television (CATV). Bond the established ground, the intended ground termination, or the outer conductive shield of a CATV coaxial cable in the same manner as other telecommunications cables to help limit potential differences between these systems and other metallic systems (*NEC* Section 820.33).

Water pipes. Historically, the first choice for a grounding electrode has been a metallic water pipe connected to a utility water distribution system. This may no longer be a good strategy. There is an increased use of nonmetallic pipe in buildings, and electrical systems should no longer rely on plumbing systems for grounding. Water pipes must be bonded to another electrode type to ensure proper protection.

For a similar reason, caution must be exercised when water pipe is used as an intersystem bonding conductor. *NEC* Sections 250.50 and 250.104 cover such usage; however, avoid this practice and use a minimum 6 AWG (4.1 mm [0.16 in]) copper bonding conductor.

System practices

There is a wide range of differing (but current) practices concerning system grounding. Each manufacturer, service provider, inspector, or customer may have unique requirements.

Small systems.   In small ERs and entrance facilities, the ground connection terminals on the installed equipment (e.g., protector panels, PBX cabinets) are usually directly connected to the closest approved grounds (typically made available by electrical service personnel). See Figure 8.9 for details.

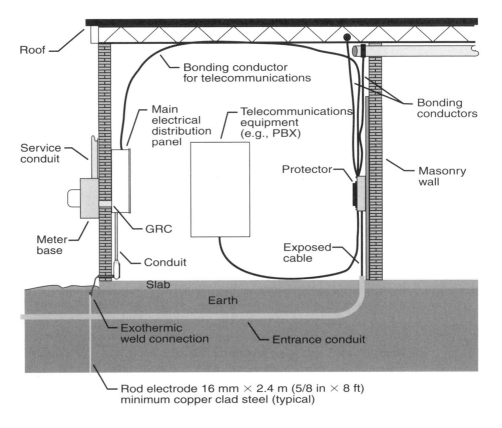

DD = Distribution device
GRC = Galvanized rigid conduit

Figure 8.9   Example of multi-dwelling unit small equipment room.

All exposed metallic cable shall be directly terminated at the associated protectors.

Any exposed metallic cable shields are bonded directly to the closest:

- Protector.

- Protector's ground terminal.

- Approved protector ground.

Beyond this, equipment manufacturers' guidelines must be followed. Typically, a bonding conductor is installed directly between the protector's ground terminal and the PBX (or other equipment) ground terminal.

If the protector and PBX are colocated and rely on the same approved ground, then the bonding conductor may not be needed.

Large systems. A grounding busbar or multiple busbars are used for sites with several ERs and separate entrance facilities where there are usually many connections. Requirements and guidelines for this situation are detailed in the ANSI/TIA/EIA-607 standard.

Busbars should be:

- Positioned so that telecommunications cabling will generally follow the related bonding conductors.

- Positioned near associated equipment.

- Insulated from their support.

The busbar designated for protectors, the TMGB, must safely carry lightning and power-fault currents. The TMGB is directly bonded to the electrical service ground. It should be positioned adjacent to the protectors and directly between the protectors and the approved building ground for protector operation. The minimum dimensions of the TMGB are 6.3 mm (0.25 in) thick, 100 mm (4 in) wide, and variable in length.

A TBB conductor is connected from the TMGB to the TGB in TRs within the building. The minimum dimensions of the TGB are 6.3 mm (0.25 in) thick, 50 mm (2 in) wide, and variable in length.

A TBB is a conductor that interconnects all the TGBs with the TMGB. The TBB is designed to interconnect busbars and is not intended to have equipment bonding conductors spliced on it. The minimum TBB size shall be a 6 AWG (4.1 mm [0.16 in]) and could be as large as a 3/0 AWG (10 mm [0.39 in]).

Each TBB should be a continuous conductor from the TMGB out to the farthest TGB. Intermediate TGBs should be tapped onto the TBB with a short bonding conductor.

The TBB shall be connected to the busbars with a two-bolt attachment. This provides a secure connection as per ANSI/TIA/EIA-607.

Larger TRs may require additional TGBs. Multiple TGBs within a TR are allowed, provided they are all bonded together.

In larger multistory buildings that have multiple TRs on each floor, the potentials between TRs need to be equalized. When applicable, a TBBIBC is installed between TGBs to interconnect all the TRs on the same floor. The TBBIBC is not required for every floor, but is installed on every third floor and the top floor.

To facilitate identification, busbar connections should be grouped with protector, busbar bonding, and approved building grounding conductors toward one end and equipment grounding conductors should be grouped toward the other end.

Telecommunications rooms (TRs).   In a TR, suitable ground options include:

- Building structural steel.

- An electrical receptacle box or approved conduit connection.

- A combination of the above that is accessible.

- An already established telecommunications ground.

For TRs requiring a large number of connections or requiring flexibility for equipment changes, a grounding busbar should be installed.

Telecommunications bonding principles

Most buildings have low overall impedance. Many, in fact, are designed as part of a lightning protection system to safely conduct lightning strikes to earth. However, significant differences in ground potential can exist throughout a building. Additional bonding conductors are an effective way to improve marginal situations, especially in buildings that lack an overall bonded structure.

If continuous structural steel exists along the same path, as is likely to be the case in large high-rise buildings, there may be little actual improvement from installing a separate telecommunications bonding and grounding infrastructure. Even here, a certain assurance is gained by having specific bonding conductors that can be verified and inspected by the residential cabling installer.

Three specific principles behind the use of telecommunications bonding conductors are:

- Equalization—Potentials between different ground points are very dependent on the impedance between them. Ground equalization is improved because the additional bonding lowers the impedance between different ground points. The shortest and most direct path using large conductors provides a low impedance (both resistive and inductive). Multiple conductors or wide straps will provide a lower impedance.

- Diversion—Because the bonding conductor follows the telecommunications cable and is directly connected to system grounds at each end, electrical transients that are forced down the cable path may be diverted (carried) by the bonding conductor and are less likely to influence the telecommunications conductors.

- Coupling—The closer the bonding conductor is to the telecommunications metallic cable, the greater the mutual electromagnetic coupling. During electrical transients, this coupling tends to cancel the transient partially when it reaches the telecommunications equipment at the end. A tightly coupled bonding conductor or a backbone cable shield is often called a coupled bonding conductor (CBC).

Each of these three effects is achieved in varying degrees, depending on a wide range of factors. It is often difficult to predict or measure specific results, but any combination of the three is usually beneficial to telecommunications equipment.

Telecommunications bonding practices

Bonding connections. Bonding connections should be made directly to the points being bonded. Avoid unnecessary connections or splices in bonding conductors, but when necessary, use an approved connection and position it in an accessible location.

Many aluminum and stainless steel connectors are approved and available for use, but it is recommended that connectors and splices should be one of the following:

- Tin-plated copper
- Copper
- Copper alloy

Copper and copper alloy connections should be cleaned and coated with an antioxidant prior to establishing the connection.

Typical connections are made by using:

- Bolt or crimp/compression connectors, splices, clamps, or lugs—Bolt-type connectors have a tendency to loosen over time due to vibration and repeated temperature fluctuations. Listed mechanical crimp/compression connectors are virtually maintenance free and are not susceptible to vibration and temperature.

   NOTE: Use listed hardware that has been laboratory tested to eliminate most field problems.

- Exothermic welding.

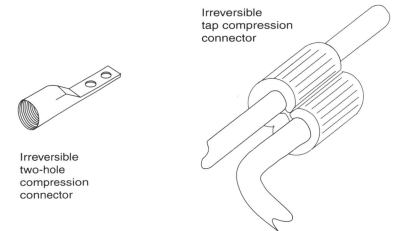

Irreversible
tap compression
connector

Irreversible
two-hole
compression
connector

Figure 8.10   Compression connectors.

The exothermic weld process uses a special heat-resistant mold to bond metallic conductors together. There are a variety of molds available to meet the requirements of:

- Conductor size and shape.
- Ground rod material (based on diameter, treated or plain).
- Application (e.g., pipe, rebar, building steel).
- Configuration (e.g., tee, tap, inline splice).

The exothermic welding process is commonly applied:

- Within the ground electrode system.

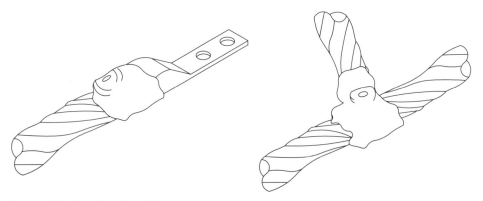

Figure 8.11   Exothermic welding.

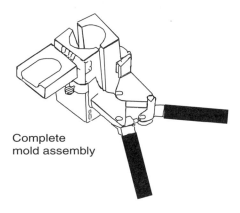

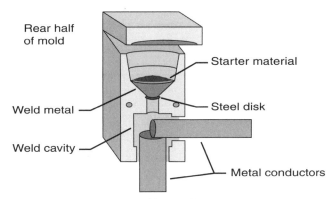

Figure 8.12   Exothermic welding mold.

▪ To parts of a grounding system that are subject to corrosion or that must reliably carry high currents.

▪ To locations requiring minimal maintenance.

Follow the steps below to complete an exothermic weld.

Step	Completing an Exothermic Weld
1	Determine what type of connection is needed.
2	Obtain the proper mold for conductors to be bonded.
3	Follow safety procedures and wear leather gloves and safety glasses.
4	Open the mold and place the metal objects into the mold, following the manufacturer's instructions.
5	Close the mold around the conductors to be bonded.
6	Place the steel disk in the base of the crucible.

Step	Completing an Exothermic Weld
	NOTE: The disk keeps the weld metal in the crucible until the disk melts and the weld metal flows down into the weld cavity.
7	Fill the crucible with the weld metal.
	NOTE: Each canister of weld metal is sized for a specific mold.
8	Sprinkle the starter material over the weld metal.
	NOTE: The starter material has a very low flash point to start the exothermic process.
9	Close the cover and spark the starter material to ignite the weld.
10	Allow the mold to cool, then open and clean it for the next weld.
	NOTE: Always adhere to the manufacturer's instructions when using exothermic welds. The most common exothermic welds are for outside use only. The process described above is for outside or well-ventilated areas. Inside welds are similar but use a nontoxic welding material, a battery, and small spark igniter with a special filtering material to cut down on smoke.

All bonding connections should be made using the proper tools and following the manufacturer's guidelines.

Bonding conductors.   Bonding conductors must be made of copper. To avoid unintentional ground connections, bonding conductors within buildings should be insulated. Bonding conductors must be routed with minimum bends or changes in direction.

NEC Section 800.40(A) requires at least a 14 AWG (1.6 mm [0.063 in]) stranded or solid insulated conductor for connecting the telecommunications protectors and associated metallic cable sheaths to the selected ground. Other *NEC* ground requirements generally indicate a 6 AWG (4.1 mm [0.16 in]) minimum (for ground rod and for intersystem bonding), but a 6 AWG (4.1 mm [0.16 in]) stranded conductor should be used since this accommodates different code requirements and allows for future changes.

In most applications, the bond type selected depends on the application and the fault-current carrying capacity needed, but a minimum 6 AWG (4.1 mm [0.16 in]) copper conductor is generally used throughout typical commercial buildings. According to ANSI/TIA/EIA-607, consideration should be given to sizing conductors as large as 3/0 AWG (10 mm [0.39 in]).

An exception to the minimum 6 AWG (4.1 mm [0.16 in]) bonding conductor is in smaller entrance facilities. UL 497 requires the protection unit to determine the minimum-sized conductor based on anticipated current flow from the number of conductor pairs. Units that are designed for:

▪ One and two pairs of protection require a minimum 12 AWG (2.1 mm [0.083 in]).

- Three to six pairs of protection require a minimum 10 AWG (2.6 mm [0.10 in]).
- Greater than six pairs of protection require a minimum 6 AWG (4.1 mm [0.16 in]).

Some bonding conductors must be guarded against physical damage. The preferred physical protection is nonmetallic. If metallic conduit is used, the bonding conductor must be bonded to the conduit at both ends.

Inspection.   Ground systems require scheduled maintenance. They should be checked for tight connections annually. Critical telecommunications systems should be tested annually to ensure low-resistant connections throughout the system and to the earth.

Visual inspection can usually reveal problems, such as:

- Loose connections
- Corrosion
- Physical damage
- System modifications

During any service work, the residential cabling installer should visually inspect bonding connections. Once a ground system has been installed properly, 90 percent of all grounding troubles are the result of loose connections.

Exposed cables.   Bond and ground exposed cable shields and metallic sheath members according to manufacturer's installation instructions. These items should be grounded as close as practical to the point of entrance, and grounding conductors should be routed directly to the closest approved ground. This may also apply to optical fiber cables with metallic members.

Backbone cable protection

Cables inside a building are generally considered not exposed, but there are situations that call for protective measures. The following guidelines should be applied:

- Electrical power cabling should not be routed directly alongside telecommunications cable. Electrical cabling is usually in conduit, providing additional shielding (see *NEC* Section 800.52 for separation and exceptions).
- The telecommunications metallic cable should be routed near the middle of the building. Surrounding the cable with structural building steel or a lightning-protection system provides shielding and usually diverts transient currents.
- A lightning protection system should be installed.
- The other exposed cables that enter the building must be protected and grounded.

- Some form of bonding conductor should be installed along each backbone cable pathway (see ANSI/TIA/EIA-607).

In high-rise buildings and low-wide buildings, particularly if located in an area with high lightning activity or high soil resistivity, protective measures are vital. The same may be true for buildings that are close to an electrical substation or heavy industrial facilities.

Telecommunications Bonding Backbone (TBB). A TBB is a 6 AWG (4.1 mm [0.16 in]) or larger bonding conductor that provides direct bonding between different locations in a building, typically between ERs and TRs. A TBB is usually considered part of a grounding and bonding infrastructure but is independent of equipment or cable.

ANSI/TIA/EIA-607 provides standard requirements for TBBs and associated hardware.

Coupled Bonding Conductor (CBC). A CBC is a bonding conductor that provides equalization like a TBB. A CBC also provides a different form of protection through electromagnetic coupling (close proximity) with the telecommunications metallic cable. The CBC is generally considered part of an installed telecommunications metallic cable and not part of a grounding and bonding infrastructure. Some PBX equipment manufacturers specify a CBC between their equipment and exposed circuit protectors.

There are two basic forms of CBC:

- Cable shield

- Separate copper conductor tie wrapped at regular intervals to an unshielded cable

Typically, the CBC is specified as 10 AWG [2.6 mm (0.10 in)]. To work properly, the CBC must be connected directly to the protector ground and to the PBX ground.

Unshielded backbone cable. For unshielded metallic backbone cable, a corouted bonding conductor (TBB) should be installed as follows:

Step	Installing Corouted Bonding Conductor (TBB)
1	Route a 6 AWG (4.1 mm [0.16 in]) copper conductor along each backbone cable route. (Ensuring a minimal separation between the conductor and the cables along the entire distance may satisfy the equipment manufacturer's requirement for a CBC.)
2	Bond each end at the nearest approved ground (i.e., TMGB, TGB) in the area where the associated cables terminate or are spliced or cross-connected to other cables.

NOTE: A TBB may also be required for installation of shielded backbone cable. This depends on customer requirements. (Refer to ANSI/TIA/EIA-607.)

Backbone cable shields. Many indoor and outdoor metallic backbone cables have an overall cable shield. These shields serve as:

- Physical protection.
- A coupled bonding conductor.

They should be directly bonded following the manufacturer's guidelines, typically to the nearest approved ground at each end. (Cable shields do not satisfy the requirements for a TBB.) See Figures 8.13 and 8.14.

Shielded cabling systems. Some indoor metallic cabling systems rely on shielding as an integral factor in their signal transmission performance, most notably those with coaxial, shielded twisted-pair (STP), or screened twisted-pair (ScTP) wire. Through standard metallic cable connectors, the cable shields are typically grounded to a connector/administration panel at one end only, so that even after administration changes, cable shield continuity is maintained throughout the cabling system.

The administration panels should be bonded to the nearest approved ground with a direct, minimum length grounding conductor. At the user terminal end, these cable shields are commonly terminated by the user terminal, which relies on the nearest power plug's green wire (safety ground) instead of direct bonding. The bond is completed through the equipment's chassis ground connection.

Use manufacturers' instructions and apparatus for terminating and grounding these cable types.

Exposed cable sheath terminations. OSP cables usually have metallic sheath members or a metallic shield that can be bonded together at the point of entrance and grounded at the nearest approved ground.

Under conditions of high lightning incidence, an isolation gap is sometimes designed into the system. In this case, the cable sheath routed within the building is isolated from the cable sheath that enters the building. Both sides of the isolation gap are grounded with individual grounding conductors.

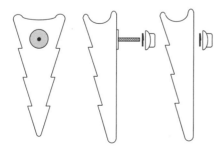

Figure 8.13   Bullet bond clamp kit.

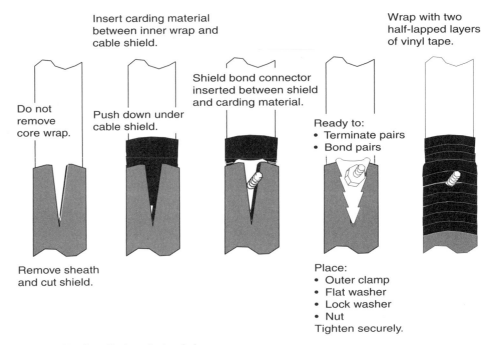

Insert carding material between inner wrap and cable shield.

Wrap with two half-lapped layers of vinyl tape.

Do not remove core wrap.

Push down under cable shield.

Shield bond connector inserted between shield and carding material.

Ready to:
• Terminate pairs
• Bond pairs

Remove sheath and cut shield.

Place:
• Outer clamp
• Flat washer
• Lock washer
• Nut
Tighten securely.

Figure 8.14 Installation of a bond clamp.

Under conditions where direct current (dc) ground currents can be expected, an insulating joint is sometimes designed into the system. An insulating gap is made in the cable sheath that is similar to the isolation gap, but the field (outside) side is not grounded directly. Instead, it is grounded through a capacitor that blocks dc current.

Overview

It is every organization's responsibility to provide safety training for its employees and ensure that safety procedures are strictly followed. Safety training should be given on a regularly scheduled basis, not as a result of an injury.

Due to the daily hazards to which residential telecommunications cabling installers are exposed, it is vital that all residential cabling installers have a complete understanding of rescue and first aid procedures.

Residential cabling installers must know their company's safety policies and practices, and follow them while working. Be aware of any local site-specific safety issues that affect a task. Pay close attention and ask questions during every company or job site safety meeting.

A safe work environment is an important part of every job. Do not depend on the efforts of others to ensure job safety. Personal safety will inevitably protect workers, homeowners, and the public.

When working, consider the possible effects of every action. This is especially important for actions that could have consequences in a remote location (e.g., turning power on or off or activating distant machinery).

This chapter is not intended to provide complete coverage of workplace safety. It is the responsibility of the residential cabling installer to be aware of regulations, standards, and local policies and to play a primary role in practicing safety.

Occupational Safety and Health Act

Passed by the U.S. Congress in 1970, the Occupational Safety and Health Act attempts to ensure a safe and healthy environment for every working person in the United States. The provision and requirements of the Occupational

Safety and Health Administration (OSHA) are set forth in the *Code of Federal Regulations (CFR)*.

While OSHA is responsible for the administrative work relating to the statute, most field work has been assigned to each state's Department of Labor. As a result, each jurisdiction is responsible for field inspections and enforcement. OSHA works closely with the individual jurisdictions to review workplace hazards, conduct inspections, and provide training.

OSHA or an approved jurisdiction is responsible for job site inspections and has the authority to shut down a job site and levy fines against individuals and companies for noncompliance with OSHA regulations. Additionally, OSHA is responsible for the development, publication, and enforcement of safety standards. The two main standards sections with which the telecommunications industry is concerned are:

- OSHA Regulations (Standards-29 CFR)–1910, *Occupational Safety and Health Standards.*

- OSHA Regulations (Standards-29 CFR)–1926, *Safety and Health Regulations for Construction.*

> **NOTE:** These may not be the only CFR regulations that apply in a given situation. Check with the relevant state agency about applicable safety regulations.

Safety Awareness

The rules regarding the specific type of safety program or training that a company must provide varies by company size, by state, and at times by city. However, there are common themes for safety among them all, including:

- Common sense.

- Training.

- Awareness.

- Cooperation and participation.

Workplace safety is not simply remembering to wear a hard hat or knowing how to secure a heavy load. The subject includes a full range of health-related issues as well as knowledge of the physical aspects of safety.

There is a significant amount of information related to safety in the workplace available to individual workers and companies. In addition to OSHA and local or state government agencies, most industry organizations and unions offer extensive resources relating to safety and safety planning. Specialized safety training is also available from private training enterprises.

The remaining sections of this chapter provide a limited overview of some common health and safety issues related to the telecommunications field. It is up to the individuals to seek additional information from their own company, trade organization, or authority having jurisdiction (AHJ).

First Aid

All residential cabling installers should take courses in and be capable of providing:

- Basic first aid.
- Cardiopulmonary resuscitation (CPR).

The American Red Cross offers local courses in standard first aid and community CPR around the country.

First aid is the emergency aid or treatment given before medical services can be obtained (e.g., Emergency Medical Services [EMS]). Training in first aid prepares individuals to act properly and help save lives in the event of an emergency.

CPR is the emergency procedure used on a person who is not breathing and whose heart has stopped beating (cardiac arrest).

First aid, cardiopulmonary resuscitation (CPR), and the law

Legally, a victim must give consent before a person trained in first aid can provide assistance. Therefore, it is important to ask a conscious victim for permission. However, the law assumes that an unconscious person would give consent. In the United States, "Good Samaritan" laws provide legal protection to rescuers who act in good faith and are not guilty of gross negligence or willful misconduct. Laws may vary from state to state and should be verified as part of first aid training.

First aid and CPR certifications should be current. Certification cards have expiration dates that require refresher courses to be renewed.

First aid kits

First aid kits and portable eyewash stations must be a part of the equipment for every job. Fresh water to rinse out debris or toxins may not be available during certain construction periods.

Check to see that first aid kits are restocked after each use and ensure eyewash stations have not passed their expiration dates. Promptly report any use of supplies from the first aid kit to the proper supervisor.

> **NOTE:** Many companies keep additional first aid kits. At the end of each month, kits that have been used on the job are swapped with fully stocked kits. The used kits are then restocked and prepared for reuse.

Figure 9.1   First aid kit.

First aid kits and eyewash stations should be accessible to all personnel on the job site. Prevent eyewash stations from freezing, since a frozen station is of little use in an emergency.

Written copies of the first aid procedures for exposure to a hazardous substance, such as Material Safety Data Sheets (MSDSs), should be brought to any job where residential cabling installers might be exposed to the hazardous substance. Review these procedures before work begins.

Emergency Rescue

Training in emergency rescue and first aid is often provided in one comprehensive course.

There are six basic steps to safely assist others without endangering yourself:

1. Survey the scene—Check for fire, toxic fumes, heavy vehicle traffic, live electrical wires, a ladder, or swift-moving water. If the victim is conscious, ask questions to get information.

2. Notify someone—It is imperative to let someone know that you need help and where you are. Direct this person to notify emergency personnel. If no help is available, extreme caution should be used so that the rescuer does not become a victim.

3. Secure the area—Make the area safe for you and the victim. Locate and secure the power to the energized circuits and turn off gas or water mains. Move the victim to a safe area only if it will not further complicate their medical condition.

   NOTE: Do not move someone with a neck or back injury unless the person is in a life-threatening situation. Remember that you cannot help anyone if you become part of the problem.

4. Primary survey of victim—Check the victim's "ABCs." "A" is for opening the victim's "airway." This is the most important action for a successful resuscitation. "B" is to check for "breathing," and "C" is to check "circulation" or pulse.

5. Telephone EMS 911 or applicable local number—Direct someone to call EMS and relay all the information you have collected in your initial surveys.

6. Secondary survey of victim—If qualified, perform CPR as needed and check for life threatening injuries (such as serious lacerations), then check for minor injuries that may have been overlooked previously.

> **CAUTION:** The above rescue techniques are a basic outline that should only be used after receiving the proper training. Some rescues require specialized training and equipment.

Communication

Communication is an important part of any safety program. Attend and pay close attention to all safety meetings and safety equipment training. Ask questions.

On the job, residential cabling installers must communicate freely and clearly with everyone affected by their work and those whose work may affect them. These people include:

- Coworkers.
- Supervisors and building management.
- Building occupants (if any).
- Other workers on site (e.g., construction and electrical utility personnel).

When work is being performed in two locations (e.g., an electrical circuit is being switched off from one location to allow a residential cabling installer to work safely in another location) the worker(s) in each location should be in radio contact. They should repeat each message, and get confirmation that it was heard correctly before acting on the message. Never assume that related tasks have been performed; always get confirmation.

> **NOTE:** Portable radios use a limited number of frequencies; therefore, it is very likely that different crews will be using the same frequencies. When using radios to communicate between two locations, workers should always confirm that they are talking to the correct person.

Be alert and read any warning signs or markings. Bring them to the attention of coworkers who may have missed them. Encourage communication by accepting repeated information politely; it is better to be notified about the same hazard several times than it is not to be notified at all.

If a residential cabling installer discovers any defective or damaged equipment or facilities, the residential cabling installer must report them

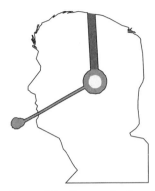

Figure 9.2 Communications headset.

promptly to a supervisor or directly to persons qualified to handle the problem. For example, report damaged electrical power lines to electrical workers on site or to a building or construction supervisor who will contact the proper electrical workers. If the defect or damage poses an immediate hazard, the residential cabling installer should do whatever is safely possible to ensure that others are not harmed by the hazard before qualified personnel arrive to fix the problem. This may involve notifying other workers in the area, putting up signs and barriers, or standing guard until qualified personnel arrive.

> **IMPORTANT:** Residential cabling installers must promptly report all accidents or injuries to their employer.

Designating and Securing Work Areas

When access by untrained personnel is a concern, always use safety cones to designate work areas and to restrict access. Yellow caution tape and folding A-frame signs may also be used. Be sure to leave enough room inside the cone perimeter to do the required work.

Consider the needs of the building occupants whenever possible. Try not to block a doorway or hallway for which there is no alternate route any longer than necessary. When working near doors or hallway corners, try to ensure that oncoming pedestrians can tell there is a work area ahead.

Do not leave open floor systems unattended. Do not leave open ceiling systems with dangling access panels or equipment unattended. Do not leave work areas cautioned off longer than required. The customer and coworkers will soon realize the work is complete and start disregarding the warnings. This could cause an accident in the future when a real danger exists.

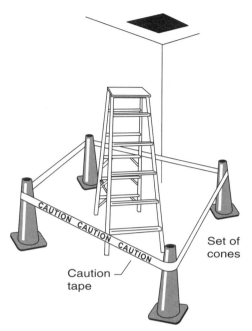

Figure 9.3   Designated work area.

Tools and Equipment

Residential cabling installers must use only those tools for which they are trained or certified to operate. They should also use the correct safety equipment as advised by manufacturers. Manufacturers require users to be certified to use certain devices such as powder-actuated tools. Never be afraid to ask questions or say that you are uncomfortable using a tool.

Use tools only for the purpose for which they are intended (e.g., use a tool designed for stripping to strip cable insulation—not a pocketknife). A screwdriver is not a scrapper, chisel, pry bar, Drywall saw, hole punch, or a drill for wood.

Examine both hand tools and power tools regularly to ensure that they are in safe working condition. Broken tools must be immediately tagged as "BROKEN" and removed from the job site. A detailed description of the problem should accompany the tag. This will prevent an injury to someone using the tool prior to its repair or replacement.

The wooden or plastic handles of hand tools must be kept free of splinters, sharp-edged cuts, or other surface damage that could injure a worker's hand. Do not use a hand tool if its handle is loose. Loose-handled tools can give way suddenly, causing injuries to both people and equipment.

When using powder-actuated tools, always verify that the area behind the work area is clear.

Figure 9.4 Tag broken tools.

Do not attempt to drive nails or other fasteners into very:

▪ Brittle or very hard materials (e.g., glazed tile, glass block, or face brick). The shattering of the material (or the fastener) can scatter dangerous shards across a wide area.

▪ Soft or easily penetrated materials. The nails or fastener can pass through the material and create a hazard for people on the other side.

> **NOTE:** These precautions apply to both manual and powder-actuated fastening devices.

Power tools should be inspected regularly to ensure that automatic cutoffs, guards, and other safety devices work properly. Follow the manufacturer's recommended maintenance schedule to ensure reliable operation.

Before each use, examine power tools to ensure that all guards are in place and securely attached. Also examine the power cord for fraying or other damage.

The *National Electrical Code® NEC®* requires a ground-fault circuit interrupter, residual current device, and/or circuit breaker when temporary wiring is used. Temporary wiring is an extension cord or even an entire building's internal wiring prior to final inspection and acceptance.

Power tools that require a three-conductor power cable must be grounded (earthed). Never use a power tool if the ground prong of the plug has been cut off. Never use a power tool with an extension cord or adapter that eliminates the ground prong before the cord reaches the outlet.

Never use a tool's power cord to lift or lower the tool. Carefully follow all manufacturer's instructions when mounting, securing, and using potentially dangerous mechanical equipment such as tuggers (for cable pulling), tension arms, cable wheels, cable brakes, and powder-actuated guns. Do not set up or operate such equipment without first receiving adequate training and having access to the manufacturer's instructions.

> **NOTE:** Keep original instructions for tools in a file at the office. Provide a photocopy of the instructions with each tool. This ensures that

a set of clean instructions is always available at the office that can be used to make additional copies for the job site.

Ensure that adequate lighting is available to safely and efficiently perform work. Proper lighting will help prevent accidents and rework. Use portable lighting when needed and make sure to keep it away from combustible materials. Always check the routing of the power cables to ensure safety.

Ladder Safety

Residential cabling installers must know how to choose a ladder, place it securely, and climb and work on it safely. The location of telecommunications cabling and equipment requires that ladders will often be used for both installation and repair work.

> **NOTE:** Laws require that manufacturers print ladder use guidelines on the ladders. Read and follow these guidelines.

OSHA does not require workers to wear fall-arresting safety equipment while working on portable ladders. However, it is a good safety practice to belt into a secure anchorage when working aloft with heavy equipment or over a prolonged period.

Use the correct type of ladder. Never use a metal ladder where there is a chance that the residential cabling installer or the ladder will touch energized electrical cables or equipment. Use ladders made of wood or nonconductive synthetics in these situations. Most construction sites will not allow metal ladders on site for safety and insurance reasons.

Figure 9.5 Ladder.

Ladders should be examined before each use. Check to ensure that:

- Joints between the steps and side rails are tight.
- Antiskid feet are secure and operating properly.
- Any moving parts operate freely.
- Rungs are free of dirt, liquids, or other substances that could cause slipping.
- Side rails are not damaged.

Choose a secure location to set up the ladder, such as flooring or ground that is solid, level, and offers adequate traction for the ladder's feet. If adequate traction is not available, the ladder must be lashed in place or held in position by another worker or workers. Never set a ladder on top of a box, furniture, or any other unstable surface.

Place an extension ladder so that both side rails are supported at the top, unless the ladder has a single support attachment at the top. For stepladders, verify that the supports that link the ladder rails to the back rails are fully extended and locked into place.

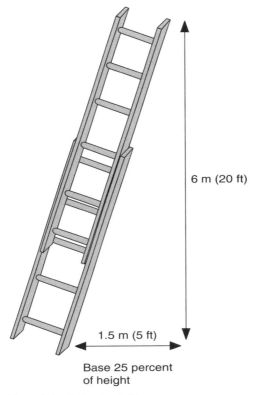

6 m (20 ft)

1.5 m (5 ft)

Base 25 percent
of height

Figure 9.6 Extension ladder.

Confirm or verify that the extension ladder is set at the proper pitch (angle). The distance from the base of the ladder to the supporting wall should be one-quarter (25 percent) of the length of the ladder. A ladder extended 6 m (20 ft) up a wall, for example, would have its base 1.5 m (5 ft) from the wall.

Extension ladders should always overlap between sections by at least three rungs. The top of the ladder should extend up to the work area and 0.91 m (3 ft) above catwalks or lofts. This allows the residential cabling installer to easily find the steps when getting back onto the ladder from the catwalk or loft.

Never paint a ladder. Doing so will hide any stress cracks or damage.

Try to place the ladder where it will be out of traffic. Use safety cones to designate a restricted area around the base of the ladder. Never set a ladder in front of a door that opens toward the ladder, unless the door is locked or can be blocked or guarded from the other side.

When using a ladder:

- Never exceed the ladder's weight rating. Most ladders are designed for one person only.

- Always face the ladder when climbing up or down.

- Never stand on the top two rungs of a ladder.

- Never leave any object, such as tools or gloves, on any rung of a ladder.

- Never straddle a ladder or stand on the rear rungs. The rear rungs are narrower than the front steps and are not designed to support weight.

- Never intentionally drop or throw down anything (e.g., tools, excess wire, or scraps) when on a ladder. Use a hand line and a "grunt sack" to raise and lower items.

- Never fasten two or more ladders together to create a longer section unless they are specifically designed for such use.

- Never move, shift, or extend a ladder while it is in use.

When the job requires a ladder, use a ladder. Do not stand on furniture, boxes, or any other ladder substitute. If a ladder is broken or stressed, tag it according to company policies with a large "DO NOT USE" sign to keep others from becoming injured. Defective equipment should be removed from the job site immediately and returned to the office for repair or disposal.

Personnel Lifts

A personnel lift is required when a ladder cannot be used safely because of the required working height or weight of personnel and equipment. Personnel should be properly trained in the use of a lift before use. There are two types of lifts:

- A bucket lift (cherry picker)—This is a fiberglass bucket mounted on the end of an extendable arm in which the user stands. The articulating arm allows the user to approach the work area from several angles and to avoid obstacles or possible safety hazards. These units are usually large and can be used in limited areas.

- A scissor lift—This is a working platform mounted on a large scissor jack. The scissor lift is very stable but not very flexible in its use. As the scissors are extended, the platform moves straight upward. If there are any obstacles above it, the platform is unable to maneuver around them.

Factors that determine if a lift is suitable include:

- The maximum working height of the lift.
- The size of the work area.
- Obstacles that may obstruct the lift.

Lifts must be secured by setting brakes and using stabilizing legs or outriggers, if the devices are so equipped.

Personnel must be surrounded by side rails or the bucket's outer walls and should wear a full body harness with two lanyards, one of which must be attached to the lift at all times.

As of January 1998, OSHA prohibited the use of safety belts and lanyards equipped with the "nonlocking type of self-closing keepers," also known as metal clips.

Personal Protective Equipment (PPE)

PPE is safety equipment worn by the residential cabling installer. When used correctly, PPE greatly decreases the residential cabling installer's risk of injury. When it is used incorrectly—or not used—it can leave the residential cabling installer exposed to a wide variety of dangers.

The PPE that a residential cabling installer is required to wear when performing a task depends on:

- The hazards of the task.
- The hazards at the work site.
- Other workers, homeowners, and the public.
- Local, state, and national safety requirements.

PPE must fit well and be as comfortable as possible. Equipment that fits properly and comfortably ensures that the residential cabling installer and the protective equipment can work at the same time.

Pay careful attention to the training for each item of PPE. Be sure to learn:

- When the equipment must be used.
- How to put on, adjust, and take off the equipment.
- What the equipment can and cannot protect against.
- Care and maintenance of the equipment.

It is important to inspect PPE each time it is used. Look for wear, cracks, tears, punctures, weak joints, or other signs that the equipment may not be able to provide protection. Report any problems to the proper supervisor. Never use defective protective equipment.

Remember that no amount of protective equipment can provide complete protection. Often the best personal protection comes from using caution, proper procedures, and common sense when working.

Headgear

Residential cabling installers must wear protective headgear (hard hats) when working in any area where there is danger of:

* Falling or flying objects.
* Electrical shock.
* Striking their heads.

Generally, the hard hats provided for residential cabling installers afford both physical and electrical protection. Residential cabling installers should ensure that their hard hats provide electrical protection before working around power lines or equipment.

The hard hat must fit securely enough to ensure that it will not slip and block the residential cabling installer's vision or fall onto the equipment the residential cabling installer is working on. Residential cabling installers may choose to use a chin strap to secure the hard hat only if the chin strap is thin enough to give way easily if the hard hat catches on something during a fall.

Before putting headgear on, inspect it for cracks, weakness of the internal support structure, or other defects.

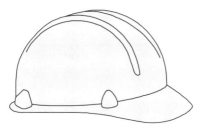

Figure 9.7   Hard hat.

Eye protection

Residential cabling installers must wear eye protection (full-face shield, goggles, or glasses) whenever there is a potential hazard to the eyes. If the residential cabling installer must wear prescription glasses to correct vision, pre-

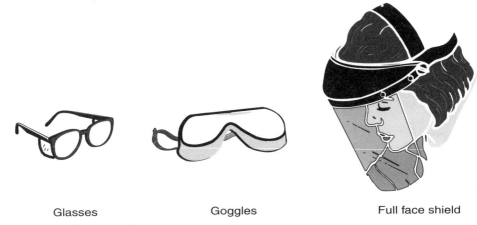

Glasses   Goggles   Full face shield

Figure 9.8 Eye protection.

scription safety glasses with side shields may be required, or goggles that fit over personal prescription glasses can be used. All eye protection must meet or exceed OSHA or AHJ requirements. A wide variety of work situations require eye protection, including:

- Working with batteries.
- Using powder-actuated tools.
- Using power tools, such as drills, which may result in flying objects.
- Working with optical fibers. (This does not include laser light protection.)
- Any situation in which the residential cabling installer is working above eye level and looking up at the work.
- Wearing a full face shield where there is danger of splashing chemicals, such as when working with batteries.

Wear protective goggles or glasses that provide side protection as well as front protection when the hazards involve flying objects. Residential cabling installers who wear prescription lenses must have eye protection that either fits over the prescription lenses or includes the prescription in the protective lenses.

Breathing protection

Residential cabling installers must wear a respirator or filter mask whenever harmful dust, gas, smoke, chemical vapor, or some other pollutant is present at the work site.

Never work without the proper breathing protection. The effects of breathing some harmful substances may not show up until hours, weeks, or years after exposure.

Filter masks are used in cases where the atmosphere is moderately hazardous. Very hazardous atmospheres require the use of gas masks or even compressed air respirators.

To provide the proper protection, each mask must seal itself to the user's face. This may require the removal of facial hair.

Residential cabling installers should not try to work while wearing a breathing protection device unless:

- They have been fully trained in how to use the device.
- The device has been carefully fitted.
- They have been found physically fit to work while wearing the device.

Although employers are required to inspect and maintain breathing protection devices, the residential cabling installer should also inspect the device every time it is used. Report any problems to the supervisor in charge of breathing protection devices.

Lifting belt

A lifting belt does not give the user any added strength. The belt is designed to be worn around the wearer's abdomen to help support the stomach muscles while encouraging proper posture.

Use correct lifting techniques when lifting any object on the job. These include lifting with the legs and not with the back, and turning with the feet and not at the waist. Whenever possible, wear a lifting belt when lifting or moving heavy objects or equipment.

When carrying or moving items, always know the path to the destination point, and be sure the items being carried do not block vision.

Protective footwear

Wear protective footwear on work sites where feet could be injured by falling objects, rolling carts, or by stepping on sharp objects. A good pair of shoes will protect feet from injury and fatigue. Steel toes will keep feet from getting smashed, while steel or fiberglass shanks will offer protection from stepping on sharp objects. Steel or fiberglass shanks also help distribute weight across the base of the shoe. This reduces foot fatigue while standing on the thin rungs of an extension ladder. Leather-soled shoes are not advisable, since leather conducts electricity when wet. Tennis shoes are not acceptable footwear.

Leather   Rubber

Figure 9.9   Gloves.

Gloves

Wear physically protective gloves when performing any work that has the potential for hand or forearm injuries. Keep in mind that:

- Leather gloves provide protection from cuts, abrasions, and extreme temperatures.
- Rubber gloves provide protection from harmful chemicals.
- Rubber and leather gloves are not for high-voltage use. All high-voltage situations should be referred to qualified persons.

Hearing protection

Hearing loss is one of the most frequent injuries encountered in the construction trades. The victim does not feel any pain, but after years of exposure to high levels of construction noise, varying frequencies of hearing may be lost.

Wear hearing protection while working in the vicinity of loud noises. Even the sound of a hammer striking a metal clamp onto red-hot iron requires hearing protection.

There are three major types of hearing protection:

- Disposable, foam plugs—These plugs can be rolled between the fingers and slipped into the ear canal.
- Reusable rubber earplugs—These may be on a breakaway cord or individually housed in a pocket-sized plastic container. They are convenient because they can be attached to a hard hat or around the residential cabling installer's neck and tucked inside the shirt. It is vital that they be on a breakaway cord to prevent strangling.
- Aural—These types resemble earmuffs. They are available in passive or active models. When wearing the active models, normal conversation can be heard; however, when a loud noise occurs, the protection automatically dampens the louder sound.

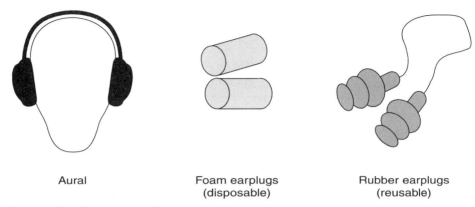

| Aural | Foam earplugs (disposable) | Rubber earplugs (reusable) |

Figure 9.10 Hearing protection.

When working in a noisy work site (with or without earplugs), be careful not to rely on hearing to detect the location of machinery, coworkers, or other hazards.

Safety harness

To prevent falls, residential cabling installers must wear a full-body safety harness and two lanyards any time they use an elevating device (scissor lift or bucket lift) to reach their work. Using two lanyards allows the wearer to always have one lanyard attached to a safe support. If the user has to move along a catwalk, but runs into an obstacle, he or she simply attaches one lanyard beyond the obstacle prior to disconnecting the second one, allowing the person to be safely attached to the structure at all times.

Safety lanyards are available as simple nylon ropes with self-closing and locking keepers (metal safety hooks) on each end, or they can incorporate a shock absorber into the line. If a fall occurs, the shock absorber reduces the force of the sudden stop to the victim.

Many work sites do not allow the use of safety belts. They do not provide the same level of protection as the full-body harness.

Before ascending or descending:

- Ensure the body harness fits properly. A harness that is too large or too small will not properly protect the wearer and may cause injury.

- Inspect the harness and its hardware carefully for signs of wear or damage.

- Ensure that the harness is properly secured to the elevating device's anchoring point, never the supporting guard rail or platform.

Whenever securing the lanyard, always check the connection. The residential cabling installer must ensure the metal clip has captured the lanyard's rope or the equipment's safety ring and is securely fastened.

Clothing

Work clothing should be reasonably snug but must allow the residential cabling installer to move freely. Do not wear dangling or floppy clothing that may get caught on tools or obstacles. Keep shirttails tucked in, and cuffs (if any) buttoned or neatly rolled up. This is especially important when the residential cabling installer is working in a confined space, on an elevating device, or near operating machinery.

Do not wear metal jewelry or metal watchbands when working on telecommunications circuits or equipment. Field-damaged clothing should be repaired immediately or the individual should leave the job site.

Grooming

Long hair can be extremely dangerous when working around operating machinery and while working aloft. Hair can easily be pulled into machinery or become caught on ceiling grids. Pulling the hair back in a ponytail usually provides adequate protection, while allowing the worker to wear safety equipment.

NOTE: The safest way to ensure that hair does not catch on something is to completely tuck it under a hat.

Hazardous Environments

Although there are hazards involved in any telecommunications cabling installation work, some indoor situations are especially hazardous. These situations require extra safety precautions and often require extra protective equipment.

A complete study of hazardous indoor environments is beyond the scope of this manual. The following sections give a brief overview of several commonplace hazardous environments and note some of the extra precautions they require.

IMPORTANT: Carefully follow all safety procedures when working in a hazardous environment.

Electrical hazards

The presence of electrical power cabling and electrical equipment is probably the most common environmental hazard faced by residential cabling installers. Like telecommunications cables, electrical power cables run in walls, under floors, and over ceilings. Power may also be required for telecommunications equipment.

All electrical systems are potential killers; therefore, all personnel should be aware of the dangers and have electrical safety training. Use power tools and equipment only for the purposes for which they were made. Use tools only according to the manufacturer's instructions.

Wear rubber-soled safety shoes and remove all metallic jewelry. Most jewelry is made of gold or silver, which are two extremely good conductors of electricity.

Never intentionally expose yourself to an electrical shock. That is, do not run your finger down a termination block to check for ringing current. Physical effects of current (milliampere [mA]) are as follows:

- 2–3 mA—Produces a tingling of the skin.

- 10 mA—Produces a painful shock and the muscles cannot release the contact.

- 50–100 mA—Breathing becomes difficult.

- 100 mA—Ventricular fibrillation occurs, causing the heart to repeatedly relax and violently clamp shut. This action destroys the heart and usually results in death.

- 200 mA and above—The heart clamps shut, severe burns occur, and a sickening smell is produced as the skin and hair burn away. At this level, the damage to the heart may actually be less than at 100 mA, allowing the victim to survive if medical treatment is given in time.

Treat all electrical circuits as if they were live (energized). Even after the circuits have been turned off and tested to ensure that they are off, treat electrical circuits as if they were likely to become live again at any moment. Always lock out all electrical circuits that have been turned off. Continue to maintain the proper clearances, wear the proper PPE, and take all the proper precautions.

Residential cabling installers must be especially careful in situations where electrical circuits or equipment may be contacted blindly (e.g., when drilling into walls or fishing conduits).

> **WARNING:** Do not use a metal fish tape in a conduit if the exit point is unknown.

Avoid working in standing water. If residential cabling installers must work in standing water (e.g., in a basement tunnel), take extra care to ensure that there are no electrical power circuits near the water or the work area.

Never cut the ground prong off a power tool plug. Removing the ground prong creates a serious possibility of severe electrical shock for the worker using the tool.

Avoid working on energized equipment. If the residential cabling installer must work on an energized circuit, such as when performing an alignment in a microwave radio or troubleshooting a telephone system, have a qualified safety person standing by. A qualified safety person must know:

- Where and how to secure electrical circuits.
- First aid and CPR.
- Where and how to get help.

Some equipment rooms (ERs) may be outfitted with an emergency electrical safety board. This board is mounted to the wall and may have a:

- First aid kit.
- Static grounding wrist strap—This is used to take the built-up static charge from the residential cabling installer to the ground and avoids damaging sensitive circuits.
- Safety grounding wand—This is an insulated handle with a metal tip that is connected to an insulated 1.8 m (6 ft) cable with a large metal clip on the opposite end. The clip is connected first to a ground source. The metal tip is used to short any transient voltages left on a de-energized circuit.
- Pair of high-voltage rubber gloves and protective leather outer shells—The rubber gloves should be inspected regularly for holes and cracks, while the leather outer shell must show no signs of wear. Only residential cabling installers trained in high-voltage rescues should use these gloves.
- Wooden cane—The cane should be lacquer free so as to be a nonconductive rescue device. It may be used to pull live wires off a victim or to pull the victim to safety.
- Class C fire extinguisher—The Class C extinguisher does not have any chemicals with conductive properties and is used for electrical fires.

Figure 9.11   Fire extinguisher.

In the case of an electrical fire, it is most important to protect people. Protection involves four steps easily remembered by the word RACE. This acronym stands for:

- Rescue—Get people out of danger.

- Alarm—Sound the alarm, call for help.

- Confine—De-energize all electrical circuits involved with the fire. Close windows and doors and deactivate the heating, ventilating, and air conditioning (HVAC) system.

- Extinguish—Control the fire with the correct type of firefighting equipment.

In multiple dwelling units, all multi-dwelling unit telecommunication rooms (MDU-TRs) and ERs should have access to a fire extinguisher designed to fight an electrical fire. The extinguisher must be Class C.

When working in MDU-TRs, make it a habit to check for an approved fire extinguisher. If the "C" rating appears on the label, it is approved. For example, "ABC," "AC," "BC," or "C" are approved. "A" is for combustibles: paper, wood, or anything that leaves ashes. "B" is for liquids, gas, oil, or alcohol. "C" is for electrical fires. These letter designations are important.

Lightning hazards

Although lightning is generally thought of as an outside plant (OSP) hazard, it can also endanger indoor workers. This is especially true during construction or renovation, when protective systems may be incomplete or disconnected. Exercise caution during an electrical storm when working on premises cables that are electrically connected to OSP cables.

Crawl space hazards

Telecommunications cabling often runs over suspended ceilings, below raised floors, and in other spaces where cabling installers cannot stand upright. These areas are called crawl spaces.

It is a good idea to wear protective headgear (hard hat) when working in crawl spaces, especially when electrical wiring is present. The hard hat will also protect the residential cabling installer's head from hard surfaces and sharp edges that may be found on supporting hardware for the floor or ceiling system.

Ensure that lighting is adequate to see the work clearly. If not, use a flashlight or other work light for extra lighting.

Before beginning work in any crawl space, take the time to locate and identify any other facilities that are routed through the crawl space (e.g., electrical power wiring, pipes, or HVAC ducts, etc.). Identifying surrounding hazards can keep the residential cabling installer from accidentally damaging another system or endangering himself/herself.

A filter mask or other breathing protection may be required if dust, fibrous insulation, or other breathing hazards are present in a crawl space. Check with employers and the building management to determine the nature of the hazard and the protection required.

When moving through a crawl space, walk or crawl only on surfaces designated to support movement. The residential cabling installer should never put weight on ceiling support hardware that is not designed to support human weight. Before putting full weight on a walk or crawl surface, the residential cabling installer should ensure that the surface is strong enough to bear weight. Never put weight on cable support devices (e.g., cable trays, etc.).

Never intentionally drop or throw anything (e.g., tools, excess wire, and scraps) from a crawl space above a suspended ceiling. Do not drop, place, or throw anything on top of ceiling tiles.

Crawl spaces may be considered as confined spaces that require additional precautions from hazards, such as:

- High or extreme heat.

- Animals (e.g., snakes, spiders, skunks, rodents, etc.).

- Electrical loads.

Confined spaces

According to OSHA 1910.146, *Permit-Required Confined Spaces*, "a confined space:

1. Is large enough and so configured that an employee can bodily enter and perform assigned work; and

2. Has limited or restrictive means for entry or exit (for example, tanks, vessels, silos, storage bins, hoppers, vaults, and pits are spaces that may have limited means of entry); and

3. Is not designed for continuous employee occupancy."

Maintenance holes, splice pits, crawl spaces, and attics can fall under the OSHA definition of a confined space.

Confined spaces may require testing for a hazardous atmosphere because they may contain:

- Unsafe oxygen levels below 19.5 to 23.5 percent.

- Flammable gas, vapor, or mist.

- Combustible dust.

- Any toxic substance in a concentration greater than deemed safe by OSHA standards. Nonlethal or incapacitating toxins are not covered by this provision.

- Any other atmospheric condition that is immediately dangerous to life or health.

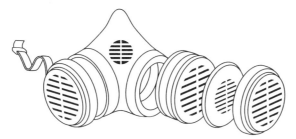

Figure 9.12   Breathing respirator.

When toxins, gases, or combustibles are detected, the residential cabling installer shall:

- Provide continuous forced air ventilation to purge the contaminants from the space.
- Periodically monitor the air quality within the space.
- Evacuate the space immediately if contaminants return.
- Determine why the contaminants returned and take corrective action prior to re-entering the space.

Certain confined spaces may require:

- Breathing apparatus.
- Protective clothing.
- A trained safety person stationed outside the space.
- A lifeline attached to the worker inside the space.

Hazards from toxic or flammable gas are rare when working inside a building. MDU-TRs and ERs should be completely free of gas hazards. However, there may be some situations where a sealed vault-type structure (e.g., an entry facility or splice pit) may accumulate gases.

Always use caution when entering any room or work area that is marked with warning signs that prohibit open flames or indicate other potential gas hazards.

When opening a vault-type structure, treat the vault like a maintenance hole. Before entering the vault, use a gas detector to determine whether any dangerous gases are present. If dangerous gases are detected, the gases must be cleared from the vault before any worker enters it. Testing a maintenance hole for gases and exhausting gases out of such a facility are usually considered OSP procedures and are outside the scope of this manual. Detailed descriptions of these procedures are available in OSHA 1910.146, and 1910.146, Appendixes A–D.

Optical fiber hazards

Optical fiber systems involve some hazards. Most of these hazards involve the optical fiber itself or the transmission light source.

Optical fibers are very thin but surprisingly strong. Small scraps can easily penetrate skin, causing irritation or infection. Ensure that cleanup is thorough after any optical fiber splicing or termination. Many workers use a loop of sticky tape or a container to collect fiber scraps after each cut. This ensures that all scraps can be disposed of properly at the end of the job.

Never throw bare fiber scraps into community trash containers. Always seal fiber scraps in a container, tape it closed, and mark it as optical fiber glass scraps. Take the container directly to the Dumpster® to avoid accidents to the unsuspecting customer.

> **WARNING:** Never look into the end of an optical fiber cable. Most optical fiber transmission light is invisible but can still burn the retina of the eye before the residential cabling installer realizes that the light is present. Light sources for test equipment may be just as hazardous as the regular system's light source.

Always wear eye protection when handling exposed fibers. Small fragments of optical fiber can easily fly into the eyes during cleaving. Exposed fiber ends can also injure the eyes when cables twist, flip, or fall.

Battery hazards

Working with or around vented lead-acid (wet) cell or nickel cadmium (NICd) batteries requires:

- An eyewash station.
- Training in handling electrolytes.
- Full face shield protection (front and side).
- Acid-resistant gloves and apron.
- Training in emergency procedures for spills.

Always use care when working around batteries. Batteries are always live. Batteries also release hydrogen gas as they are charged. Hydrogen is extremely combustible (four percent by volume in air) and must be vented outdoors. Most batteries are vented into the room, and an exhaust fan is used to pull the hydrogen and oxygen outdoors. If the fan fails to operate, the gases will build up and create a potential hazard.

Neutralize small acid spills with baking soda and clean up with damp rags.

Flush electrolyte burns to the skin with large quantities of fresh water. Apply a salve such as boric acid or zinc ointment to the skin and seek medical attention.

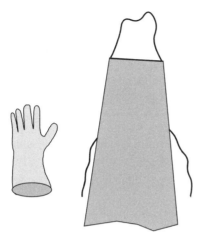

Figure 9.13   Rubber gloves and apron.

Asbestos hazards

Asbestos is a fibrous mineral substance that was used in many buildings as an insulation material between the mid-1940s and 1978. Asbestos has been widely used in acoustical ceilings, wall and ceiling insulation, fireproofing for structural steel, and pipe and boiler wrapping. The material often looks like plaster and cloth tape wrapped around pipes or the expanding insulation that is sprayed on boilers and structural steel. The difficult part of identifying asbestos is that it resembles these other forms of insulation and must be checked in a laboratory to verify its presence.

The use of asbestos was banned in 1978, after it was learned that breathing asbestos fibers could cause cancer of the lungs, stomach, colon, esophagus, and other organs.

Schools and other public buildings were required to locate asbestos-containing materials and determine the threat they posed. Then the asbestos-containing materials were either:

- Removed.
- Cleaned and sealed to prevent fiber releases.
- Labeled as asbestos and left intact.

The mere presence of an asbestos-containing substance is not hazardous, as long as it is not releasing fibers into the air. However, disturbing a substance that contains asbestos (by sawing, drilling, breaking, rubbing, or other construction procedures) may cause it to release fibers and create a serious breathing hazard.

Working in an asbestos atmosphere requires full breathing protection (respirators), protective suits, specialized training, and other special precautions.

Laboratory tests are required to determine whether a substance contains asbestos. If residential cabling installers are working in a building that was built or renovated between 1945 and 1978, and they encounter a substance that may contain asbestos, they should check with building management and maintenance personnel before disturbing the substance. Building managers and owners are required to maintain records of any known or suspected asbestos-containing substances in their buildings.

If residential cabling installers encounter a labeled asbestos-containing substance and cannot perform the job without disturbing it, they should stop work immediately and consult their employer about alternative plans.

Chemical hazards

Many products used in the telecommunications industry contain chemicals that can be hazardous to people and the environment. To help the residential cabling installer work safely with commercial products, manufacturers are required to provide MSDSs.

MSDSs are provided for all products used in a commercial environment that can be absorbed through the skin, inhaled, ingested, or require special handling for disposal.

MSDSs must be readily available at the job site where the products are being used. The sheets should be indexed and kept in a three-ring binder in alphabetical order by product name.

Chemicals and chemical vapors can be very dangerous. Without proper handling and ventilation:

- Toxic vapors can overwhelm the user, causing immediate and long-term effects. Fumes may cause nausea, headache, or vomiting. Prolonged exposure can cause disease to internal organs.

- Vapors may be flammable and create a fire when exposed to a spark.

- Vapors may be explosive when concentrated in a confined space.

- Lead-based paint hazards may be present in existing residences.

When working with products that produce toxic or flammable vapors, it is always best to use them outdoors. When this is not practical:

- Notify other workers and have a safety person check periodically.

- Open windows.

- Restrict air flow to other areas where people are working.

- Blow toxic air directly outdoors, paying attention not to send the vapors into an unsuspecting building next door.

- Blow fresh air into the work area.

- Take frequent breaks to keep the levels of toxin low.

Outside Plant (OSP) hazards

Hazardous outdoor working environments (e.g., maintenance holes, tunnels, ditches, and aerial facilities) are outside the scope of this manual. Special OSP training is required for working in these environments.

Safety Planning

Overview

Most large companies already have established safety plans and training following federal or state guidelines. However, many small firms do not. Some rely on materials provided by general contractors or specific industry organizations. Although it is not the intent of this manual to provide any formal safety plan, this section of the manual provides an overview of what such a plan should contain, identifies sources for this information, and includes a sample checklist for safety practices.

Per OSHA 1926 Subpart C, a safety and health program is designed to "institute and maintain a program that provides adequate systematic policies, procedures, and practices to protect their employees from and allow them to recognize, job-related safety and health hazards."

The following are the major elements of an effective program.

Management commitment and employee involvement

Employees generally will not fully implement a program not actively supported by management. Conversely, without the support of every employee, even a good plan is doomed to failure. At a minimum, the following actions are needed:

- Clearly state the company policy on health and safety.
- Establish a clear goal for the program.
- Provide visible top management support.
- Encourage employee participation.
- Provide adequate authority and resources.
- Hold everyone accountable.
- Review and update the program.

Work site analysis

Each work site must be examined to identify existing and potential hazards and to anticipate and prevent problems. The following measures are recommended:

- Conduct baseline work site surveys.
- Analyze planned and new facilities.
- Perform routine hazard analyses.
- Conduct risk assessments on workers' tasks.
- Conduct regular inspections of the site.
- Provide a reliable system of problem notification.
- Investigate accidents and incidents promptly.
- Analyze trends over time to detect patterns.
- Utilize OSHA material to review case studies pertinent to the work site.

Hazard prevention and control

The goal is to eliminate hazards in the workplace or, if that is not feasible, to control the hazards. To accomplish this, the following procedures must be in place.

- Utilize specific safety-defined engineering job techniques.
- Establish safe work practices through training and enforcement.
- Provide (or require) PPE.
- Maintain the site in a manner to reduce risk.
- Plan and prepare for problems.
- Establish a medical program that includes first aid.

Health and safety training

Training is a major portion of any effective program and must include the following components.

- Employee training.
- Supervisory training.
- Management support.

For the complete original text of these nonmandatory guidelines, check the Federal Register 54 (18):3094-3916, January 26, 1989. A full range of OSHA standards and directives is available on CD-ROM from the U.S. Government Printing Office. Portions may be found on the web at http://www.osha-slc.gov/Publications.

The following is an example of a job site inspection checklist.

Telecommunications Job Site General Inspection

Company Name: _____   Job Site Address: _____

Superintendent: _____   Date and Time: _____

Inspector(s): _____

Yes	No	N/A	Date Correct	Area/Issue Inspected
				1. Project site safety analysis complete?
				2. Hazardous materials identified and planned for or removed from site?
				3. Safety activities and protective equipment identified?
				4. Site emergency response and evacuation plan prepared?
				5. Monitoring plan prepared?
				6. Subcontractor conformance accepted?
				7. Employee and manager training completed?
				8. Posters, barricades, and safety signs in place?
				9. Safety meetings scheduled?
				10. First aid kit stocked and available?
				11. Emergency contract telephone numbers displayed?
				12. Emergency communications in place?
				13. Site safety supervisor assigned?
				14. Site recordkeeping system in place?
				15. Adequate toilets and washing facilities provided?
				16. Adequate work area illumination provided?
				17. Electrical service and grounding adequate?
				18. Support equipment (ladders, lifts, lights, hand tools) checked and serviceable?
				19. Work site cleaned daily?

Example 9.1   Job site inspection checklist.

> **NOTE:** This general safety inspection checklist is not designed to supercede existing safety inspection checklists. It should be used only as a general guideline.

10

Alternative Infrastructure Technologies

Introduction

The pattern of ever-increasing bandwidth requirements experienced in the commercial networking market over the past 10–15 years is occurring within the home market. These requirements are driven by video, multimedia, and similar applications. Bandwidth requirements have outpaced 56 kilobit per second (kb/s) dial-up Internet access and brought about the need for digital subscriber line, cable modem, and other broadband delivery systems. The residential structured cabling system (RSCS) promises the best hope of extending the useful life of the home telecommunications system.

The residential market has one major drawback associated with the recabling of existing homes. The cost of retrofitting a comprehensive RSCS may prove to be costly for most homeowners. While there are more than 1.5 million new homes built each year in the United States, there are over 100 million existing homes.

The telecommunications industry is faced with the need to develop and use telecommunications media within existing homes at a reduced cost and with minimal inconvenience to the homeowner. While many homeowners may choose to hard wire an RSCS to maximize their home network's performance, vendors are offering alternative solutions. These solutions include:

- HomePNA™ systems.
- Power-line systems.
- Wireless systems.

A properly designed and installed RSCS provides both superior bandwidth and access to high-volume, low-cost networking components that have commonly been used in the commercial environment (e.g., Ethernet hubs and network interface cards). An RSCS is always preferred for use with these systems. However, if the cost of installing the cabling media required by such systems is prohibitive, a number of alternatives are available.

HomePNA™ Systems

The first place to look for an adequate alternative telecommunications pathway in an existing home is the medium that most closely resembles an RSCS. It is a medium already installed in most residences—the telephone cabling system.

HomePNA (Home Phoneline Networking Alliance) systems are designed to use existing telephone cabling to transmit data. In addition to providing multiple access points for the networking system, HomePNA solutions also benefit from the experience most homeowners already have using telephone jacks for telephone, facsimile (fax), and dial-up Internet telecommunications.

HomePNA is an association with over 120 member companies worldwide. The group's 1.0 specification defines equipment that can transmit 1 megabit per second (Mb/s) data rates over existing telephone lines without disrupting standard telephone service. New-generation specifications are adding additional bandwidth to this technology and moving the performance of HomePNA systems into the range of the most basic networking technologies.

The main benefit of telephone-line networking is the limited amount of recabling required to make the system work. However, HomePNA may not be as accessible as it seems. In many homes, there are only a few telephone outlets. To allow for adequate coverage in all the areas of the home in which networking is required, additional cabling must be installed.

HomePNA technology uses frequency-division multiplexing (FDM) to transmit data signals at the same time that voice signals are being sent. This method allocates channels of specified bandwidth and frequency to each signal.

Implementation of a HomePNA system is relatively straightforward. HomePNA hardware is connected to the personal computer (PC) or similar networked device. A patch cord connects this device to the telephone outlet, and software is installed and configured in the computer.

Problems with a HomePNA installation occur most often due to noise being generated on the cabling and interfering with the data signal. The noise may be from either analog communication devices (e.g., fax machines and telephones that are transmitting on the same wires) or the interference may be from the electrical power used to supply a fax machine or similar device. To overcome this difficulty, low-pass filters can be added to the system. Low-pass filters transmit the lower frequencies used by the telecommunications system while blocking other interfering frequency ranges.

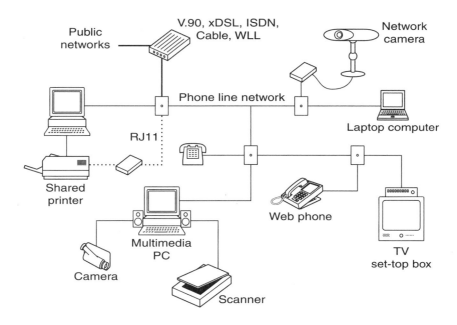

Figure 10.1   HomePNA system design.

ISDN = Integrated services digital network
  PC = Personal computer
  RJ = Registered jack
  TV = Television
WLL = Wireless local loop
xDSL = x digital subscriber line

Design criteria is important when considering a HomePNA system.

- Only 25 networking devices can be on the system, and cable lengths between devices cannot exceed 150 m (500 ft).

- HomePNA technology is not recommended for homes with over 1000 m^2 (10,000 ft^2) of floor space.

- HomePNA systems may not work properly in residences that use private branch exchange-based telephone systems.

Power-Line Carrier (PLC) Systems

PLC solutions use the existing electrical cabling as a telecommunications media throughout the home. In most homes, power receptacles are available in multiple locations within each room. Since virtually every device in the home (e.g., lights, appliances) draws power from the electrical system, these devices can be communicated with and controlled through the power system.

Many telecommunications requirements are well-suited to PLC solutions. Home control systems, for example, are typically bursty telecommunications systems, whereas voice, data, and video telecommunications are more continuous and transmit larger files. In burst mode, bandwidth requirements are low and the system is less demanding in terms of reliability. Missed instructions can simply be resent by the control mechanism to the individual control devices.

> **NOTE:** Control and monitoring functions were among the first to be addressed by the two major power-line systems, X-10 and Consumer Electronics Bus (CEBus®).

Typical applications for PLC systems are:

- Lighting control.
- Appliance control.
- Telephone distribution.
- Energy management.
- Heating, ventilating, and air conditioning.
- Computer networking distribution.
- Security.
- Audio/video distribution.

Some drawbacks of PLC systems may include:

- Limited bandwidth.
- Distance limitation of communicating devices.
- Lack of security between neighboring properties.
- Interference from neighboring systems.
- Reliability issues due to electrical disturbances (e.g., power surges, lightning strikes).
- Power usage in the home that can impact performance of the system.

Although the performance of PLC technologies may not be sufficient for residential computer networks, PLC systems are evolving to meet these challenges.

Typical two-phase electrical systems can also block telecommunications between devices on different phases of the same power grid. This is particularly troubling for larger homes that may be served by multiple electrical panels. Careful design practices can largely eliminate the functional impact of this issue.

Surge protectors are problematic for these systems, therefore, a whole-home surge suppression system may be appropriate. Some types of lighting systems may not work with PLCs (e.g., fluorescent, halogen, and sodium-vapor lighting).

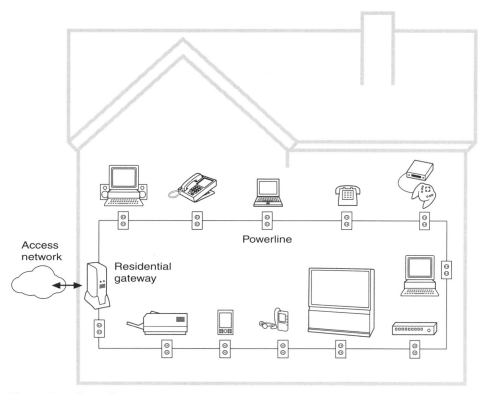

Figure 10.2   Power-line carrier system design.

Wireless Systems

Wireless technologies are meant to be an extension of the cabled RSCS. Wireless solutions are a practical alternative in hard-to-cable areas of an existing home, or for applications where the portability of computing devices is the critical issue.

In many applications, wireless may be a better solution than a communications system employing an RSCS. This is the case, for instance, where portability is critical or where areas in the home simply cannot be reached. When either issue exists and bandwidth requirements are not critical (and are likely to remain so), wireless technology is a good choice.

Some advantages of wireless systems are:

- Portability.
- Ease of installation.
- Accessing unreachable areas.

Some drawbacks of wireless systems may include:

- Limited bandwidth.
- Security.
- Interference.
- Distance limitations.
- Proprietary protocols.

Although the performance of wireless technologies may not be sufficient for residential computer networks, wireless systems are evolving to meet these challenges.

There are several types of wireless telecommunications technologies on the market today. The most widely accepted include the Institute of Electrical and Electronics Engineers, Inc.® (IEEE®) 802.11, *Wireless Networking* (Wi-Fi), HomeRF, and Bluetooth.

IEEE 802.11, *Wireless Networking*

This telecommunications solution is defined by the same committee of the IEEE that developed Ethernet, token ring, and other networking standards. IEEE 802.11 systems are designed to serve as a commercial-grade wireless local area network technology, but they also have application in the residential environment.

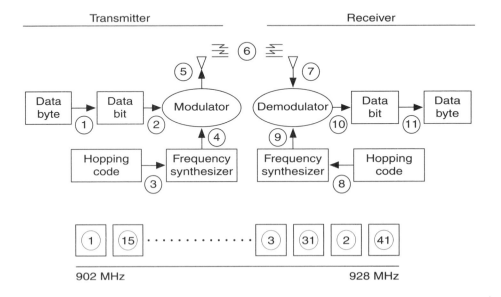

MHz = Megahertz

Figure 10.3   IEEE 802.11 wireless solution.

HomeRF networking

HomeRF is a radio frequency (RF) networking technology. HomeRF systems use a protocol called shared wireless access protocol, which includes six duplex voice channels in addition to the IEEE 802.11 Ethernet specification for data transmission. While up to 127 devices can be installed in a HomeRF network, typical transmission distances are 23–37.5 m (75–125 ft), and extending up to 45 m (150 ft) indoors.

Bluetooth

Bluetooth is a short-distance RF communications technology that operates for distances of up to 10 m (33 ft). Bluetooth is intended to be an RF solution in the home to replace the infrared remote controls used by many devices

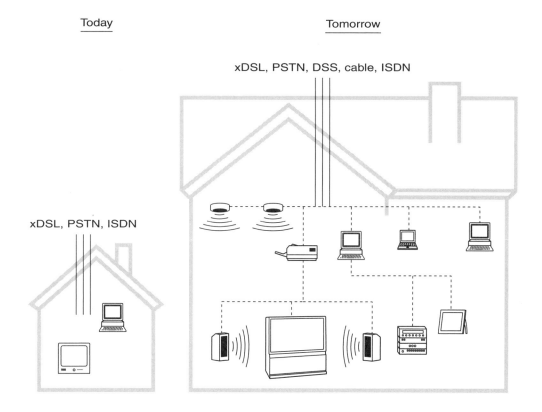

DSS = Digital satellite system
ISDN = Integrated services digital network
PSTN = Public switched telephone network
xDSL = x digital subscriber line

Figure 10.4 HomeRF wireless.

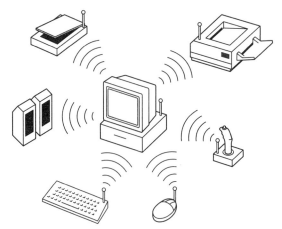

Figure 10.5   Bluetooth wireless computer usage model.

(e.g., televisions, video cassette recorders, etc.). These RF systems are designed to automatically connect devices into a small telecommunications network. Some examples of electronic devices that will benefit from Bluetooth technology include computer peripherals, cell phones, personal digital assistants, and headphones.

Broadband Coaxial Cabling

Another media already installed in many homes is the coaxial cabling used to provide community antenna television (CATV) service. Coaxial cabling is used by many providers of broadband services to transmit video programming, telephone service, and high-speed broadband Internet services (cable modems) to residential customers.

High-speed broadband services can be transmitted over the same coaxial cabling as television programming because each telecommunications service transmits and receives signals at a different frequency. Using the same principle, vendors are working on products that can be used to distribute voice and data (computer networking) information over the same cabling infrastructure, again using a different frequency range.

The Society of Cable Telecommunications Engineers (SCTE) is working with the Cable Television Laboratories, Inc. (CableLabs®), a nonprofit cable CATV industry organization, to develop an open cable initiative providing interoperability for advanced terminal devices, including set-top CATV boxes. The initiative is slated to:

- Deliver a blueprint for delivering advanced telecommunications services to consumers.

- Define a family of digital cable devices that provides these services.

IEEE 1394, *Standard for a High Performance Serial Bus—Firewire™*

The IEEE 1394 specification, also known as FireWire™, is now an open-architecture standard supported by the IEEE and managed by its IEEE 1394 Working Group. The standard is designed to support high-bandwidth requirements of devices (e.g., digital video equipment and high-capacity mass storage).

The objective of IEEE 1394 is to replace the various input and output connectors used in consumer electronics equipment and connectivity at the back panel of the PC using a scalable, high-speed serial interface. IEEE 1394 technology allows different digital signals (e.g., video, audio, and device control commands) to be multiplexed onto the same twisted-pair conductors. It uses an arbitration approach to ensure that all devices with data to send have a fair chance to transmit.

IEEE 1394 was developed to meet the need for high data transfer rates between PC components and between the PC and connected devices. The objective is to develop a single connection type that is able to move large amounts of data into and out of PCs or peripheral devices—typically from consumer electronics devices (e.g., video cameras). The current standard for point-to-point cable connections supports data rates of 100, 200, and 400 Mb/s, with higher data rates in development.

Some of the advantages associated with IEEE 1394 include the following:

- It provides a standard serial connector type on PCs and associated peripheral devices.
- IEEE 1394 serial cable is less bulky than parallel cables.
- A high data transfer rate accommodates multimedia applications.
- It has the ability to connect and disconnect devices without disrupting computer operations.
- Devices can be connected to a PC without requiring terminators.

The IEEE 1394 standard is a specification for a layered transport system. The standard defines three layers, as follows:

- Physical layer specifications for digital signaling
- Link layer specifications for formatting data into packets before transmission over a 1394 serial cable
- Transaction layer specifications for presenting link-layer packets to the application

IEEE 1394 supports both asynchronous and isochronous communications. Asynchronous communications guarantees data delivery, since the receiving device transmits an acknowledgment of receipt to the sending device. However, the delivery latency of data cannot be predicted, since it depends on the amount of traffic at the time of communications.

The isochrono-communications mode guarantees a time slot for data transfer between devices, ensuring the fixed data rate that is required by time-sensitive applications. Ongoing isochronous communications between two or more devices is referred to as a channel. Once such a channel has been established, the requesting device is guaranteed use of the time slot.

Providing both asynchronous and isochronous capabilities on the same interface allows both nonreal-time applications such as printing and scanning to operate over the same link as real-time applications (e.g., video conferencing).

To achieve the high speeds associated with IEEE 1394, the cable used to connect two devices cannot be more than 4.5 m (14.8 ft) in length. However, a maximum of 17 devices can be daisy-chained to each other over 16 links, for a maximum end-to-end distance of 72 m (236 ft). Following a tree topology, a maximum of 63 devices (due to addressing limitations) can be interconnected using IEEE 1394 technology.

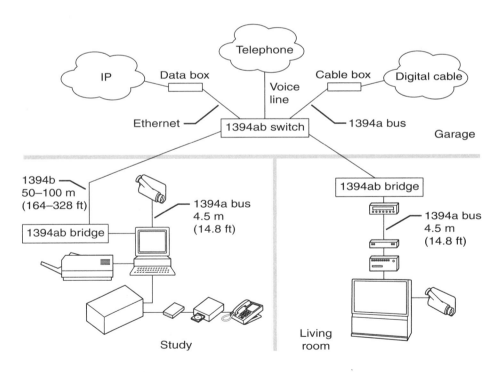

IP = Internet protocol

Figure 10.6   IEEE 1394 home network.

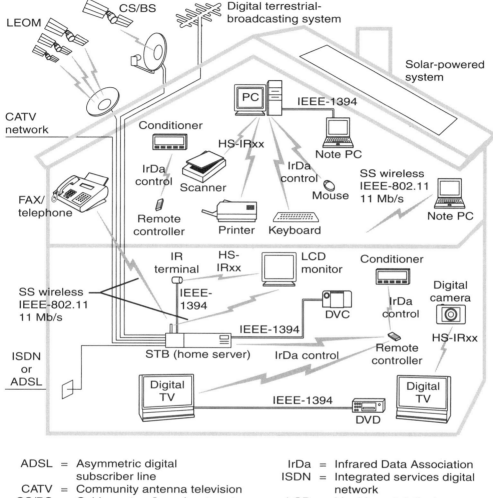

Figure 10.7   IEEE 1394 Firewire residential network with plastic optical fiber.

ADSL	=	Asymmetric digital subscriber line	IrDa	= Infrared Data Association
CATV	=	Community antenna television	ISDN	= Integrated services digital network
CS/BS	=	Cable system/broadcast system		
DVC	=	Desktop video conference	LCD	= Liquid crystal display
DVD	=	Digital versatile disc	LEOM	= Low earth orbit microsatellite
fax	=	Facsimile	Mb/s	= Megabits per second
HS-IRxx	=	High-speed infrared	PC	= Personal computer
IEEE®	=	Institute of Electrical and Electronics Engineers, Inc.®	SS	= Spread spectrum
			STB	= Set top box
IR	=	Infrared	TV	= Television

In an IEEE 1394 network topology, devices can have one or more 1394 ports. A one-port device is referred to as a leaf device, since it can be connected only to the end of a branch on the network. To create a daisy-chained topology, two-port devices are also required. In addition, to construct a full 63-device network, devices with three or more IEEE 1394 ports must be used. In such a network, each device acts as a repeater.

Plastic Optical Fiber (POF)

POF has recently become an alternative patch cabling media used for the transmission of information primarily between audio devices. Connectors are now typically being installed on digital versatile disc players and other audio/video products.

POF technology has long intrigued the computer networking industry. Certain experimental varieties of POF can transmit data at rates nearly as high as that of singlemode glass optical fiber; however, they are not used in the home environment. POF also enjoys the same immunity as its glass counterpart to electromagnetic interference, radio frequency interference, and other types of interference, an especially important factor in the typically electrical noisy residential environment.

What separates POF from glass optical fiber and copper cabling in the residential environment is the ease with which the media can be terminated. The core of the fiber, through which the information-carrying light travels, is large enough to allow do-it-yourself homeowners to install and maintain it with a specialized tool kit.

However, a drawback to POF is its limited range of transmission—typically under 50 m (164 ft), including patch cords. This distance limitation has proved a major stumbling block in the commercial networking environment, where many computers are up to 100 m (328 ft) from the networking hub. This nonstandard media does not lend itself to the RSCS deployed by other media described in this manual or ANSI/TIA/EIA-570-A, *Residential Telecommunications Cabling Standard*.

11

Other Residential System Applications and Implementations

Introduction

This manual describes the residential structured cabling requirements for the distribution of voice, data, and video (VDV) applications. The flexibility, reliability, and longevity (in terms of extended bandwidth) of these residential structured cabling systems (RSCSs) make them an attractive technology, particularly in new construction where the cost of installation is minimized.

Several of the alternative residential telecommunications and control media (wireless, power line, etc.) have already been described. The main benefit being sought by the supporters of these technologies is the use of media that is already installed in the home to distribute information more cost-effectively. The assumption with these technologies is that a portion of the homeowner public will trade performance, and potentially higher hardware costs, for portability and the ability to use the residence's existing infrastructure.

RSCSs serving applications other than VDV are described in this chapter. These systems provide residential services such as:

- Satellite telecommunications.
- Security.
- Whole home audio.

Some of these systems can be deployed using alternative media described in Chapter 10: Alternative Infrastructure Technologies. Wireless, for instance, is a strong contender for home security applications. This chapter, however, will largely focus on wired options.

Direct Broadcast Satellite (DBS)

DBS systems are becoming popular for a variety of reasons. In new construction and in areas where community antenna television (CATV) is either unavailable or unreliable, the quality of DBS reception and its ease of installation provide significant benefits. The variety of programming and the system's ability to be customized to meet personal tastes are also well-accepted features of this technology. The availability of broadband speed Internet services using the same system is an increasingly important feature of DBS.

Whoever the vendor for the DBS service (e.g., DishNetwork, DirecTV, World Satellite Network, Inc. [WSNet], etc.), the basic layout of the system is the same. There is a satellite dish or dishes mounted on the exterior of the home. Satellite grade coaxial cable runs from the satellite dish to the home's distribution device (DD). Once the RSCS distributes the signal throughout the home, each television (TV) requires a satellite receiver or set-top box.

The satellite dish should be mounted high on the exterior of the home to provide an unobstructed, long-term line of sight to the satellite. In aligning the satellite dish with the satellite, it is critical to avoid:

- Trees (which will grow over time).
- Utility poles.
- Other homes.
- Similar obstructions.

Mounting the satellite dish high on the home may make repairing the system difficult, but mounting it low to the ground may invite vandalism and other types of damage.

It is recommended to mount satellite dishes in the same location (preferably an unobtrusive one), on multiple homes in the same subdevelopment, in order to maintain the aesthetic appeal of the homes (see Figure 11.1).

Once the satellite dish is mounted, the residential cabling installer or homeowner should contact the satellite service provider (SP) or visit the company's Web site to determine the proper direction and angle for the satellite dish. By entering the home on an interactive Web site display, the residential cabling installer can easily determine the direction and angle at which to set the satellite dish to guarantee proper reception. The direction and angle of reception are more sensitive with digital satellite TV than with analog off-air TV signals received by an omni-directional antenna. Use of a field test instrument is advisable since it is easier to determine direction and angle and adjust the system while the residential cabling installer is still on the roof (see Figure 11.2).

The number of cables running from the satellite dish into the home may be one, two, or more, depending on the SP and the programming and services being offered (e.g., foreign language channels, Internet service, etc.).

If DBS is prewired during construction, it is advisable to run cable into the attic area (if one exists) to hide the cable from view. Cable can also be hidden

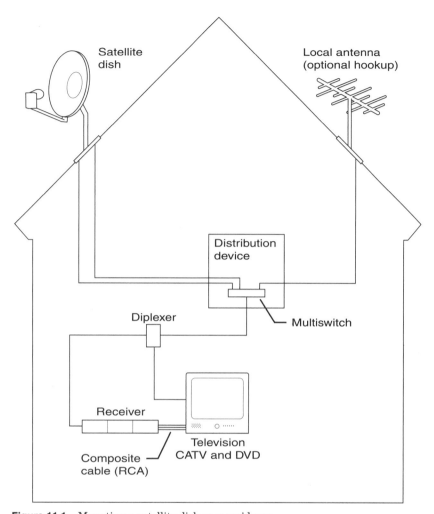

Figure 11.1   Mounting a satellite dish on a residence.

behind chimneys, water drains, and other building structures. Make sure a long length of cable is looped in the attic for future attachment to the satellite dish.

The cable should be high-frequency (minimum 2.2 gigahertz) Series 6 quad-shielded coaxial cable. The cable should be terminated in the DD, along with the home's internal coaxial cabling, using F connectors.

DBS satellite service cannot be distributed in the same way as traditional coaxial cable services. A multi-switch must be used when more connections are required than are supplied by the satellite dish low noise block.

A diplexer is a device that combines the digital DBS signal and the analog CATV signal in the DD, allowing both services to be transmitted along the

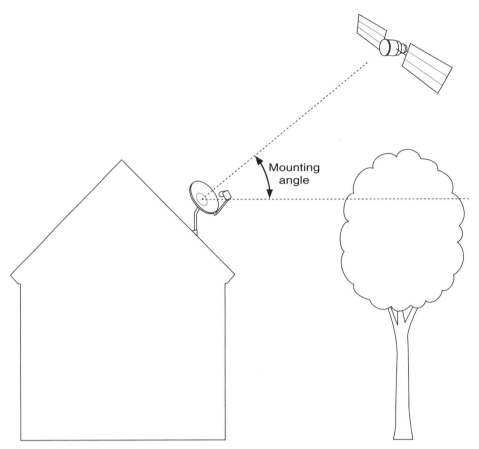

Figure 11.2   Satellite dish mounting angle.

same coaxial cable. Since different frequencies are used for each service, a single outlet cable can be used to provide the two separate services. An additional diplexer is needed at the outlet to separate the combined signal. The appropriate device is then connected.

Alternatively, the output from one satellite receiver can be sent back through the home's cabling system and viewed on multiple TV sets, but each set in this case must necessarily show the same program. Multiple receivers allow multiple TV sets or home theaters to be used to independently control the video source being displayed.

A well-designed RSCS should include a twisted-pair cable beside each video cable. The twisted-pair cable provides telephone access to the set-top box so that the homeowner can more efficiently order pay-per-view programming through the satellite system.

The homeowner often has a choice concerning how to receive off-air broadcasts provided by major local network affiliates. Most satellite SPs offer local

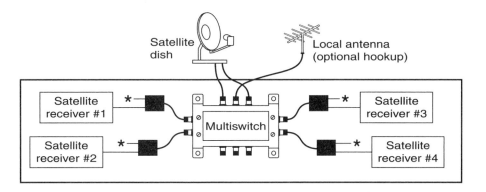

= Diplexer, if hooking up local antenna to multiswitch

★ = To local antenna feed on television or receiver

Figure 11.3 Typical direct broadcast satellite signal distribution.

> **NOTE:** All cables in Figure 11.3 are Series 6 quad-shielded coaxial cables.

channel options for an additional monthly fee. Alternatively, the homeowner can draw signals from a standard off-air antenna, mounted in the attic or on the roof, and feed video signals via coaxial cabling into the home's telecommunications system.

Home Theaters and Entertainment Systems

The market

For homeowners, homebuilders, and integrated services professionals, the growing demand for prewired, in-the-wall entertainment systems is a significant trend. In-the-wall entertainment systems range from a simple system consisting of a pair of speakers in a room to a complete, cinema-quality home theater. The main benefits of prewired entertainment solutions in new construction are:

- More efficient use of floor space.
- Improved aesthetics.
- Elimination of cordage.
- Unobstructed audio service for higher quality sound.

- Sharing expensive audio components to reduce costs.

- Whole-home, multizone control.

The number of home entertainment system vendors and the broad range of prices and features they offer allow the homeowner of even the most modest home the opportunity to take advantage of the benefits listed above. By installing a home entertainment system during construction along with an RSCS, security system, and home automation solution, the homeowner may be able to cut labor costs significantly.

The importance of these systems to the homebuilder is that they are mounted within the walls or ceilings and become part of the home. Besides minimizing installation costs, the integration of these systems during new construction also allows the homebuilder, homeowner, and integrator to determine if any structural design changes are needed to accommodate optimal acoustic and visual performance.

Another important trend driving this market is the desire of residents living in high-rise multi-dwelling buildings in metropolitan areas to gain the additional usable floor space made available by the installation of these in-wall audio/video (A/V) systems. Residents in these situations are often paying a premium of several hundred dollars per square foot for floor space. This makes the installation of in-wall, A/V systems, including the new generation of flat TVs, extremely attractive and cost-effective.

A critical issue to keep in mind is that many multi-dwelling buildings are built using concrete slabs or other materials that make cabling after construction difficult, if not impossible. The desires of the eventual homeowner must, therefore, be determined before construction.

A/V cabling comprises a low-voltage system fully within the scope of the structured cabling professional or homeowner. Since much of the video portion of A/V cabling (i.e., coaxial cabling) is terminated within the DD as part of the ANSI/TIA/EIA-570-A, *Residential Telecommunications Cabling Standard*, this section focuses on the design and installation of audio systems.

Both audio and video signals can be transmitted via a number of media. Plastic optical fiber (POF) may be used for video systems in the future. POF ports have already made their way into many A/V components in the market. POF is mostly used today to transport high-data-rate signals over short distances between devices colocated, for example, within an entertainment center. Twisted-pair technologies are also being deployed for the distribution of audio and video throughout the residence. Even wireless systems can deliver A/V services in some circumstances.

Audio Residential Structured Cabling Systems (RSCSs)

Audio cabling should be designed into the home and installed as another structured telecommunications cabling subsystem. As an RSCS, the topology of the audio cabling system should be designed as a generic (i.e., nonvendor- and nontechnology-specific) system intended to stay with the home over time and to meet the various needs of multiple owners.

It is important that considerations are given to the installation of this type of system as early in the construction or remodel process as possible. The homeowner or builder will walk through the home with the building plans to discuss how each room is expected to be used. Speaker and volume control placement issues should be resolved before beginning the prewire phase.

The main components in distributed home audio today are:

- Audio source—Typically, a device that converts recorded media to line level signal (e.g., multi-disc compact disc [CD] player, audio from a computer, satellite music source, etc.).

- Amplifier—A device that amplifies line level audio from the audio source to speaker level output.

- Distribution system—The cabling and any bridging or impedance-matching equipment. The number of zones that a distribution system can support depends directly on the impedance-matching system used. For standard impedance-matching analog volume controls, the best practice is to limit the number of zones to eight. However, up to 32 zones could be installed on one amplifier, with some penalties in overall system performance. Large distribution systems suffer from reduced power to the speakers proportional to the size of the system.

- Volume controls—The volume in each room (zone). Every zone should have a means of adjusting the volume independently from the other zones in the residence. Many volume controls incorporate impedance matching.

- Speakers—The electromechanical device that converts electronic audio signals to sound. When selecting speakers, keep the following criteria in mind:

  - Aesthetics—The physical appearance of a speaker in the room is perhaps the most important consideration. The speakers must fit the overall décor and use of the room.

  - Dispersion—Select a speaker that can be installed in such a way as to give the homeowner maximum stereo sound coverage. Ceiling speakers offer the best coverage in most applications. Wall-mounted speakers, either flush or surface, can work well in areas with low ceilings or rooms with major ceiling features that draw attention upward (e.g., chandelier).

  - Sensitivity and room size—Balance is the key to a quality sound system installation. Large rooms (e.g., living rooms) typically have challenging acoustical problems. These rooms have soft, sound-absorbing furniture and carpets that effectively attenuate music volume. The opposite is true of small spaces (e.g., a bathroom). There a residential cabling installer is likely to find hard surfaces everywhere, including the floor. Music sounds much louder in a space like a bathroom or utility room. When installing

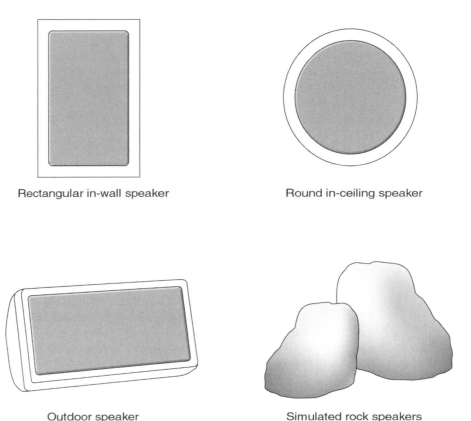

Rectangular in-wall speaker   Round in-ceiling speaker

Outdoor speaker   Simulated rock speakers

Figure 11.4   Speakers.

speakers, keep in mind that spaces that are considerably larger than other spaces in the home may require additional speakers to offer smooth coverage and balanced sound levels.

Speaker and volume control location considerations.   There is a certain science to the placement and installation of speakers for multi-room stereo systems. It is usually best to place the speakers in the room in such a way as to provide stereo left and right imaging when facing the main feature of the room. This feature could be a large picture window or fireplace in a general use space. In a bedroom, the speakers are best placed over the foot of the bed. Speakers installed in the ceiling usually provide the best dispersion in any space. Generally, speakers should face the typical sitting location of the listeners. Once the best scenario for speaker mounting has been selected, the rule-of-thumb is to mentally divide the room into three equal sections. The imaginary lines where any two sections meet are the best locations for the speakers (after taking into account furnishings, fireplaces, windows, and light fixtures). This

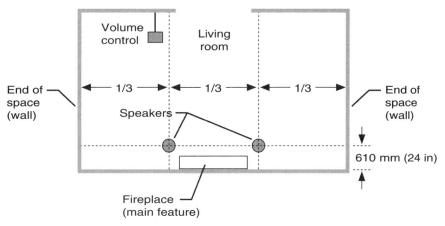

Figure 11.5   Example of speaker placement.

method places the speakers roughly equidistant from the room corners (called boundaries) and from each other.

The wall/ceiling interface, however, is another room boundary. Placing a speaker too close to a room boundary can result in "doubling," or an excessively booming sound. Today's sophisticated A/V receivers and audio processors can help correct some room speaker interaction problems, but it is better to start from as sonically correct a placement as possible.

Keep speakers at least 600 mm (24 in) from any room boundary, including the ceiling (see Figure 11.5). An exception is soffit locations, where there might be a protrusion of a Drywall space into the room. Because soffits represent extruded room boundaries, they can sometimes be used to the residential cabling installer's advantage.

Audio Residential Structured Cabling System (RSCS) design

The following wiring method for home audio is intended to cover the majority of current and near-future technologies. There are many other possibilities designed around specific manufacturers and their equipment. This design gives the residential cabling installer and homeowner the widest choice possible in selecting equipment, keeping the optional future upgrades open.

There are three distinct cabling runs in this system, with multiples of each to create a layout to suit the home. Figure 11.6 shows an example of this layout in a typical residence.

The three runs are:

- From the amplifier location(s) to the DD—From a suitable wall plate located at the amplifiers and source equipment location, run six conductors of speaker wire and one Category 5e cable. Four of the speaker wire conductors are to deliver left and right speaker signal to the DD, the remaining two

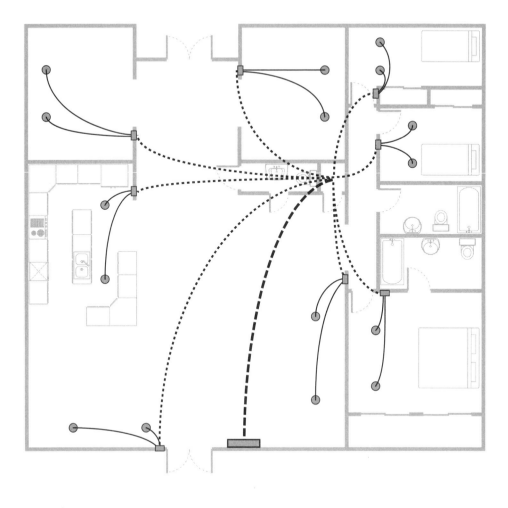

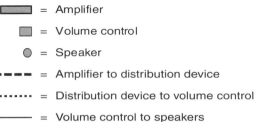

━━━ = Amplifier

▨ = Volume control

● = Speaker

━ ━ ━ = Amplifier to distribution device

• • • • • • = Distribution device to volume control

──── = Volume control to speakers

Figure 11.6   Example of an audio residential structured cabling system.

conductors may be used to deliver direct current (dc) voltage to the DD for switching or control. The Category 5e cable will support other communications relevant to audio distribution (e.g., infrared [IR], data communications, A/V distribution).

- From the DD to the volume control locations—Each audio zone should have at least one volume control. Some systems can have more than one. From the DD to each volume control location, pull four conductors of speaker wire and one Category 5e cable to each volume control location. It is important to know that most volume controls are large devices that do not fit well into the average single-gang electrical box. Low-voltage mounting brackets are highly recommended if allowed by local code. The other alternative is to use the largest, deepest four square box available with the appropriate square drawn cover (commonly known as a mud ring) attached.

- From the volume control locations to the speakers—Each speaker requires two conductors of speaker wire from the speaker to its associated volume control at a minimum. It is advised to install one Category 5e cable for future applications such as IR repeating or the installation of amplified speakers.

Although home theaters can be tied into the home's A/V system and share the high-end components usually placed within the theater (e.g., digital versatile disc players, CD carousels, etc.), in most cases the home theater is developed as a completely isolated subsystem. This is particularly helpful in the setup of the home theater's precise audio speaker system.

While many low-voltage specialists may be comfortable cabling for preinstalled speaker systems, some may find the entire discussion of specifying and providing speakers and other A/V components troubling. It is not enough to run cabling to a point behind the walls. A true rough-in of an audio system may include the speaker mounting brackets, which must be sized to match the manufacturer and make of the speakers to be used. In many cases, the low-voltage specialist must obtain a significant level of expertise in A/V solutions. Also, the development of home theaters often requires unique woodwork, acoustic tiling and walls, special theater seats, and other products, which may or may not be a task assigned to the low-voltage specialist. Cabling for whole-home audio systems is specified in ANSI/TIA/EIA-570-A-3.

Home theaters

Home theaters are becoming an increasingly popular feature of today's homes. They also offer unique design and installation challenges for the integrated services professional.

Audio. Many vendors provide theater quality in-wall and in-ceiling speakers. Typically, a sophisticated sound system is composed of five audio channels plus low frequency (subwoofer):

- Right front
- Left front
- Center

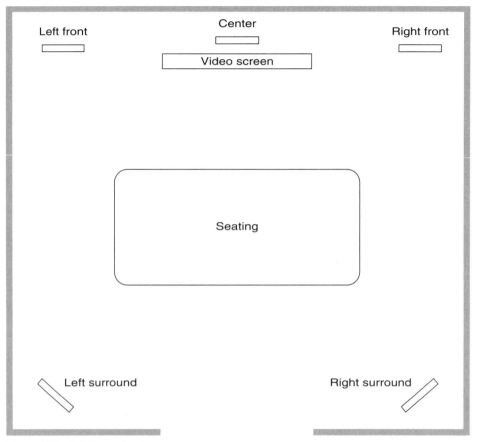

Figure 11.7   5.1 Home theater.

- Left surround
- Right surround

> **NOTE:** All front channels speakers should be installed at the same height from the floor.

Although there are more advanced systems that may have seven channels, including both mid- and rear-surround sound speakers, they are beyond the scope of this manual.

The size of the room is measured in terms of cubic footage, which is then compared to the speaker manufacturer's product specifications.

Video.   An important component in a home theater is the video projector or TV. There are currently three main types of device to choose from.

High-definition plasma monitor   High-definition projection TV

Front projection

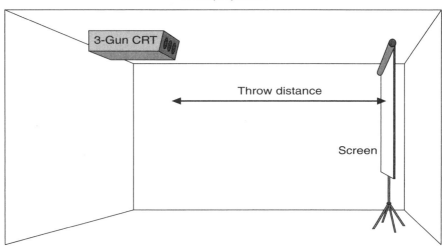

CRT = Cathode ray tube
 TV = Television

Figure 11.8   Home theater video.

- Plasma TVs
- Rear projection TVs
- Front projectors

   The projection systems employ a projection unit and a separate screen. The screens can easily exceed 2500 mm (100 in) and are ideal for dedicated home theater rooms. The room must be made dark for optimal viewing. There are three main types of projection TVs.

- Liquid crystal display (LCD) projectors
- Digital light processor (DLP)
- Three-gun cathode-ray tube (CRT)

Home Automation Systems

The following sections describe three important residential telecommunications technologies:

- Security
- Intercom
- Lighting and heating, ventilating, and air-conditioning (HVAC) control

These systems are similar to the wide range of products and systems developed as part of building automation systems (BAS) for commercial buildings. BAS provide monitoring and control of many aspects of a commercial building. Additional information concerning BAS is in Chapter 21 of the BICSI *Telecommunications Distribution Methods Manual*, 9th Edition.

BAS in commercial buildings include fire alarms, security and access control (including closed-circuit TV), HVAC, and energy management systems (including lighting control). Many of these systems are available to the residential homeowner and are additional services that fit well within the business of the residential structured cabling installer. Some of the vendors of VDV structured cabling products integrate these home automation systems into their DDs.

Home automation services require the availability of reliable telecommunications, but differ from VDV in many respects. Data rates for such systems are typically less than one megabit per second (1 Mb/s). They operate in burst mode instead of sending a continuous or semi-continuous stream of data, and they are often connected through various media as separate communication subsystems.

Designing the home automation system to work in conjunction with, or upon, the RSCS intended for use as a distribution method for VDV provides many benefits, particularly when installed during new construction. Implementing all of these systems together allows a single project team, or one telecommunications residential cabling installer, to manage the entire project, reducing costs and eliminating contention between various trades. The integrated SP supplies a single point of contact for builders and homeowners, providing flexibility to make system upgrades and rearrangements when needed.

Installation costs are minimized and aesthetic issues are reduced when implemented during new construction, since multiple telecommunications requirements can be handled over the same cabling system, or the residential cabling installer can implement multiple cabling systems along the same cable paths with a single cable pull. The potential for damage to the residence during installation is also reduced, as is interference between multiple cable pathways and other residential utilities such as HVAC ducts, electrical power, and plumbing.

Once the installation is complete, a DD can be used to house and power multiple networking and automation subsystems. The use of a single DD will often reduce:

- Costs.
- Required wall or floor space.
- Long-term maintenance costs.

The critical issue in determining whether an ANSI/TIA/EIA-570-A-style of RSCS can be used for a particular home automation solution is the distance limitations imposed on the system by the use of twisted-pair cabling.

Twisted-pair 24 AWG (0.51 mm [0.020 in]) cable can handle 1 amp of current draw per conductor to a maximum of 3.3 amperes for a 4-pair cable. The voltage required to drive an automation system is defined by the system hardware. The voltage crossing a twisted-pair cable is reduced due to the resistance of the cable itself. A typical 24 AWG (0.51 mm [0.020 in]) twisted-pair cable has 0.1876 Ω of resistance per meter (.0572 ohms per foot).

To calculate the maximum distance an automation signal can travel down a twisted-pair cable, consult the system manufacturer.

Security and Monitoring

Residential security and monitoring is one home automation system that is often preinstalled into new construction or added to existing residences. These systems rely on a low-voltage (or wireless) telecommunications media and, therefore, can be a natural fit for today's integrated SP. Local and state codes may require licensing for the low-voltage firm to provide security services.

Today's security systems are significantly more advanced than previous versions. They can be wireless systems, making installation into existing residences simpler and more straightforward. They are often designed to be monitored through the same DD that the residence's RSCS uses for VDV telecommunications. The security system may also be designed to allow the user to manage it remotely through dial-up or Internet connections.

As with the audio cabling subsystem, a prewired security and monitoring system is typically star-wired from each device location to the central control panel, although some level of daisy-chaining can be allowed for multiple devices on the same security zone. Most systems allow the integrator to program in multiple security zones to better track down where an alarm has been triggered. Each security device can be an individual zone. For example, multiple windows can be managed as a single zone with cabling daisy-chained between the various monitoring devices (e.g., window contacts and glass break detectors). However, a home run star topology is much more reliable, easier to troubleshoot, and can be easily reconfigured.

Home security system components

The critical devices in a home security system are:

- Control panel—This is the command center for all programming and provides a connection to the incoming telephone service. It may be a stand-alone product, part of an RSCS DD, or integrated into a security system keypad. The control panel should be mounted close to, or within, the DD so that a connection can be inserted into the telephone service in order to capture the phone line to make outgoing emergency calls to the monitoring service. Local and national codes apply. Make sure that the intended DD is Underwriters Laboratories Inc.® approved for this service.

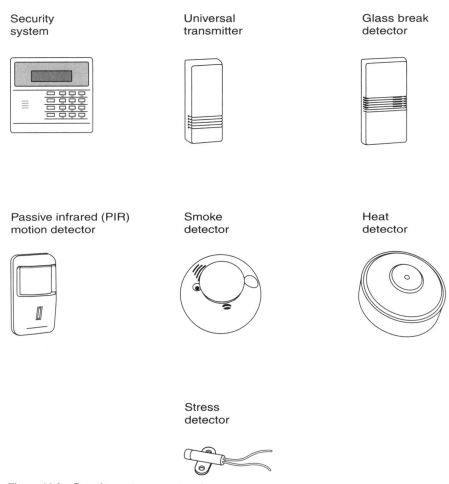

Security
system

Universal
transmitter

Glass break
detector

Passive infrared (PIR)
motion detector

Smoke
detector

Heat
detector

Stress
detector

Figure 11.9 Security system components.

- Keypad—The keypad provides the homeowner's primary interface with the security system. Keypads allow the homeowner to activate and deactivate the security system with a numeric code. Most keypads also provide the means to identify which security zones have been violated should an alarm be triggered. Keypad functions can also be incorporated into the automation system interface.

There are various sensors available for security systems, including:

- Door and window contacts.
- Glass break sensors.
- Motion detectors.
- Stress detectors/pressure pads.
- Carbon monoxide detectors.
- Smoke detectors.
- Heat detectors.
- Water level/flood detectors.
- Electricity detectors/temperature sensors.
- Panic buttons.

Installation

The components common to most security systems include the master control panel, keypad(s), and the warning sirens or sounders. Keypads are mounted to the wall within close proximity of the doors used for entering and leaving the residence. This makes setting the system and turning it off easier for the homeowner. Additional keypads can be placed in other locations (e.g., the master bedroom, office and bedroom hallway, etc.) to allow the homeowner to control the system more conveniently.

The master control panel (which may be integrated with a system keypad) requires power and a connection to a telephone jack. The use of a RJ-31X jack in the DD and connected to the master control panel allows the system to seize the telecommunications lines in order to call out an alert to the monitoring service.

Most hard-wired security systems can be prewired by running cable between the master control panel and the other devices.

The alarm-reporting telephone circuit and cabling shall adhere to ANSI/TIA/EIA and Federal Communications Systems voice transmission standards. For precabling an RJ-31X, do not use alarm circuit cabling. Category 3 (minimum) twisted-pair cable must be used. When special services are provided on the same telephone line as the alarm panel, refer to the security system manufacturer for instructions. Wireless devices may be used to expand an existing system.

Intercom/Paging Systems

Paging systems are available in two common styles. The first and more traditional paging system combines speakers at the entryways (typically front and rear doors) and speakers within the residence for communications. The second type of intercom, the broadcast system, is routed into the telephone system. This type of system rings the homeowner's telephones with a distinctive ring when the external intercom, doorbells, and other triggers are activated.

These systems can also become part of a more sophisticated residence telephone system. These advanced telephone systems offer much more functionality than commonly found in the office environment.

The cabling for the traditional intercom system should be run in a star topology, as is the case with other communications systems in the residence. In a broadcast-style intercom system, the various intercoms are run back to the master unit. When the master unit is used, all the intercoms receive the same communications. The single master control panel provides power to the system components and often provides some additional functionality beyond what's found in the other devices.

In a selective intercom system, many, if not all, of the intercoms (see Figure 11.10) can be used to communicate with each other on a one-to-one basis.

> **NOTE:** Many intercom systems can also provide music and, in some cases, video through the intercoms. These systems may require additional cabling. The music, for example, can be from an AM/FM radio source or from additional devices such as CD players. Video sources are typically small closed-circuit cameras mounted near the front and rear doors.

For an intercom/telephone system combination, the residence's voice and data RSCS (twisted-pair) would be used. For simple systems that only ring the telephones when the doorbell is rung, a single twisted-pair cable run from the doorbell to the DD is sufficient. The intercom's central control unit is mounted within, or near, the DD.

Other common functions available with the use of an intercom are a connection to a standard door chime, and a door lock controller, which allows the person answering the intercom to remotely open the door lock through the telephone keypad. Prior to installing a door lock controller, the door must be visually inspected to guarantee the appropriate dimensions to ensure successful installation of the electronic door latch and required cabling.

Lighting Control and Energy Management Systems

Today's automation system is most commonly thought of as lighting and HVAC control devices. Many manufacturers, however, have brought additional offerings to the market, allowing a single unit to control a large number of devices and services in the residence. These applications include lighting and HVAC

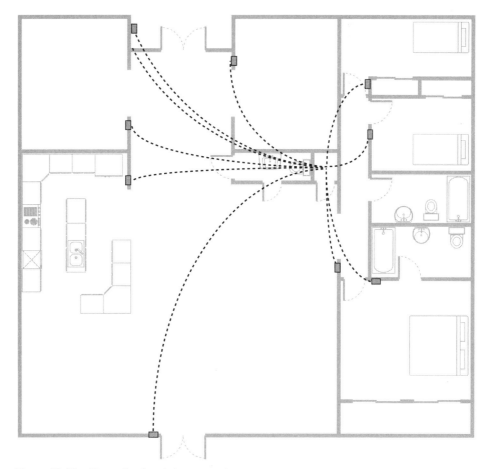

Figure 11.10   Example of an intercom system.

control, but also security and monitoring, sprinkler control, energy management, window treatment control, and others.

Control of the residence's HVAC system offers the following benefits:

- A more comfortable living environment
- Greater energy efficiency
- Protection of the residence against freezing pipes and other hazards

The provision of lighting control provides the following benefits:

- Improved energy efficiency and bulb longevity
- The safety of external night lighting outdoors
- The convenience of controlling multiple lights from a single location

- Improved aesthetics and the beautification of the residence
- Enhanced safety and security
- Setting the proper mood for entertaining

Hard-wired systems

Hard-wired systems are those systems that typically are not terminated on connectors be it plug-in or screw-on wire termination. Hard-wired systems are generally systems that have a wire connected to a device that must be spliced to a wire that extends back to a controller.

There are hard-wired solutions to the functional requirements of both lighting and HVAC control systems. Hard-wired control systems are often used in new construction, due to the quality of service available through the dedicated cabling and the lower component costs compared to wireless solutions.

Hard-wired systems emanate from a central control station—much like the home security systems previously described. Twisted-pair cabling run from the central control station of the home automation system to each device is adequate for most applications. Refer to equipment manufacturer's guidelines for specific information.

The technician or homeowner, through a computer interface, alters the software commands to define what each device in the control system will do under different circumstances. A useful feature of many of these systems is that they automatically adjust to sunrise and sunset, as well as to daylight savings time. This makes the exterior lighting virtually hands-free.

Other solutions allow for the manual setting of light levels and scenes without the need of software-based programming. Most of the other features of this system are similar to the wireless or power line system.

Installation Practices

Planning and Implementation

Introduction

Residential cabling responsibilities have shifted from the access provider (AP [e.g., telephone company, community antenna television (CATV) company]) to the homebuilder or homeowner. Many APs provide access cabling to the minimum point of presence of a single-family home or multi-dwelling building. Often, a network interface device (NID) is mounted on the exterior of a home or within an apartment complex (or within the apartment itself). The homeowner or homebuilder is responsible for the design and installation of the internal residential telecommunications cabling system.

This chapter provides residential telecommunications cabling design guidance. Traditional telecommunications cabling within homes provided minimal connection to telephone and CATV services. Homes today are providing a wide range of communications, entertainment, security, and automation to the family. The universal nature of the required residential telecommunications cabling systems and the high-bandwidth requirements of the homeowner require a significantly more capable and flexible system design.

This chapter describes the planning and administration of a residential cabling system installation. Individual residential units within a multi-dwelling building shall follow the basic in-home residential cabling system design and installation practices as a single-family home.

In the multi-dwelling environment, the builder/owner defines and controls the installation of the telecommunications backbone cabling system. The requirements for the design and installation of these systems are best planned and executed using a combination of the information in this manual (particularly Chapter 17: Multi-Dwelling Unit Structured Cabling Systems and Chapter 18: Multi-Dwelling Unit Structured Cabling Installation) and BICSI's commercial building cabling manuals, the *Telecommunications Distribution*

Methods Manual (TDMM) and the *Telecommunications Cabling Installation Manual (TCIM)*.

In many multi-dwelling unit (MDU) applications, the common areas of the building may contain commercial and office space. Residential structured cabling systems (RSCSs) within these areas shall comply with the standards defined by ANSI/TIA/EIA-568-B.1, *Commercial Building Telecommunications Cabling Standards, Part 1: General Requirements*. The common areas may also require various low-voltage solutions including fire alarm, closed circuit security camera, emergency call buttons, and in-ceiling speakers. There may be many locations within the common areas of this type of structure that will need voice, data, and video services in common with the services provided to the residents. The cabling system for these services shall be defined by ANSI/TIA/EIA-570-A, *Residential Telecommunications Cabling Standard*, as described in this manual.

MDU properties include townhouses, apartment buildings, condominiums and assisted-living facilities, among others. The design of the RSCS for each individual unit may be controlled by the building owner, or by the individual residential unit owners. The planning and design of the unit cabling systems may occur many months prior to the rough-in period as many units are presold prior to construction.

Townhouses closely resemble single-family residences and the RSCS should provide the same level of flexibility as the system designed for a stand-alone home.

Condominiums are similar to townhouses and require the same RSCS considerations. Condominium cabling should be designed and implemented in the manner described by ANSI/TIA/EIA-570-A, with each unit being a separate system. Due to space constraints, the distribution device (DD) is usually mounted within the living space, preferably flush-mounted within a closet.

In condominiums and apartment buildings, backbone telecommunications cabling often passes through common pathways from the entrance facility (see Chapter 17: Multi-Dwelling Unit Structured Cabling Systems) through one residence and into another. To provide for long-term performance and the possibility for maintenance, however, cables should be kept in the building's common areas until the point at which they enter the particular residence that they serve.

Assisted-living facilities often require a higher level of cabling within the common areas to support a number of applications. The requirements of these systems are determined through an interview with the building owner.

Design for Flexibility

It is difficult and costly to alter an RSCS once the walls of the structure are completed. The RSCS should be designed to:

- Meet the resident's anticipated long-term grade-of-service requirements.
- Maximize system flexibility.
- Provide convenience to the owner.

Where future outlets or expansions of the structure are anticipated, a means of extending the cabling system to these locations should be provided. For example, if a new home has a finished basement that could be converted into a recreational area or home theater, cabling should be preinstalled to outlet locations, or conduit of sufficient size should be placed within the walls during construction from a DD to potential outlet areas. This level of preplanning is especially critical in environments where the building structure and materials (i.e., concrete floors) make future additions particularly difficult.

Keep in mind the initial, and future, intended uses of each living space. A spare bedroom today may be a baby's room within a year. The homeowner may want to prewire to support the addition of a camera to monitor the infant. The same room may later require data and video support for a television and high-speed Internet connection. Still later, the room may be used as an office and may require multiple telephone lines and data networking connections. The potential uses of each living space should be reviewed prior to the final design of the home's cabling system. Living spaces include unfinished basements, front and back porches, sunrooms, garages, and other areas that are often overlooked in the planning stage.

Plans and Specifications

Introduction

Plans and specifications provide the information required to implement a customer's design. Plans are sometimes referred to as blueprints, building plans, or drawings. They give a pictorial representation of a project. Specifications provide a detailed written description of a project. They identify the materials to be used, as well as applicable standards, performance criteria, and testing methods. The following sections give an overview of project plans and specifications and identify the components that impact the residential telecommunications cabling installer.

Drawings

To understand project blueprints, it is necessary to understand the design used in creating them. The overall design of the structure(s) in new construction and most renovation projects requires site, architectural, mechanical, electrical, telecommunications, and structural drawings. Normally, the architect is responsible for:

- Preparing the architectural plans and general information drawing sheets.
- Hiring engineers and consultants, as appropriate, to prepare the remaining plans.
- Producing all plans and compiling them in a drawing set.

The drawing set, sometimes referred to as architectural, mechanical, electrical (AME) or architectural, mechanical, electrical, structural (AMES), is used in the bidding process and during construction. These drawing set segments are described in the following subsections.

General information. The drawing package generally contains one or more sheets that can help the residential cabling installer understand how the overall project is expected to come together. These sheets usually note any special requirements or limitations and indicate who is responsible for installing the telecommunications pathways. The sheets include:

- Title sheet—Provides general information, including the site address, owner's name, architect's name, and consulting engineers' names.
- Index sheet—Lists the drawings and any revisions.
- Legend sheet—Provides a list of standard symbols and abbreviations used throughout the particular drawing set.

Site drawings. Many projects include site drawings, also known as the site plan. These are prepared by a civil engineer and:

- Indicate the relationship of the various buildings on the project.
- Depict:
  - The location of the building on the site.
  - External features such as parking areas.
  - The entry point of the utilities, including telecommunications.

Architectural drawings. Architectural drawings, usually coded as the "A" series, provide floor plan layouts of the building or residence depicting rooms, doors, walls, and other structure-related items. Other A drawings provide elevations of the building exterior and define the relationship of the building's walls, floors, foundation, and roof.

Architectural drawings also identify specific construction items (e.g., wall finishes, fire ratings, ceiling heights, door types, windows, and floor coverings) and include construction details that define how the various components are to be built. This may include specialty items such as window treatments or particular wall fixtures. However, only items that are considered part of the building itself are typically identified on the A series drawings. Movable items such as furniture, if depicted at all, are generally provided by a specialty firm.

Most of these drawings are created to a scale that is noted on each print or, in some cases, as part of an individual detail drawing.

In some cases, the residential cabling installer will need to reference several drawings to obtain an accurate picture of the work to be undertaken. As an example, if specialty cabinetwork with built-in outlets is to be constructed, the architect will generally provide details of the woodwork, layout, and materials, but the electrical and telecommunications prints will identify the pathways to be used by residential cabling installers.

Mechanical drawings.   Mechanical drawings, frequently identified as "M" or "ME" drawings, identify the heating, ventilating, and air conditioning (HVAC) systems. These drawings include floor plan layouts, mechanical space elevations, and general installation guidelines the HVAC contractor must follow. On some projects, generally smaller in scope, this series of drawings may also include plumbing and fire sprinkler system layouts.

The M drawings are a good place to start when attempting to define a potential pathway for cable trays or cable placement. Sometimes, but not always, these drawings provide a planning elevation of crowded ceiling spaces that identifies the positions of ductwork, fire sprinkler pipes (if applicable), plumbing, electrical conduits, structural members, and telecommunications trays.

Electrical drawings.   Electrical drawings are normally completed by the electrical engineer and show the requirements for lighting, power, alarms, special electrical systems, and related services. The electrical drawings include:

- Power and, sometimes, telecommunications backbone (riser) diagrams.
- Symbols list (legend).
- Schematic diagrams and load tables.
- Large-scale details (where necessary).

Telecommunications drawings.   Historically, telecommunications requirements were shown on the electrical drawings. Because of technological advances and the complexity of today's integrated networks, some engineers now create separate drawings for telecommunications. When separate electrical and telecommunications drawings are provided, the telecommunications sheets will often show the pathways, multi-dwelling unit telecommunications room (MDU-TR) layouts, and media required for the project.

SYMBOL	DESCRIPTION
◀BA	Building automation outlet
◀	Drop location
◀XTW	Existing location to be rewired
◀P	Public phone
◀W	Wall phone
◀SP	Secure wall phone
◀S	Security panel
◀E	Emergency phone
◀B	Drop location with blank plate
◀FP	Future phone location
◀C	Ceiling mounted phone location
◀F	Fire alarm panel
◀D	Security or fire alarm dialers
◀OF	Horizontal fiber to the desktop
◼◀	Floor mounted

Figure 12.1 Sample symbols and abbreviations.

Sometimes, a Registered Communications Distribution Designer (RCDD®), or equivalent, will prepare drawings for the telecommunications portion of a project. These drawings may include a plot or site plan showing the entry points for all electrical and telecommunications services.

Under pending Division 17 (described later in this chapter), Figure 12.1 illustrates the symbols and codes that can be used on a T2 drawing to indicate specific drop location details, including pathways, intended devices, unique identification for each cable, and type of faceplate required. Required information that cannot be clearly illustrated (i.e., as a symbol) should be described with a drawing note.

Structural drawings. Structural drawings provide the detail necessary to construct the structural support components of the building—foundations, footings, floors, ceilings, and load-bearing structures. These drawings are based on engineering guidelines set up for vertical loads, lateral stress, etc. Structural drawings may be included with the architectural drawings or in a separate section within the drawing set.

The residential cabling installer needs to reference these drawings to identify any potential limitation on the placement of vertical or horizontal pathways. In most cases, the designer will have coordinated with the structural engineer to define entrance conduit methods and other structural penetrations. Any penetration through a structural or load-bearing wall or floor must be reviewed and approved by the structural engineer.

Considerations for retrofit projects

Blueprints for new construction are generally an accurate representation of the actual site. However, during the construction process, changes may occur in the location of:

- MDU-TRs.
- Telecommunications equipment.
- Work area outlets.
- Electrical equipment.
- HVAC equipment and ducting.
- Pathways.
- Utilities.

Because changes may not be noted during the construction process, plans available for retrofits may not reflect the site as-built. They should be carefully checked for accuracy. A site survey is a critical part of planning. The survey provides the opportunity to match the information on the as-built blueprints with observed:

- Pathways and spaces.
- Work area outlets.
- MDU-TRs.
- Entrance facilities.

The information on the blueprint, the elements of the job plan, and the physical design of the building should combine into a complete concept for the project.

Construction specifications

The drawing set shows the dimensions and the relationship between components but does not provide sufficient detail to complete the project. Specifications provide the additional detail required to complete the project. Specifications are a written description of the work to be performed and the responsibilities and duties required for the architect, engineer, and owner. Together with the drawings, these specifications form the basis of the contractual requirements.

For new construction or when the telecommunications systems are part of the general contract, the telecommunications specifications will be included in the overall project specifications.

There are three documents widely used in the construction industry in North America that standardize the format of specifications: PageFormat™, SectionFormat™, and MasterFormat™. All three are jointly produced by the Construction Specifications Institute (CSI) and Construction Specifications Canada (CSC). The American Institute of Architects (AIA) uses these documents to define the requirements of a specific part of the construction work on the project. The following chart shows each division, its title, and the specific area it covers.

NOTE: Division 17 is not in the current CSI MasterFormat. Telecommunications is currently included under Division 16.

TABLE 12.1   MasterFormat™

Division	Title	Description
1	General Requirements	A summary of the work, alternatives, schedule of the project, project meetings, submittals, quality control, temporary facilities, products, and the project closeout.
2	Site Work	Foundations, fill, drains, underground utilities, and other areas outside the building.
3	Concrete	All concrete work. Maintenance holes (MHs) are covered in this section.
4	Masonry	All external and internal masonry.
5	Metals	Structural, metal joists, decking, framing, fabrication, and other uses.
6	Carpentry	All wood products and their usage within the project. Also includes plastics if a written specification is provided.
7	Thermal and Moisture Protection	Covers such items as waterproofing, insulation, shingles and roofing tiles, preformed roofing and siding, sheet-metal work, wall flashing, roof accessories, and sealants.
8	Doors and Windows	All doors and windows, hardware and specialties, glazing, and window/curtain walls.
9	Finishes	All finishes for floors, walls, and ceilings.
10	Specialties	Special items, such as chalkboards and tackboards, modular furniture, louvers and vents for the HVAC system, access flooring, and protective covers are part of this division.
11	Equipment	Vaults, darkrooms, food service, industrial, laboratory equipment, and other equipment.
12	Furnishing	Furniture, fabrics, rugs and mats, seating, and other furnishings.
13	Special Construction	Air-supported structures, incinerators, and other special items.
14	Conveying Systems	All types of conveying systems.
15	Mechanical	All work of the HVAC and plumbing contractor.
16	Electrical	All of the interior and exterior electrical work.
		NOTE: Includes all telecommunications until Division 17 is adopted.
17	Telecommunications	All cabling components for the project. In the past, requirements for new or upgraded voice, data, and video systems were often bid separately from the construction phase. The specifications and designs included in these bids did not match construction industry documents or computer-aided design (CAD) drawing standards. Cabling requirements, therefore, are not effectively addressed in the current CSI specification format.

Planning the Residential Premises Cabling System

The telecommunications distribution designer must understand the homeowner or builder's immediate requirements, while attempting to provide a cabling system capable of cost-effective migration to the desires of their future telecommunications systems.

The key to effective and efficient residential telecommunications cabling is careful planning. Telecommunications system planning and adherence to local building code requirements determines the:

- Materials required for the project.

- Type and amount of labor needed.

There are four groups of people who require a detailed knowledge of the planning process. They are the:

- Residential cabling designer and installer.

- Builder.

- Architect or remodeler.

- Homeowner.

The following sections provide a step-by-step guide that addresses the most common working conditions each of these four specifiers will encounter.

Residential Telecommunications Cabling Step-by-Step

Residential cabling designer and installer

The residential cabling designer and installer should follow the steps below to plan a residential telecommunications cabling system.

Step	Planning a Residential Telecommunications Cabling System
1	Determine if the decision maker is a builder, architect, or homeowner.
2	Determine whether the installation is a retrofit to an existing property or a new system for new construction.
3	If a retrofit, obtain or create floor plans: ▪ Review the specific telecommunications hardware desired for use by the homeowner, if any. Ensure that all specific manufacturer requirements are met. ▪ Ensure that the HVAC, power, and plumbing routes (and other potential obstructions) are identified. Use the building plans to review the pathways of the system and to avoid obstacles such as power cabling, water pipes, and heating ducts that will affect cabling system performance or installation. ▪ Perform a site survey. Observe the locations within the home where existing walls are open or being replaced, and where new walls are being constructed. The open nature of these areas makes them ideal locations for running new structured cabling facilities.

Step	Planning a Residential Telecommunications Cabling System
	• Advise the decision maker of potential areas in which walls or other structures may have to be damaged to install the desired cabling. Obtain permission to do so, or determine new locations for the telecommunications cabling.
	• Choose an appropriate location for the placement of the DD. Ensure that power is available within or near the DD location.
	• Visually observe potential cabling paths to ensure that they are free of obstruction. If cabling is required in the upper floor of a home, choose an appropriate place for the installation of a vertical pathway from the DD location to the attic. Avoid running cabling within exterior walls wherever possible to minimize difficulties with installed insulation. In addition, be aware of the direction and slope of roof lines, since some exterior wall locations will not be accessible due to cramped workspaces.
	• Ensure that all lower-floor locations can be serviced from below, if the DD is to be located in the basement.
	• Determine if the number and type of auxiliary disconnect outlet (ADO) cables will be appropriate to handle the incoming telecommunications cabling requirements. If not, design in the appropriate capacity.
4	For new construction, work with the AP to determine the proper location for the new NID. The AP will install and maintain the network interface as authorized by the authority having jurisdiction (AHJ). Do not disturb, remove, modify, or connect anything to the protectors, grounding systems, or any other attachments placed by the AP.
5	Locate the ADO and DD centrally. This minimizes the length of outlet cable runs and ensures the homeowner:
	• Convenience.
	• Security.
	• Ease of administration.
	• Space for future growth.
	• Premises equipment space requirements.
	• Cost-effectiveness.
	Consider using a utility closet for this segment of the system. This location may be used for residential gateways or multimedia equipment. Also, locate the DD within 1.5 m (5 ft) of an electrical outlet, but not near possible sources of electromagnetic interference (EMI [e.g., motors or power transformers]).
6	Identify the type and number of cables to be installed between the ADO and NID.
7	Provide a conduit of sufficient size between these locations to support the future addition of necessary cables.
8	Consider the installation of coaxial cabling from the DD to the attic to support future satellite telecommunications requirements.
9	Plan to install at least one outlet run in each room (e.g., kitchen, living room, dining room, den, study, bedrooms, family room, great room, etc.). Other possible locations for outlets include the:
	• Basement.
	• Garage.
	• Patio.

Step	Planning a Residential Telecommunications Cabling System
	▪ Porch.
	▪ Laundry.
	▪ Bathrooms.
	▪ Security system (alarm service).
10	When a large room serves the purpose of two functional areas, plan an outlet for each functional area of the room.
11	A sufficient number of telecommunications outlet/connector locations should be planned to prevent the need for long extension cords, as follows:
	▪ An outlet location should be provided in each room and additional outlet locations provided within unbroken wall spaces of 3.7 m (12 ft) or more.
	▪ Additional outlet locations should be provided so that no point along the floor line in any wall space is more than 7.6 m (25 ft), measured horizontally, from an outlet location in that space.
	▪ To reduce noise transfer between rooms, do not place telecommunications outlet/connectors back-to-back within a wall area.
	▪ Outlet mounting heights shall be in accordance with applicable codes.
	▪ Outlet orientation, both vertical and horizontal, shall match that used locally for electrical outlets.
12	Determine whether each outlet will provide cabling compatible with the requirements of a Grade 1 or Grade 2 system. Grade 1 cabling minimum requirements consist of one 4-pair unshielded twisted-pair cable that meets or exceeds the requirements for Category 3 and one 75 Ω coaxial cable. Grade 2 cabling minimum requirements consist of two 4-pair unshielded twisted-pair cables that meet or exceed the requirements for Category 5, two 75 Ω coaxial cables, and optionally, two multimode optical fibers.
13	Plan outlet cable runs from the DD to each telecommunications outlet/connector using a star topology. Avoid running cables in the exterior walls and attic, where possible, to minimize effects of heat on the cable and the possibility of damage when siding is installed. Avoid plumbing used for hot water. Maintain as much distance as possible from high-voltage cabling and conduit.
14	Determine the proper size of the DD. The DD shall be capable of enclosing the cabling specified upon initial installation of the RSCS, and should be sized to provide for a growth factor of 50 percent.
15	Determine the proper power, grounding, ventilation, lighting, access, space, and clearance requirements for the DD (see Chapter 2: Residential Structured Cabling Systems).
16	Plan precabling so that it occurs after installation of the electrical power and HVAC systems, but before installation of the thermal insulation and wallboard.
17	Plan subsequent visits after the walls are finished to install the telecommunications outlet/connectors associated with all of the cabling. (Make recommendations to the homeowner or the homeowner's agent to add capacity during the rough-in stage, when the capabilities of the planned cabling system are perceived to be insufficient to meet the homeowner's needs.)
18	Provide required testing and certification documentation.

Home builder

The home builder should follow the steps below to plan a residential telecommunications cabling system.

Step	Planning a Residential Telecommunications Cabling System
1	Develop floor plans for the home. Ensure that the power, HVAC, and plumbing routes (and other potential obstructions) are identified.
2	Determine the scope of the system to be preinstalled. This decision will be influenced by the: ■ Sale price of the home. ■ Demographics of the potential home-buying market. ■ Target construction cost of the home and of the RSCS. ■ Scope of systems installed by builders in the area. ■ Local codes, if applicable.
3	Locate on the floor plans the desired location of the NID and DD. Locate the DD centrally. This minimizes the length of outlet cable runs and ensures the homeowner: ■ Convenience. ■ Security. ■ Ease of administration. ■ Space for future growth. ■ Premises equipment space requirements. ■ Cost-effectiveness. Consider using a utility closet for this segment of the system. This location may be used for residential gateways or multimedia equipment. Also, locate the DD within 1.5 m (5 ft) of an electrical outlet, but not near possible sources of EMI (e.g., motors or power transformers).
4	Identify the type and number of cables to be installed between the ADO and NID.
5	Provide a conduit between these locations to support the future addition of necessary cables.
6	Consider the installation of coaxial cabling from the DD to the attic to support future satellite telecommunications requirements.
7	Plan to install at least one outlet run in each room (e.g., kitchen, living room, dining room, den, study, bedrooms, family room, great room, etc.). Other possible locations for outlets include the: ■ Basement. ■ Garage. ■ Patio. ■ Porch. ■ Laundry. ■ Bathrooms. ■ Security system (alarm service).

Step	Planning a Residential Telecommunications Cabling System
8	When a large room serves the purpose of two functional areas, plan an outlet for each functional area of the room.
9	A sufficient number of outlet locations should be planned to prevent the need for long extension cords, as follows:
	▪ An outlet location should be provided in each room and additional outlet locations provided within unbroken wall spaces of 3.7 m (12 ft) or more.
	▪ Additional outlet locations should be provided so that no point along the floor line in any wall space is more than 7.6 m (25 ft), measured horizontally, from an outlet location in that space.
	▪ To reduce noise transfer between rooms, do not place outlets back-to-back within a wall area.
	▪ Outlet mounting heights shall be in accordance with applicable codes.
	▪ Outlet orientation, both vertical and horizontal, shall match that used locally for electrical outlets.
10	Determine whether each outlet will provide cabling compatible with the requirements of a Grade 1 or Grade 2 system. Grade 1 cabling minimum requirements consist of one 4-pair unshielded twisted-pair cable that meets or exceeds the requirements for Category 3 and one 75 Ω coaxial cable. Grade 2 cabling minimum requirements consist of two 4-pair unshielded twisted-pair cables that meet or exceed the requirements for Category 5, and two 75 Ω coaxial cables, and optionally, two multimode optical fibers.
11	Plan outlet cable runs from the DD to each telecommunications outlet/connector using a star topology. Avoid running cables in the exterior walls and attic, where possible, to minimize effects of heat on the cable and the possibility of damage when siding is installed. Avoid plumbing used for hot water. Maintain as much distance as possible from high-voltage cabling and conduit.
12	Determine the proper size of the DD. The DD shall be capable of enclosing the cabling specified upon initial installation of the RSCS, and should be sized to provide for a growth factor of 50 percent.
13	Determine the proper power, grounding, ventilation, lighting, access, space, and clearance requirements for the DD (see Chapter 2: Residential Structured Cabling Systems).
14	Determine the desire for an optional cable chase from the DD location to the attic and/or crawl space for future expansion, and position and size the cable chase accordingly.
15	Schedule the cabling installer to precable so that work occurs:
	▪ After installation of the electrical power and HVAC systems, but
	▪ Before installation of the thermal insulation and wallboard.
16	Schedule the cabling installer for a subsequent visit after the walls are finished to install the telecommunications outlet/connectors associated with all of the cabling.
17	Provide power to the home prior to rough-in.
18	Prepare and issue bid documents. Specify preferred or required cable and connectivity vendors, if any. Specify required testing and documentation.
19	Verify that all cables/components have been properly labeled.
20	Require as-built documentation from the cabling installer.

Architect/remodeler

The architect/remodeler should follow the steps below to plan residential telecommunications cabling systems.

Step	Planning a Residential Telecommunications Cabling System
1	Develop floor plans for the home. Ensure that new and existing power, HVAC, and plumbing routes (and other potential obstructions) are identified. Identify all current and future locations of appliances, woodwork, and other items that may affect the location of outlets and other RSCS devices.
2	Determine the scope of the system to be preinstalled. This decision will be influenced by the: ■ Needs of the homeowner. ■ Budget. ■ Local codes, if applicable.
3	Locate on the floor plans the desired location of the NID and DD. Locate the DD centrally. This minimizes the length of outlet cable runs and ensures the homeowner: ■ Convenience. ■ Security. ■ Ease of administration. ■ Space for future growth. ■ Premises equipment space requirements. ■ Cost-effectiveness. Consider using a utility closet for this segment of the system. This location may be used for a residential gateway or multimedia equipment. Also, locate the DD within 1.5 m (5 ft) of an electrical outlet, but not near possible sources of EMI (e.g., motors or power transformers).
4	Identify the type and number of cables to be installed between the ADO and NID.
5	Provide a conduit between these locations to support the future addition of necessary cables.
6	Consider the installation of coaxial cabling from the DD to the attic to support future satellite telecommunications requirements.
7	Plan to install at least one outlet run in each room (e.g., kitchen, living room, dining room, den, study, bedrooms, family room, great room, etc.). Other possible locations for outlets include the: ■ Basement. ■ Garage. ■ Patio. ■ Porch. ■ Laundry. ■ Bathrooms. ■ Security system (alarm service).
8	When a large room serves the purpose of two functional areas, plan an outlet for each functional area of the room.

Step	Planning a Residential Telecommunications Cabling System
9	A sufficient number of outlet locations should be planned to prevent the need for long extension cords, as follows:

- An outlet location should be provided in each room and additional outlet locations provided within unbroken wall spaces of 3.7 m (12 ft) or more.

- Additional outlet locations should be provided so that no point along the floor line in any wall space is more than 7.6 m (25 ft), measured horizontally, from an outlet location in that space.

- To reduce noise transfer between rooms, do not place outlets back-to-back within a wall area.

- Outlet mounting heights shall be in accordance with applicable codes.

- Outlet orientation, both vertical and horizontal, shall match that used locally for electrical outlets.

Step	
10	Determine whether each outlet will provide cabling compatible with the requirements of a Grade 1 or Grade 2 system.
11	Plan outlet cable runs from the DD to each telecommunications outlet/connector using a star topology. Avoid running cables in the exterior walls and attic, where possible, to minimize effects of heat on the cable and the possibility of damage when siding is installed. Avoid plumbing used for hot water. Maintain as much distance as possible from high-voltage cabling and conduit.
12	Determine the proper size of the DD. The DD shall be capable of enclosing the cabling specified upon initial installation of the RSCS, and should be sized to provide for a growth factor of 50 percent.
13	Determine the proper power, grounding, ventilation, lighting, access, space, and clearance requirements for the DD (see Chapter 2: Residential Structured Cabling Systems).
14	Determine the desire for an optional cable chase from the DD location to the attic/crawl space for future expansion, and position and size the cable chase accordingly.
15	Schedule the cabling installer to precable so that work occurs:

- After installation of the electrical power and HVAC systems, but

- Before installation of the thermal insulation and wallboard.

Step	
16	Schedule the cabling installer for a subsequent visit after the walls are finished to install the telecommunications outlet/connectors associated with all of the cabling.
17	Provide power to the home prior to rough-in.
18	Prepare and issue bid documents. Specify preferred or required cable and connectivity vendors, if any.
19	Specify testing and as-built documentation requirements.

Homeowner

The homeowner should follow the steps below to plan a residential telecommunications cabling system.

Step	Planning a Residential Telecommunications Cabling System
1	Develop floor plans for the home. Ensure that new and existing power, HVAC, and plumbing routes (and other potential obstructions) are identified.
2	Determine the scope of the system to be preinstalled. This decision will be influenced by the: ▪ Needs for telecommunications services and the use of each room. ▪ Budget. ▪ Local codes, if applicable.
3	Locate on the floor plans the desired location of the NID and DD. Locate the DD centrally. This minimizes the length of outlet cable runs and ensures: ▪ Convenience. ▪ Security. ▪ Ease of administration. ▪ Space for future growth. ▪ Premises equipment space requirements. ▪ Cost-effectiveness. Consider using a utility closet for this segment of the system. This location may be used for a residential gateway or multimedia equipment. Also, locate the DD within 1.5 m (5 ft) of an electrical outlet, but not near possible sources of EMI (e.g., motors or power transformers).
4	Identify the type and number of cables to be installed between the ADO and NID.
5	Plan for a conduit between these locations to support the future addition of necessary cables.
6	Consider the installation of coaxial cabling from the DD to the attic to support future satellite telecommunications requirements.
7	Plan to install at least one outlet run in each room (e.g., kitchen, living room, dining room, den, study, bedrooms, family room, great room, etc.). Other possible locations for outlets include the: ▪ Basement. ▪ Garage. ▪ Patio. ▪ Porch. ▪ Laundry. ▪ Bathrooms. ▪ Security system (alarm service).
8	When a large room serves the purpose of two functional areas, plan an outlet for each functional area of the room.

Step	Planning a Residential Telecommunications Cabling System
9	A sufficient number of outlet locations should be planned to prevent the need for long extension cords, as follows:
	▪ An outlet location should be provided in each room and additional outlet locations provided within unbroken wall spaces of 3.7 m (12 ft) or more.
	▪ Additional outlet locations should be provided so that no point along the floor line in any wall space is more than 7.6 m (25 ft), measured horizontally, from an outlet location in that space.
	▪ To reduce noise transfer between rooms, do not place outlets back-to-back within a wall area.
	▪ Outlet mounting heights shall be in accordance with applicable codes.
	▪ Outlet orientation, both vertical and horizontal, shall match that used locally for electrical outlets.
10	Determine whether each outlet will provide cabling compatible with the requirements of a Grade 1 or Grade 2 system.
11	Plan outlet cable runs from the DD to each outlet/connector using a star topology. Avoid running cables in the exterior walls and attic, where possible, to minimize effects of heat on the cable and the possibility of damage when siding is installed. Avoid plumbing used for hot water. Maintain as much distance as possible from high-voltage cabling and conduit.
12	Determine the proper size of the DD. The DD shall be capable of enclosing the cabling specified upon initial installation of the RSCS and should be sized to provide for a growth factor of 50 percent.
13	Determine the desire for an optional cable chase from the DD location to the attic/crawl space for future expansion, and position and size the cable chase accordingly. Determine the proper power, grounding, ventilation, lighting, access, space, and clearance requirements for the DD (see Chapter 2: Residential Structured Cabling Systems).
14	From the final plans, determine the work to be completed by outside vendors. Prepare and issue bid documents. Specify preferred or required vendors for components, if any. Perform a site survey with the chosen contractor.
15	Schedule the cabling installer to perform work during desired work hours. **NOTE:** For existing homes, the prewire and trim work may occur without pause for other trades to complete their work.
16	Review the completed system with the contractor, if one is used. Request, review, and verify test data and documentation. Verify that all cables have been labeled. Review the proper use of the system.

Construction Planning, Documentation, and Administration

Contract

A contract is written to document the understanding between the customer and the contractor. Some customers generate a purchase order that refers to the other documents associated with the project. If a contract is available, ensure that any documents listed therein are available. Contracts may also list penalties associated with not completing the work or delays in completion. Pay particular attention to liquidated damages. Performance bonds and insurance may be required as part of the contract.

Project schedule

Companies use a variety of different project management styles and software. Various charts and graphs are employed that allow the residential telecommunications installation team to track materials receipts and disbursements, labor items completed, and the overall status of the project on a day-to-day basis. Manually generated project schedules can also be employed, especially when the project is small and simple.

Examples of project time line and project schedule documents are shown in Examples 12.1 and 12.2.

Site survey

Once all of the initial project documents are obtained, a site survey is performed. For retrofits, a site survey should be completed prior to issuing a final quotation to the customer. For new construction, a site survey is most often performed after the initial framing of the home is completed, but prior to the beginning of the telecommunications cabling contractor's rough-in work.

A member or members of the residential telecommunications installation team will visit the place of installation. While there, the team should observe all of the various locations where installation work will be performed. When making this site survey, identify on the site's floor plans the locations of the various components specified for the RSCS. The drawings may indicate hidden obstacles not visible from floor level.

A checklist should be employed on each project to ensure that all items of concern are addressed during the site survey. Plans can be formulated to overcome any problems discovered while still on site, rather than having to return to the site.

All information gathered during the site survey should be placed into the project file. This information will become invaluable later, especially if new team members are assigned to the cabling installation after it starts.

The first stop at a job site should be at the general contractor's site office. While at the office:

- Introductions can be made.

- The work to be performed for the customer can be explained.

- Other contractors working for the general contractor can be identified and the impact of their work discussed.

- A copy of the general contractor's construction progress schedule can be obtained. This document can be used to determine how the installation schedule can be coordinated with the contractors working on the project.

Name	Month starting July 1, 2001				Month starting August 1, 2001				
	7/2/01	7/9/01	7/16/01	7/23/01	7/30/01	8/6/01	8/13/01	8/20/01	8/27/01
Contract award									
Site survey									
Installation team meeting									
Initial construction meeting									
Project schedule compiled									
Materials ordered									
Materials shipped									
Materials received									
Materials stored/staged									
Install project infrastructure									
Install backbone cables									
Install horizontal wires/cables									
Installation progress meeting									
Install backbone termination hardware									
Install horizontal wire/cable termination hardware									
Installation progress meeting									
Terminate backbone cables									
Terminate horizontal wires/cables									
Installation progress meeting									

Example 12.1 Project time line.

Name	Start Constraint	Finish Constraint	Actual Start	Actual Finish	Percent Done
Contract awarded	✳ 7/11/01	✳ 7/11/01	✳ 7/11/01	✳ 7/11/01	100%
Site survey	✳ 7/13/01	✳ 7/16/01	✳ 7/14/01	✳ 7/16/01	100%
Installation team meeting	✳ 7/16/01	✳ 7/16/01	✳ 7/16/01	✳ 7/16/01	100%
Initial construction meeting	✳ 7/19/01	✳ 7/19/01	7/19/01	7/19/01	100%
Project schedule compiled	✳ 7/16/01	✳ 7/16/01	7/16/01	7/16/01	100%
Materials ordered	✳ 7/16/01	✳ 7/16/01	7/16/01	7/16/01	100%
Materials shipped	✳ 7/17/01	✳ 8/3/01	✳ 7/17/01	✳ 8/3/01	100%
Materials received	✳ 7/19/01	✳ 8/18/01	✳ 7/19/01	✳ 9/25/02	100%
Materials stored/staged	✳ 7/19/01	✳ 10/19/01	7/19/01	10/19/01	100%
Install project infrastructure	✳ 7/20/01	✳ 8/9/01	✳ 7/20/01	✳ 8/9/01	100%
Install backbone cables	✳ 8/10/01	✳ 9/8/02	8/10/01	9/8/02	100%
Install horizontal wires/cables	✳ 9/10/01	✳ 12/22/02	9/10/01	12/22/02	100%
Installation progress meeting	✳ 9/10/01	✳ 9/10/01	9/10/01	9/10/01	100%
Install backbone connecting hardware	✳ 1/2/02	✳ 1/18/02	1/2/02	1/18/02	100%
Install horizontal wire/cable connecting hardware	✳ 1/19/02	✳ 2/26/02	1/19/02	2/26/02	100%
Installation progress meeting	✳ 2/26/02	✳ 2/26/02	2/26/02	2/26/02	100%
Terminate backbone cables	✳ 1/19/02	✳ 2/9/02	1/19/02	2/9/02	100%
Terminate horizontal wires/cables	✳ 2/27/02	✳ 3/29/02	2/27/02	3/29/02	100%
Installation progress meeting	✳ 3/29/02	✳ 3/29/02	3/29/02	3/29/02	100%
Label all facilities as per ANSI/TIA/EIA-606	✳ 3/22/02	✳ 4/5/02	3/22/02	4/5/02	100%
Test backbone cables	✳ 4/5/02	✳ 4/12/02	4/5/02	4/12/02	100%
Test horizontal wires/cables	✳ 4/15/02	✳ 5/3/02	4/15/02	5/3/02	100%
Compile all test results	✳ 5/6/02	✳ 5/8/02	5/6/02	5/8/02	100%
Installation progress meeting	✳ 5/3/02	✳ 5/3/02	5/3/02	5/3/02	100%
Punch list	✳ 5/9/02	✳ 5/10/02	5/9/02	5/10/02	100%
Correct all items on punch list	✳ 5/16/02	✳ 5/17/02	✳ 5/16/02	✳ 5/17/02	100%
Final punch list	✳ 5/16/02	✳ 5/16/02	5/16/02	5/16/02	100%
Customer punch list	✳ 5/17/02	✳ 5/17/02	5/17/02	5/17/02	100%
Customer acceptance	✳ 5/20/02	✳ 5/20/02	5/20/02	5/20/02	100%
Prepare as-built package	✳ 5/20/02	✳ 5/24/02	5/20/02	5/24/02	100%
Provide all project documents to customer	✳ 5/27/02	✳ 5/27/02	5/27/02	5/27/02	100%
Return surplus materials to distributor for storage	✳ 5/20/02	✳ 5/22/02	5/20/02	5/22/02	100%
Complete billing to customer	✳ 5/28/02	✳ 5/28/02	5/28/02	5/28/02	100%
Review billing with customer	✳ 5/29/02	✳ 5/29/02	5/29/02	5/29/02	100%
Receive final payment	✳ 5/30/02	✳ 5/30/02	5/30/02	5/30/02	100%
Close project	✳ 5/31/02	✳ 5/31/02	5/31/02	5/31/02	100%
Clear punch list	✳ 5/10/02	✳ 5/13/02	5/10/02	✳ 5/13/02	100%
Clear customer punch list	✳ 5/17/02		5/17/02	✳ 5/20/02	100%

✳ indicates completion

Example 12.2   Project schedule.

Determine if any pathways need to be installed, and find out what obstacles must be overcome to install them. An example of a cabling pathway in a single-family home is an optional cable chase, or pipe, connecting the DD in the basement to the attic. This chase allows for future system upgrades by permitting additional cabling to be run from the attic down to locations on the top floor. It is important to specify a vertical pathway of sufficient size and that is free of obstructions (e.g., "plumbing" and HVAC).

It is important to review the cabling contractor's responsibilities with the builder in order for the general contractor to understand the role the cabling company will play in the completion of the overall project. Remember, the builder owns the home until the owner accepts it.

Be sure to bring the tools required to perform a successful and thorough site survey. These tools might include: personal protective equipment such as hard hat and safety glasses, leather gloves, leather boots, and hearing protection. Additional items that may prove useful are a ladder, flashlight, measuring wheel, handheld tape recorder, digital camera, or video camera.

Determine the physical location of the DD(s), their size, type of construction, configuration of other utilities within the confines of their walls, and responsibilities and requirements for cooperating with other trades.

Locate the existing utility pathways that have been or will be constructed by the builder or its subcontractors. Determine their state of completion. Ask, as a minimum, the following questions:

- Have the subcontractors adhered to the architect's and designer's drawings and specifications?

- Are there any change orders that will affect the pathways and spaces? If so, how do they affect the project?

- How is the building's grounding infrastructure installed?

- Does the grounding infrastructure comply with the *National Electrical Code®* *(NEC®)?*

- When will the DD be completed?

- When will the pathways be completed?

- When will the building inspector and fire marshal be on site to perform the certificate-of-occupancy inspection?

- When will the project be turned over to the owner or resident?

The answers to these and other questions will determine how to plan and implement the project. See Examples 12.3 and 12.4 for site survey checklists.

If the project is a retrofit, identify all of the existing pathways and spaces being used for telecommunications, their size, capacities, usability, congestion, and compliance with code. Some questions to be answered are:

- Is a new telecommunications location required?

- Are new pathways required?

- Are any existing pathways vacant?
- Do any existing pathways have usable space?
- Can existing facilities be used to assist in installing the new cable?
- Will the new hardware fit within the space confines of the DD?

Issue	Responsibility	Date Scheduled	Date Ready
1. Owner's telecommunications needs defined and documented.			
2. Telecommunications design complete and up to standards.			
3. Modifications (if required) submitted and approved.			
4. Drawings and specifications distributed.			
5. Building construction complete for: • Telecommunications spaces — Backboards — Lighting — Electrical — Grounding — HVAC — Fire suppression — Alarms and controls • Backbone pathways • Horizontal pathways • Work area spaces			
6. Secure storage identified.			
7. Work site conditions defined. • Safety plan posted • Permits approved • Safety equipment in place • Office and meeting area defined • Elevators available • Loading dock/access			
8. Cable placement needs defined.			
9. Access to all floors for reels.			
10. Secure anchor points identified.			
11. Start work approval date.			
12. Firestopping complete.			
13. Inspection date.			
14. Completion date.			

Example 12.3   Checklist for site survey—new construction.

Issue	Responsibility	Date Scheduled	Date Ready
1. Owner's telecommunications needs defined and documented.			
2. Telecommunications design complete and up to standards.			
3. Modifications (if required) submitted and approved.			
4. Drawings and specifications distributed.			
5. Plan defined for moving or working around building occupants.			
6. Modifications to existing rooms or construction of new spaces complete for: • Telecommunications spaces – Backboards – Lighting – Electrical – Grounding – HVAC – Fire suppression – Alarms and controls • Backbone pathways • Horizontal pathways • Work area spaces			
7. Secure storage identified.			
8. Work site conditions defined. • Safety plan posted • Permits approved • Safety equipment in place • Office and meeting area defined • Elevators available • Loading dock/access			
9. Cable placement needs defined.			
10. Access to all floors for reels.			
11. Secure anchor points identified.			
12. Start work approval date.			
13. Firestopping complete.			
14. Inspection date.			
15. Cutover/transition date.			
16. Completion date.			

Example 12.4 Checklist for site survey—retrofit construction.

- Is power available within the vicinity of the DD?

- Does any existing cable, hardware, or equipment need to be removed?

- Do any hazards exist (e.g., inadequate or nonexistent firestops)?

Answers to many of these questions should come from the materials contained in the customer's and the designer's documents. Do not leave anything to chance. Always review all requirements of the project before concluding additional pathways and spaces are not required. If they are needed, and they have not been included in the original plans and specifications, job change orders may be required. At the very least, it may change the approach to the installation.

Initial construction meeting

For larger single-family home projects, and for most MDU projects, the residential telecommunications cabling project manager should hold an initial meeting with the entire residential telecommunications cabling installation team. At this meeting, the project manager and team leader should lay out the responsibilities of the parties involved. This ensures that there will be minimal confusion during the implementation of the project. The project plan can be reviewed and updated as necessary. Questions can be asked and answered.

Communication between all personnel involved in the project is critical to its success, and should be established at this early stage. Minutes of this and all other meetings should be maintained and a printed copy provided to each person attending, as well as to additional people having responsibility for the project.

Ordering and staging materials

Project materials may be ordered on a project-by-project basis or may be taken from a company's in-house stock. Either way, there are typically two steps in ordering products for an RSCS—the rough-in step and the trim step.

The rough-in phase typically requires the following materials:

- Bulk twisted-pair, coaxial, and optical fiber cable

- Low-voltage mounting brackets and electrical back boxes (single-gang, double-gang)

- Cable management hardware (e.g., cable ties, bridle rings, etc.)

- Nail plates

- Materials for cable pathways

If using cable spools to hold the cable itself, make sure the spool size is small enough to fit within the company's delivery vehicles and light enough to be carried by the installation staff on-site. Ensure also that there is a facility available for mounting the spools on a stable platform during cable pulling.

Do not simply order the proper amount of cable in terms of overall footage required on the job site. Instead, determine the maximum number of cables that can be pulled between two locations at the same time, and order enough spools or boxes of cable to meet that requirement. Although more than enough cable may be ordered in this way, leaving excess inventory at the end of the residential cabling installation, the savings in labor time gained by pulling multiple cables simultaneously can offset the extra cost.

During the trim phase, the following materials will be required:

- DD enclosure and backboard
- DD cable termination components (e.g., video splitters, twisted-pair termination blocks, etc.)
- Outlets and faceplates
- Twisted-pair cabling jacks
- Optical fiber connectors
- F adapters
- F connectors
- Patch cords, if required
- Labels

NOTES: Additional products used for other types of cabling systems (e.g., home entertainment, security, etc.) may be required during rough-in and trim-out (e.g., in-wall speakers and speaker brackets).

Some of the items listed for trim-out can also be installed during the rough-in phase, or between rough-in and trim-out if desired. For example, the DD can be mounted whenever practicable.

Make sure to use twisted-pair cabling jacks meeting the same or higher performance requirements as the outlet cable to which it is being attached.

See Example 12.5 for a materials list and Example 12.6 for a labor list.

Project Number: XXXXXXXXXXXX
Project Name: Anywhere Elementary School
Address: Anywhere, USA

Item	Material Description	Catalog Manufacturer	Catalog Number	Unit Each	Total Price	Price
1	Wire, 4 pair, 24 AWG, UTP, Cat 5, CMR	Mfg. A	530141-TP	54000′	0	$0.00
2	Surface raceway, small	Mfg. B	MT1ABC5	350′	0	$0.00
3	Surface mount box, single gang	Mfg. C	WT12DB	56	0	$0.00
4	Cat 5 relay rack E/W wire management	Mfg. D	55053-703	2	0	$0.00
5	Vertical wire management hardware	Mfg. E	11374-703	2	0	$0.00
6	Patch panel, 96 port, T568A	Mfg. F	49485-C96	4	0	$0.00
7	Patch panel, 24 port, T568A	Mfg. G	49485-C24	2	0	$0.00
8	Faceplate, single gang, dual port	Mfg. H	41080-2AP	9	0	$0.00
9	Faceplate, single gang, quad port	Mfg. I	41080-4AP	47	0	$0.00
10	Modular jack, Cat 5, orange, T568A	Mfg. J	41108-RO5	208	0	$0.00
11	Horizontal wire management panel	Mfg. K	49253-BCM	16	0	$0.00
12	Rear cable bars	Mfg. L	49258-TWB	36	0	$0.00
13	Firestop compound	Mfg. M	AA529	10	0	$0.00
14	Electrical Metallic Tubing, 3/4"	Mfg. N	34EMT	150′	0	$0.00
15	Electrical Metallic Tubing, 2"	Mfg. O	2EMT	160′	0	$0.00
16	Cable, fiber optic, 6 str., OFNR	Mfg. P	PDRCB3510/15	300′	0	$0.00
17	Fiber distribution panel	Mfg. Q	4R130-OTA	2	0	$0.00
18	Connector panels	Mfg. R	4F100-6TM	2	0	$0.00
19	ST connectors	Mfg. S	95-100-01R	12	0	$0.00
20	10BASE-T 24-port hubs	Mfg. T	AT3624TR-15	1	0	$0.00
21	10BASE-T 24-port hubs	Mfg. U	AT3624TRS-15	3	0	$0.00
22	Chassis	Mfg. V	AT-36C3	2	0	$0.00
23	Fiber optic transceivers	Mfg. W	AT-MX26F-05	2	0	$0.00
24	Duplex, STST, 3 m F/O jumpers	Mfg. X	STST3M	2	0	$0.00
25	Data patch cords	Mfg. Y	42454-03O	104	0	$0.00
26	Data patch cords	Mfg. Z	42454-050	104	0	$0.00
27	Data line cords	Mfg. AA	42454-10O	208	0	$0.00
28						
29						
30						
31	Exempt materials	various	various	1 lot	NA	$0.00
32	Transportation					$0.00
Total Materials						**$0.00**

Example 12.5   Materials list.

Project Number: XXXXXXXXXXXXX
Project Name: Anywhere Elementary School
Address: Anywhere, USA

Item	Material Description	Units/Quantity	Unit Rate	Total Price
1	Installing horizontal wires (2 per run)	9	0	$0.00
2	Installing horizontal wires (4 per run)	47	0	$0.00
3	Installing faceplates and jacks	208	0	$0.00
4	Installing relay racks	2	0	$0.00
5	Installing patch panels	6	0	$0.00
6	Terminating wires at patch panels	208	0	$0.00
7	Certifying Cat 5 wires	208	0	$0.00
8	Installing surface raceway	56	0	$0.00
9	Installing surface mount boxes	56	0	$0.00
10	Installing backbone fiber optic cables	300	0	$0.00
11	Installing F/O connecting hardware	4	0	$0.00
12	Terminating F/O cables	12	0	$0.00
13	Testing F/O cables	6	0	$0.00
14	Install fire-/smoke-rated partition penetrations	59	0	$0.00
15	Mount hubs on relay racks	4	0	$0.00
16	Installing backbone conduit from main building to portables	120	0	$0.00
17	Installing horizontal conduit between backbone conduit and portables	48	0	$0.00
18				
19				
20				
Total Labor				**$0.00**

Total cost			**$0.00**
Materials markup	**50%**		**$0.00**
Labor markup	**50%**		**$0.00**
State sales tax on materials	**6%**		**$0.00**
Total price to customer			**$0.00**

Example 12.6 Labor list.

Testing, labeling, and documentation

The final step in the delivery of a functional residential cabling system to the owner is the provision of testing results, labeling, and documentation. This step is very important as these tests confirm the suitability of the installed cabling and can be used to mitigate disputes after the residential cabling installation is complete.

The testing required for each twisted-pair cabling link must include all required tests to the category specified in the contract. According to ANSI/TIA/EIA-570-A, "For field testing of Categories 3, 5, 5e, and higher cabling, test instruments shall meet or exceed the applicable requirements in ANSI/TIA/EIA TSB67. For field testing of twisted-pair cabling, test instruments shall meet or exceed the applicable requirements in ANSI/TIA/EIA-568-B.1."

The twisted-pair cabling tests shall include:

- Wire map (showing opens, shorts, and crossed pairs).

- Cable length.

- Cable insertion loss (formerly known as attenuation, this is a measure of lost signal over the length of the cable).

- Near-end crosstalk (NEXT) loss.

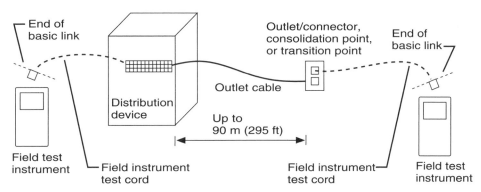

Figure 12.2   Basic link test configuration.

NOTE: Test cords are up to 2 m (79 in) in length.

Category 5e and higher twisted-pair cabling tests add the following:

- Power-sum near-end crosstalk (PSNEXT) loss
- Equal-level far-end crosstalk (ELFEXT) loss
- Power-sum equal-level far-end crosstalk (PSELFEXT) loss
- Return loss
- Propagation delay
- Delay skew

While this level of testing is far in excess of what has traditionally been done in the residential environment, networking hardware is designed under the assumption that the cabling system being used meets or exceeds the minimal specified performance levels in each of these areas.

Many manufacturers require complete testing to be done prior to issuing a system warranty. If so, the contract should specify the requirements for such a warranty. The desired length of warranty—10 to 20 years is typical—should also be determined, along with the form of warranty coverage being provided. The testing documentation required at the end of the residential cabling installation may also be noted in the warranty, and may differ from and exceed that included in the residential standard.

Coaxial cable must be tested for continuity to ensure proper termination. According to ANSI/TIA/EIA-570-A, additional tests shall include attenuation and cable link length.

For optical fiber cable, a simple test of attenuation and length is required.

A more complete description of testing, test equipment, and residential testing procedures is found in Chapter 16: Testing and Troubleshooting.

All cables and terminations must be labeled within the DD. Label the cable at the outlet although homeowners may find labeling the faceplate aesthetically displeasing. When not labeling the faceplate, the telecommunications outlet/connector inside the outlet should be labeled.

The final documentation given to the homeowner may be a list of outlet locations cross-referenced to the labeling system. The homeowner may also request as-built drawings, although this may be cost-prohibitive under some circumstances.

13

New Construction Rough-In

Overview

The installation of a residential telecommunications cabling system consists of two time periods:

- Rough-in—The time in which the residential cabling installer runs the cable between the various termination points (outlets, distribution device [DD], network interface device [NID], etc.). The main task during the rough-in is to complete all the work that cannot be easily accomplished once Drywall and other wall materials are installed. As described in Chapter 1: Residential Environment, the rough-in period usually begins after the rough-in of the electrical and heating, ventilating, and air-conditioning (HVAC) systems. The period usually ends prior to Drywall and insulating.

- Trim-out—Typically, one of the last steps in the home construction process. All the cables are terminated to the appropriate devices and testing is completed. The trim-out is covered in Chapter 15: Trim-Out Finish, of this manual.

In retrofits, the rough-in and trim-out periods may overlap, since there may be no need to wait for the completion of work from other trades on the job site. The specific requirements of a retrofit residential cabling installation are covered in Chapter 14: Retrofit.

Challenges of the rough-in

In new construction, particularly for single-family homes, the rough-in period poses many challenges to the residential cabling installer.

When the rough-in process is required, the home is typically not insulated. Therefore, steps must be taken to protect equipment from rain and other environmental factors. The home's HVAC system is not installed, so the temperature extremes may require special clothing, portable heaters, the availability of drinking water, and other items. Extra care should be taken when working in attics when the weather is hot to avoid heat exhaustion.

Bathroom facilities are often unavailable at the job site. Nails and other potential hazards left over from other trades may be present, making the use of proper footwear important. The noise generated from other trades may be extreme, so use of earplugs may be beneficial.

From a work efficiency standpoint, the following issues may also challenge the residential cabling installer's productivity. Electrical power may not be available, thus requiring the use of cordless drills. However, cordless drills can only provide limited power and are inadequate for many required tasks (e.g., drilling large holes through flooring for vertical cable runs and the installation of the cabling chase). A portable electric generator is highly desirable to provide power for larger drills.

For homes with basements, the lack of a working sump pump may lead to standing water in the work area. A solid work surface may be helpful. Also, at the rough-in stage, some builders have not yet installed stairs into the basement, presenting the need for the residential cabling installer to bring ladders.

Select Device Locations

The rough-in process begins with identifying the various device locations. A residential structured cabling system (RSCS) should follow ANSI/TIA/EIA-570-A, *Residential Telecommunications Cabling Standard*. Criteria for the distribution of voice, data, and video (VDV) application devices includes the DD and the telecommunications outlet/connectors. The locations for vertical and horizontal pathways must be established, including the position for a chase, if desired.

For new construction, the telecommunications outlet/connector and DD locations can usually be determined from reviewing the home's blueprints or floor plans. The locations marked on the plans determine the wall and general locations where a device is best located. In most instances, the exact locations are not determined until the residential cabling installer is on the job site. This is necessary because the pathways and stud bays filled by the plumbing, electrical, and HVAC systems are not known until visibly inspected.

The locations of the chase and other vertical pathways are highly dependent on obstructions from other mechanical systems. In addition, it is often difficult to determine viable vertical pathways from a multi-story floor plan because stub bays on subsequent floors may not line up.

Visible inspections at the beginning of the rough-in period are necessary. If there is no direct vertical pathway, the chase may be placed along an exterior wall, although this may be difficult due to the construction of the walls (e.g., multiple headers, etc.). Placing the chase in an exterior wall may interfere with the proper working of the home's insulation.

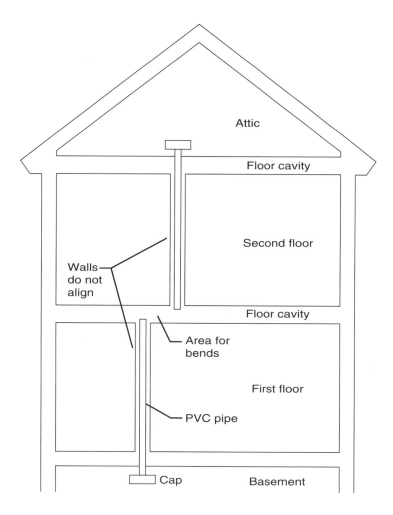

PVC = Polyvinyl chloride

Figure 13.1   Installation of a cable chase.

Alternatively, a pair of bends can be inserted into the chase in order to link two portions of the chase, each run vertically on a single floor (see Figure 13.1). It is recommended to sweep the bends as smoothly as possible and never use more than two 90-degree bends.

Mark Device Locations

In new construction, the residential cabling installer's first task is to mark the specific locations for the various devices, guided by the customer's general directions listed on the floor plans. A large marker should be used to identify the locations and the type of outlet being placed within a specific stud bay.

In addition, the type and number of cables that will be needed at each location should be identified at the device location and marked clearly on the floor plans. This reduces confusion between installing a low-mounted telecommunications outlet/connector versus a standing height wall telephone outlet. Identifying the number and type of installed cables also helps to determine the size of the holes that will be required.

When choosing the specific location of a telecommunications outlet/connector, keep the following issues in mind:

- The direction of swing for an adjacent door—Avoid placing an outlet behind an open door (see Figure 13.2).

- The location of electrical outlets—Keep adequate spacing between the telecommunications outlet/connector and electrical outlets (those electrical outlets that serve both the same room and the room sharing the wall). It is a good idea to keep the telecommunications outlet/connector and electrical outlets within a few meters (feet) of each other, so power is available for computers, televisions, and other devices served through the telecommunications outlet/connector. Keep in mind that the separation of telecommunications cable and power cable should be maintained.

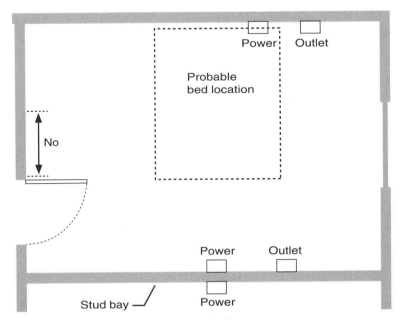

Figure 13.2   Proper outlet locations in a bedroom.

NOTE: Outlets must be ANSI/TIA/EIA-570-A Grade 1 or Grade 2.

- Ability to route cable—The cable should be able to enter the stud bay from directly above or directly below it in order to minimize the labor needed to drill horizontally through studs. Check both directions for obstructions.

- Integrity of the studs—The integrity of the studs affects the strength of the square drawn cover (e.g., mud ring) or electrical box. At times, a desired stud location cannot be used because the stud is damaged. Keep in mind the proper firestopping requirements when an outlet is placed in a fire-rated wall.

- Probable layout of the room—Consider the use of the room (i.e., bedroom, office, etc.) and the most likely position of furniture before deciding on the final outlet locations. These should be decided with the customer on-site, along with potential future room layouts. See Figure 13.2 for proper outlet locations.

Cable numbering

The cable locations should be numbered in a consistent manner. Each outlet can be given a unique outlet number, which may or may not include a room or floor identifier. Then, each cable is numbered sequentially for each outlet.

If each outlet can have up to six cables, then the numbering sequence may be set as number one through number six. Numbering always starts in the upper left corner, then goes down the columns and across the rows, regardless of how many cables are run to any individual outlet (see Figure 13.3).

Other labeling sequences may be used, as long as the system is consistently followed.

Mount Rough-In Devices

The rough-in for a telecommunications outlet/connector begins with a square drawn cover, which is better known as a mud ring. A mud ring will become the backing into which the outlet faceplate will be attached.

In some locations, a back box may be required or preferred. This is the case especially where conduit is run. While conduit is not usually needed for low-voltage cabling, some homes include conduit as an effective path for feeding

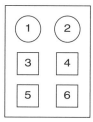

Figure 13.3 Possible cable numbering system at the outlet.

cable through cement slab floors and to make future outlet changes. Back boxes can also be used when the wall cavity is shallow (e.g., when the walls are cinder block or brick) or when studs are not available, since the box itself can be screwed into the wall.

Mud rings screw into wood or metal studs. They are mounted in one of the following locations:

- Low to the wall, at a vertical level to match the height of the electrical outlets.

- At the same height as a light switch (e.g., in the case of a wall telephone).

Some outlets may be required at other heights, particularly in kitchens and offices, to fit within openings between cabinetry and countertops.

In the same way, mud rings are installed in an audio cabling system for the future mounting of components. Volume controls are mounted at the same height as light switches, while outlets are mounted low on the wall near an entertainment center and used to terminate the speaker cable onto binding posts. Mud rings can also be installed and later covered with a blank wall plate. This provides easy access to cables that will later be used for speakers and other devices.

Mud rings and outlets used for telecommunications cabling should remain consistent in orientation compared with the electrical outlets. In some geographic areas, electrical and telecommunications outlet/connectors are mounted vertically; in other areas, they are mounted horizontally. While the performance of the system will not be affected with either mounting direction, it is more aesthetically pleasing to maintain consistency.

Residential cabling installers should do the following when mounting a mud ring.

Step	Mounting a Mud Ring
1	Install the mud ring at a height consistent with the electrical outlets (see Figure 13.4).
2	Screw one screw into the stud to hold the mud ring.
3	Use a level to ensure the proper angle of the outlet (do not simply "eye-ball" the mud ring versus the stud, the stud may not be precisely vertical).
4	Once the mud ring is level, screw in the other screw and tighten it down.

This same process is used when installing any low-voltage bracket during rough-in. The speaker locations are identified and the bracket is attached to an adjacent stud. It is critical to determine if there are other architectural elements (e.g., recessed lighting cans) that need to be lined up with the speaker bracket.

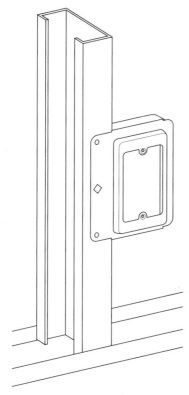

Figure 13.4 Mounting a mud ring.

In addition, it is important to mount and use rough-in low-voltage brackets at a proper distance from each other and in the desired location versus doorways, windows, and other elements of the room. See Chapter 11: Other Residential System Applications and Implementations, for further information on in-wall or in-ceiling speaker installation.

Rough-In Distribution Devices (DD)

DDs may be flush mounted within a stud bay. If so, the DD needs to be installed in position during rough-in prior to the installation of Drywall and insulation.

Another common practice is to mount the DD directly on a basement (or other available) wall. In this case, it is advisable to mount a plywood backboard first. The DD is then mounted to this backboard.

If the DD is not to be flush mounted, then the DD can be installed during the trim-out phase of the residential cabling installation (see Chapter 15: Trim-Out Finish).

The DD should be mounted within a climate-controlled location. This protects the sensitive electronics that may be stored within the DD from damage. In addition, the DD must be located near a dedicated power source in order to supply data and video electronics.

Many DDs are designed with covers that screw on and off. Other designs include hinged doors, which can be helpful in avoiding the loss of the door. However, in this case, appropriate room must be allowed for the door to swing open and shut.

The DD can be mounted at standing height in order to facilitate moves, adds, or changes (MACs). Alternatively, the DD can be mounted higher on the wall to keep it out of reach and away from potential damage.

The rough-in of the flush mounted DD is a straightforward process. The back portion of the box is screwed to two adjacent studs so the face of the box will be flush with the Drywall once installed (see Figure 13.5).

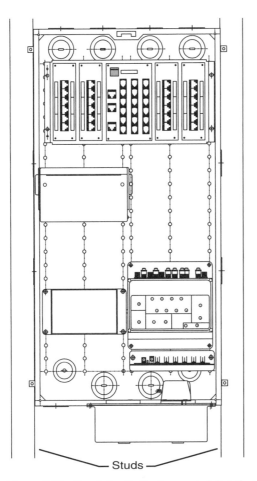

Studs

Figure 13.5   Rough-in of a flush-mounted distribution device.

Most DDs are designed to work well within standard stud openings; however, many factors can lead a builder to deviate from standard stud separation. In addition, the stud bay must be free from obstacles (e.g., electrical cabling).

Once the back portion of the DD is installed within the stud bay, the telecommunications cables (twisted-pair, coaxial, fiber, and audio) can be routed through the openings in the DD and coiled within the enclosure for future termination. Avoid leaving the cables exposed outside of the enclosure to prevent possible damage.

If the cover of the DD is to be screwed onto the back of the enclosure, the cover can either be attached at this point (a process which further limits the possibility for damage or theft) or removed from the premises until trim-out.

Ensure that the DD is properly grounded according to the vendor's instructions. Most vendors provide a grounding post within their enclosures.

Install Future Access Pathway or Chase

It is advisable to run a chase of at least 50 mm (2 in) in diameter to accommodate future cable system expansion. Check with the authority having jurisdiction (AHJ) on the amount of penetration allowed in a wall cavity. It is also advisable to install a pull string through the chase to make future pulls more efficient. Code allowing, the chase may be standard polyvinyl chloride (PVC) piping, or it can be made out of the more flexible piping used for central vacuum systems.

Multiple chases can be installed within a home if a large amount of future system expansions is expected.

Secure Cable Pulling Area

It is most efficient to run all the cables from the central point(s) of the cable system (e.g., DD for voice, data, and video; entertainment center for audio; control panel for security, etc.). It is advisable to secure the area so that neither the homeowner nor those working in other trades get in the way and risk injury.

If the rough-in period takes more than one day, all cables should be raised off the floor and removed from all walkways at the end of each day. This will protect the cable from damage and protects other persons on the job site. Once the cable rough-in is complete, the residential cabling installer should secure the cable.

Drill required holes

As one residential cabling installer prepares and labels the cables for pulling, another can complete the drilling of the required holes for the cable runs. Figure 13.6 shows cable routes for a typical two-story home with basement.

In this case, the cables serving the first floor telecommunications outlet/connectors will come up from the basement through holes drilled in the first floor's flooring. The second floor will be served by cables run through a

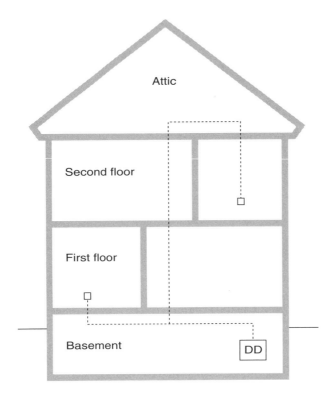

DD = Distribution device

Figure 13.6 Typical cable routes.

vertical pathway, crossing the first and second floors and entering the attic space. Holes at the top of the stud cavities are then drilled for the cables serving the second floor telecommunications outlet/connectors.

It is advisable to have both cordless and power drills available. A drill with a right-angle drilling direction is valuable for getting within cramped stud cavities and maintaining straight holes.

If possible, the holes need to be sized so that the cables passing through fill no more than 50–75 percent of the opening. In all cases, check with local regulations concerning any limitations on the size of the holes.

After the hole is drilled, clean off splinters from both edges to ensure the smooth passage of the cables. Always try to line subsequent holes up both vertically and horizontally. Maintain a straight cable path to make long cable pulls easier and limit the potential for damage.

Long drill bits are necessary in many cases, particularly when drilling through large headers. Always remember to wear safety glasses and hearing protection when drilling.

In some forms of construction, the studs are not wood, but are comprised of a sheet-metal frame. In some cases, the frame has preformed holes that can be used by the residential cabling installer for routing cables. In other cases, holes must be punched or drilled through the metal. It is advisable to install grommets to protect the cable from the sharp metal edges. Carefully pull cables to avoid pinching in tight corners. Avoid pulling around more than one corner or the pulling tension may exceed recommended maximums (refer to ANSI/TIA/EIA/570-A).

Pull Cable to the Network Interface Device (NID) and Outlets

Cable can be set up at the DD location and pulled in the direction of the NID and outlets. This allows cables going to multiple outlets in the same area to be pulled through the majority of the path together, thus lowering cabling installation time. By pulling the cable from the DD location toward the outlets, the cable reels do not have to be moved from location to location. In addition, the amount of cable that needs to be fed back to the DD (a difficult process to judge and one that can lead to excessive wasted cable) from an intermediate pulling location will be minimized.

In many cases, the initial cable run will be long, going from the DD location to the point at which the cables become vertical. To ease the pulling tension across this long run, bridle rings or J-hooks may be installed, particularly at points where the cables make 90-degree bends.

Step	Mounting J-Hooks in Structures
1	Determine the J-hook size required in each cable path and lay them out along the cable route.
2	Identify the location of the first J-hook to be installed.
3	Position the J-hook at its proper location and, using a pencil, mark the holes for the anchors.
4	Predrill the holes and install the correct anchors for the type of structure. A masonry structure requires one type of anchor while metal structures or Drywall require others.
5	Install the anchors.
6	Reposition the J-hook at the desired location.
7	Using the correct screwdriver, install the appropriate screw through the hole in the J-hook and into the anchor, securing the J-hook to the anchor.
8	Repeat Step 7 for the second screw.
9	Install the remaining J-hooks using the same procedures, until all J-hooks are installed.

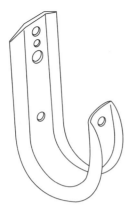

Figure 13.7 J-hook.

The cable will generally be housed within a pull box or on a spool. Pull boxes contain the cable, which is wrapped in such a way that it will not tangle when the cable is pulled out of the box. They are convenient since an additional cable-pulling apparatus is not needed. The rectangular pull boxes are convenient for transporting the cable to and from the job sites and for carrying throughout a job site. Pull boxes can be susceptible to cable tangles, particularly if excess cable from one cable pull is fed back into the pull box prior to a subsequent pull. In cold weather, cable tangling is often a significant problem, as the cable becomes cold and brittle. Keep cable boxes away from damp or wet floors and dripping water.

Cable spools, or reels, allow for relatively snag-free cable pulling, although they require additional cable-pulling apparatus. Longer cable lengths may be available on spools rather than in pull boxes, which can help the efficiency of the cable rough-in. Since the cable on a spool is visible, it is easier to judge whether there is sufficient cable left for an additional run.

Each cable end must be labeled with the same cable identifier; white electrical tape and a permanent marker work well. It is advisable to label the cable in multiple places near the end because marks can be rubbed off during the installation process. Since a few millimeters (inches) of cable are discarded at the end during the trim-out period, an additional label a few meters (feet) back from the end may be helpful. Self-laminating labels provide extra protection for the markings, particularly in wet environments. Most manufacturers will not offer a warranty on cables that have been immersed in water/damp conditions.

Always clearly label the pull box or reel with the cable-identifying number prior to the pull. It is easy to confuse the cables, which requires additional testing during the trim-out phase. At the end of each pull, cut the cables at the spool or pull box and label the ends appropriately. Clearly mark out the number identifier from the spool or pull box to eliminate confusion on subsequent pulls.

Once the cables going to a particular outlet are labeled, they can be secured together with electrical tape into a cable bundle. Multiple bundles (serving multiple outlets) can be bundled together to allow the residential cabling installer to run more cable on an individual pull and reduce installation time. However, it becomes difficult to run more than six to eight cables in a single pull. Additional turns and bends in the cable path will increase the difficulty of running large mounts of cables at once, particularly when going through the small cable openings between floors. The time lost untangling cables will quickly offset any time saved by running additional cables on each pull.

Once the pull has begun, it is important to maintain a pulling tension that will not damage the cables. For 4-pair twisted cable, this pulling tension is limited to 110 N (25 lbf). It is important to be sensitive to the stress applied to cables going around sharp bends and to maintain proper bend radii throughout the cable run. It is easy to create a friction burn on the jacket of a cable that, if serious enough, could affect the conductors in the cable.

The telecommunications cable should also consistently stay above, or below, all obstructions. If a cable is routed above one obstruction (e.g., pipe, beam, etc.) and then below the next obstruction, the cable pull will become difficult. The extra force required in the pull may damage the cable performance and break the obstruction, and the cable route will not look orderly.

Maintain as much distance between the telecommunications cables and the electrical cabling or conduit. If the paths of the two services are required to cross, do so at a 90-degree angle, if possible. Maintain adequate space (300 mm [12 in] or more) in attics between the telecommunications cables and light fixtures. Visibly inspect the final cable run to ensure that once the Drywall is installed, there will be no exposed cables.

When a wall or floor needs to be crossed by a cable run, it is advisable to use two people—one on either side of the obstruction. Vertical runs through small floor or ceiling openings represent a particular opportunity for snagging cables. With only one residential cabling installer, it is difficult to determine if the cable length has been completely pulled out or if there is a snag in the cable at the opening. Pulling overly hard on a snagged cable will cause a knot that can severely impact the cable performance.

Cables can either be fed through vertical and horizontal obstacles by pushing the cable from behind, or by installing a fish tape from the other side of the wall or ceiling. The cable is then taped to the fish tape and pulled through from the other side. If the pathway is a difficult one to cross and will be used again for further cable runs, it is advisable to attach a pull string to the cable long enough to make it through the obstruction. Tie off the pull string at both ends until a future pull through the same path is required.

In new construction, holes drilled through vertical or horizontal obstructions should be 13 mm (0.5 in) in diameter for two or fewer cables, 25 mm (1 in) for four cables, and 38–50 mm (1.5–2 in) in diameter for eight cables or more. When drilling through multiple wall studs on the same cable path, it is important to keep the same vertical height to ease the cable pull. The holes should be drilled as straight as possible through the studs for the same reason. Angled cuts direct the cable off the pulling axis and make installation more difficult.

Check the other side of the obstruction prior to drilling any hole. Ensure that all piping, cabling, and mechanical systems are away from the exit area of the drill.

Once all the cable is pulled through to the outlets, it is advisable to confirm that the number and types of cables match what is called for on the floor plan.

Secure Cable

Once the cables are in place behind the mud rings, they need to be secured. Excess cable slack can then be worked back through the cable run toward the DD. Secured cables are protected from damage during the remainder of the construction process.

For new construction, a staple may be placed into the stud behind the mud ring opening. The cables are coiled (leaving approximately 300–460 mm [12–18 in] of slack for subsequent termination) and secured to the staple with a cable tie. Ensure that the cable tie is close enough to the opening to be easily removed once the Drywall is in place.

The cable should then be secured in one or several places as it moves vertically out of the stud bay and into the attic (first floor cables simply drop through holes beneath the mud rings into the basement). Once in the attic, the cables need to be secured back to the vertical pathway, routing the cables on 90-degree angles and keeping proper distance from light fixtures and other devices. Securing the cable this way will help protect the cable during subsequent construction and maintain a professional look and feel. Make sure the cables, when dropping vertically, are not resting on a sharp edge, since weight alone over a period of time can damage the physical characteristics.

Slack is removed from the vertical pathway(s) as the cables are routed back to the DD. Secure the cables together and to the building structure every 1–1.8 m (3–6 ft) and at every turn.

Do not staple directly over a cable due to the potential for damage. Instead, staple beside the cable and use a cable tie through the staple and around the bundle of cable to complete the procedure. Screw-in cable mounts (cable tie bases) are available, which can provide additional longevity and strength.

Once dressed back to the DD, the telecommunications cables should be cut to similar lengths, with enough slack to be terminated anywhere within the DD. The cables are then fed into the DD. In most designs, the twisted-pair cable is terminated on one side, or on one vertical level, and the coaxial cable is terminated on the other side, or another vertical level. It is advisable to separate the cable types prior to entering the DD.

Protect Cable

The new construction process presents many potential dangers to installed telecommunications cables. The greatest risk is from the nails or screws used to install Drywall. Cables can be protected through the use of nail plates. Verify proper usage with the AHJ. Nail plates are to be installed across each stud that is passed through with cable. Nail plates should be installed to cover the entire hole width. Ensure that nail plates are attached properly and that subsequent nail plates do not overlap and cause a buckling, which may affect the installation of the Drywall.

According to the *NEC* 300.4, penetrations through structural members may be no closer than 32 mm ($1^1/_4$ in) from the edge of the member. Where this is impossible, a steel plate of at least 1.6 mm ($^1/_{16}$ in) thick must be installed to protect against nails or screws penetrating the electrical wiring. The *NEC* does not state the maximum size hole that can be made. However, in order to comply with the above article, the maximum size hole for a 2 × 4 would be 25 mm (1 in [89 mm ($3^1/_2$ in) by 32 mm ($1^1/_4$ in) by 32 mm ($1^1/_4$ in)]). For a 2 × 6, it is 50 mm (2 in), etc.

Secure cables to the stud within all stud bays so the cables will not freely float and get pinched between the studs and the Drywall during Drywall installation.

Both the outlets and DD can be further protected by installing a piece of cardboard across the openings. Many vendors provide cardboard with their products for this purpose.

Firestopping

Firestop as required by the AHJ. See Chapter 3: Codes, Standards, and Regulations for further information.

Perform Clean Up

The clean up process during the rough-in process helps maintain a professional look to the job and protects those who perform other trades. Since other trades will perform a number of tasks at the premises prior to the completion of the job, a "broom clean" level of tidiness is appropriate.

All excess cable should be thrown away or reused if long enough. All cut cable ties, empty boxes, and other refuse should be thrown away. Dispose of this in an environmentally friendly manner.

14

Retrofit

Introduction

The telecommunications (voice, data, and video) requirements for owners of existing homes are the same as for buyers of new homes. While some existing homeowners may choose a wireless system (see Chapter 10: Alternative Infrastructure Technologies), many others will want the full benefits (e.g., bandwidth, reliability, etc.) available only through a residential structured cabling system (RSCS). This chapter covers the design and installation of these systems.

While new construction is a relatively straightforward environment for the residential cabling installer, the number of residences that will potentially be retrofitted with residential telecommunications technologies over the next decade is high. Properly planned and executed, the revenue generated from each home can far exceed that generated with new construction. This is due to the necessarily increased amount of labor required to perform the work.

Retrofit work, while messy, also provides the residential cabling installer with the benefit of being able to work straight through on a job—from start to finish—often without the interruption of the other trade personnel always present in new construction.

The key to success is a thorough understanding of the materials and home construction methods that will be encountered during the residential cabling system installation. The most common of these materials and methods are covered in this chapter.

Determine Customer Requirements

Due to the complexity and cost of installing an RSCS within existing homes, it is critical to determine the current and future telecommunications requirements of the homeowner and family. The criteria set forth in ANSI/TIA/EIA-570-A, *Residential Telecommunications Cabling Standard*, and Chapter 2:

Residential Structured Cabling Systems, will assist in determining the scope of these requirements.

Most existing homes will have cable or satellite television services and telephone/Internet services already in place. These services will enter the home from one or more locations and connect to the existing (probably daisy-chained) cabling system to serve one or more outlets.

It is necessary to determine whether any of the existing RSCS can be reused within the new RSCS. In some cases, the position of particular outlets might be beneficial to the homeowner because new cable cannot easily be installed in the same areas. In this case, the homeowner may choose to maintain that portion of the existing cabling system.

The designer must be aware of the current cabling system. An outlet/connector within an outlet may be at the end of a long daisy-chain of cables. Disruption or removal of an earlier section of the daisy-chained cable will take the desired outlet/connector location offline as well. For example, if the telephone cable terminated on an outlet/connector earlier in the daisy-chain is to be removed from the outlet, the matching conductors from the two cables in the daisy-chain must be spliced together to allow the signal to continue down the daisy-chain to subsequent outlet/connectors (see Figure 14.1).

If any part of the existing residential cabling system is reused, the new distribution device (DD) will need to be linked to both the incoming service cables and the internal cabling system.

The designer should review the number of telephone, facsimile, and dial-up modem lines the homeowner will require in the home for family use and home

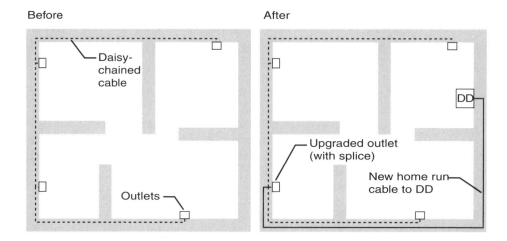

DD = Distribution device

Figure 14.1   Continuation of a daisy-chain.

office applications. The appropriate number of twisted-pair auxiliary disconnect outlet (ADO) cables should be installed to handle these requirements. The
proper number and grade of coaxial cables should also be in place (or added)
between either the incoming community antenna television (CATV) demarcation point, or the satellite dish, or both.

The ADO should be mounted:

- In an unobtrusive location.
- Within 1.5 m (5 ft) of a dedicated electrical circuit.
- Within easy reach of the desired cable routes.

The DD will most likely be mounted onto an existing wall or new plywood
backing that is screwed into an existing wall. If a basement is available, the
DD can be mounted into the concrete, although this requires special drills and
increased installation time.

The DD needs to be sized appropriately to house the desired DD modules.
Products mounted within the DD may include:

- Twisted-pair punch-down blocks.
- Coaxial adapter panels.
- Video splitters and amplifiers.
- Multi-line telephone modular blocks.
- Networking hubs.
- Residential gateways.

The use and benefit of each device should be explained to the customer. The
final design should be laid out on paper to ensure that there is adequate space
within the chosen DD enclosure. Experience suggests that the DD should be
sized to accommodate at least 50 percent more cabling and electronics than is
initially installed in order to handle future growth. The residential cabling
installer can choose a location where there is space directly below the DD in
order to add additional enclosures in the future, if needed.

The residential cabling installer needs to determine the desired service
availability in each room in the home. Refer to Chapter 2: Residential
Structured Cabling Systems as a guide. The following needs to be determined:

- Outlet cable requirements in each bedroom, living room, family room,
  kitchen, office/den, and other locations.
- Quantity and location of outlets in each room.
- Any special-use cabling requirements. Keep in mind not only the initially
  planned use for each room in the home, but also the potential use(s) in the
  future.

Identify Pathways and Outlet Locations

Visually inspect each proposed outlet location to determine if a usable cable pathway is readily available. The first floor (where a basement is available) can be served most easily from below. The second floor (if an attic exists above it) can be reached most readily from above.

The preferred choice is to maintain all cabling within existing wall cavities. Interior walls are preferred over external walls. Exterior walls may subject the cable to greater environmental extremes. In addition, exterior walls are often difficult to penetrate due to insulation, as well as headers for windows, doors, and other features of the home. Entering an external wall from above through an attic may not work if the slope of the roof line makes the working space above the wall unreachable. In the same way, exterior walls are often hard to penetrate from below through a basement.

If cable cannot be installed within the walls, channel molding can be mounted to interior walls or in the corners of the rooms to provide cable access from the ceiling or floor. Cable raceways of this type can be installed within telecommunications rooms through multiple floors to provide an access method from a basement to an attic area. Use channel molding that can be painted and (with fittings) can provide proper telecommunications cable bend radius control (see Figure 14.2).

Another option is to route cables on the exterior of the home and then drill through the exterior walls to the outlet locations. Exterior-rated cable should be used. The cable may be run within an exterior cable raceway to maintain the aesthetic beauty of the home and further protect the cable. If the exterior wall is concrete or brick, use a hammer drill to make the required holes for screwing in the mounts and brackets to hold on the cabling raceway. All metallic cables and containments on the exterior of the building or the home must be bonded and grounded (earthed) to protect against lightning strikes and voltage potential difference.

Once a main cable route is determined, the residential cabling installer needs to identify the exact location of the new telecommunications outlet. Use

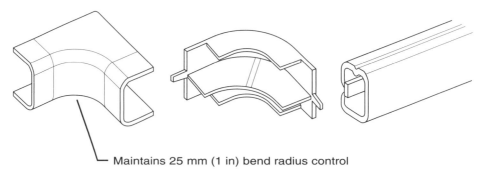

Maintains 25 mm (1 in) bend radius control

Figure 14.2 Channel molding with bend radius control.

a stud finder to determine a location with sufficient clearance on all sides. One way to determine the location of studs is to remove the faceplates and receptacles in electrical outlets and light switches that reside on the same wall as the desired telecommunications outlet locations. In most cases, the electrical outlet box is mounted on a wall stud.

> **NOTE:** Prior to removing any faceplates, follow proper safety procedures.

Determine the presence of electrical wiring, if possible. This will help determine which stud bays are available for the mounting of the telecommunications outlet. Confirm the location and direction of other mechanical apparatus (e.g., heating ducts and plumbing piping [see Figure 14.3]).

For retrofits, a low-voltage mounting bracket is used. Identify the location and height of the desired outlet opening. The low-voltage mounting bracket is first used as a template to cut a proper opening in the wall (using a utility knife or Drywall saw). Remember to match the orientation (vertical or horizontal) used in the area for the electrical and other telecommunications outlet. Be careful to keep enough space between the low-voltage mounting bracket and the wall studs and other mechanical devices to allow the low-voltage mounting bracket to be properly installed. Ensure the opening is level and then cut out this section of Drywall with a knife or saw. Do not install the low-voltage mounting bracket at this time because sharp edges may damage the cable being installed.

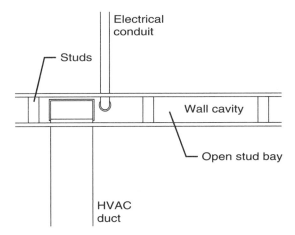

HVAC = Heating, ventilating, and air conditioning

Figure 14.3   Identify potential obstructions.

If the outlet is on an exterior wall, determine the point of penetration through an external wall. Always measure to the desired outlet location from a window (see Figure 14.4) or other point that can be easily referenced from both the interior and exterior of the home. Drill a small pilot hole from the outside to verify position. This allows the residential cabling installer to better position the hole compared with the external wall materials and design.

For brick exteriors, the residential cabling installer should drill through the mortar with a concrete drill bit. If the residential cabling installer attempts to drill out from the inside of the home, the middle of a brick may be hit, which could increase the time and expense of the installation.

The next important pathway to consider is the location for the vertical cables being run from the DD to the attic to service the upper floor. To support future additions to the cabling system, and to assist in the pulling of the cables, a space large enough for a 38–50 mm (1.5–2 in) chase is desirable. The best place to run this chase is near plumbing or heating, ventilating, and air conditioning pipes that run vertically from the DD area into and through the attic. These vents and pipes are contained within the wall structure and may have enough room around them to run the chase. Verify that the attic is properly ventilated to avoid the possibility of excessive temperature buildup. It is important to ensure that the vents run vertically through the home and do not take a bend within a floor or wall cavity.

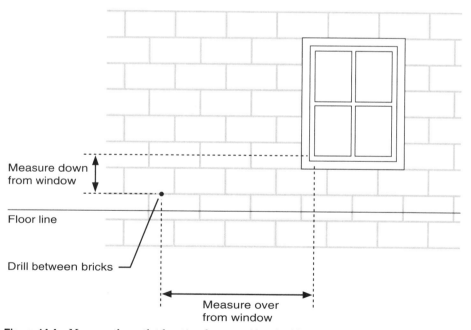

Figure 14.4   Measure the outlet location from outside a building.

Old Cable

In some cases, it is desirable to remove old cable and hardware. If existing outlets will be disconnected from service but not removed from the wall, it is highly desirable to make it clear to the homeowner that those particular facilities can no longer be used. This can limit the number of false maintenance calls and reduce the homeowner's difficulty in using the new telecommunications services in the home.

Existing outlets can be removed from the wall. Once removed by simply unscrewing the faceplate and removing the cable(s) from the termination point(s), the cutout in the wall needs to be filled in. This can be accomplished either by replacing the outlet with a new, blank outlet faceplate or by installing new Drywall over the hole and painting to match the room.

The former option is much less costly and provides the homeowner and residential cabling installer with a ready location to add to the RSCS in the future. However, some homeowners will not like the look of a blank faceplate and will opt to refinish the wall.

The cable terminated to the old outlet can be left in the walls or removed. Unless there is a compelling reason, it is preferred to leave the cable within the walls. First, this limits the potential for damage caused through the removal of the cable. Second, the old cable may be usable in the future as a pull string to route future cables through the wall to the outlet location. The cable may also be attached to the wall studs, making it extremely difficult to remove.

Old cabling that is exposed (e.g., in an open basement location) is more likely to be removed during the installation of the new cabling system. Prior to cutting or moving any cable, however, confirm that the cable will not be needed in the new telecommunications system.

When removing old cable that has penetrated an exterior wall, ensure that the remaining holes in the walls are properly sealed against the elements.

Fish Cable into Place

Installing cabling in retrofit applications often requires the placement of cables in blind areas behind walls. The primary way of accomplishing this task is to drill holes at both ends of the proposed cable run and feed a fish tape through the hole from one end to another. The fish tape is available in various lengths that are in coiled, handheld reels. The tape is semirigid so it can be pushed through these long, blind areas, yet flexible enough to make it past various obstructions.

It is often necessary to use two people when fishing a wall—one to push the fish tape into place and the other to catch the tape as it passes the other end of the cable run. The second person can also hear the fish tape as it makes progress along the run, as well as determine if any obstructions are in the way. A set of portable walkie-talkies is useful in this instance.

Once the tape has been caught and fed through the hole at the far end of the pathway, a pull string can be tied onto the fish tape and pulled back through

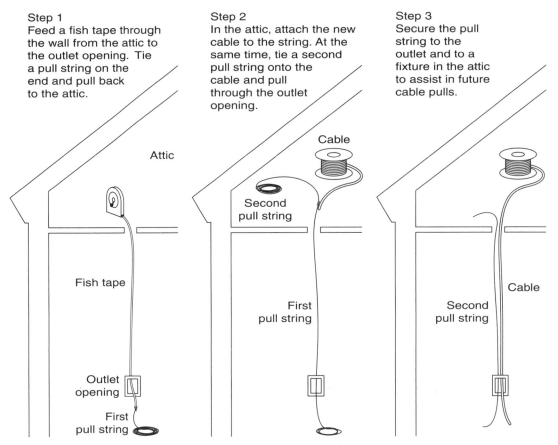

Step 1
Feed a fish tape through the wall from the attic to the outlet opening. Tie a pull string on the end and pull back to the attic.

Step 2
In the attic, attach the new cable to the string. At the same time, tie a second pull string onto the cable and pull through the outlet opening.

Step 3
Secure the pull string to the outlet and to a fixture in the attic to assist in future cable pulls.

Attic

Cable

Second pull string

Fish tape

First pull string

Second pull string

Cable

Outlet opening

First pull string

Figure 14.5   Fish tape fed through the wall from the attic to the outlet opening.

the pathway (see Figure 14.5). This allows for the future attachment of the telecommunications cable to the fish tape for pulling in either direction. Make sure the pull string is properly secured at both ends so it will not fall back into the pathway before the cable is run.

When the first cable pull is made, pull on the pull string until the ends of the cables appear. Then, pull the cables. If obstructed, do not pull the pull string with so much force that it pulls free of the cable bundle, or the pathway will need to be fished again. It is preferred to pull the cables back up (or down) to the starting point and retape them to the pull string, securing one cable at a time in order to make a smaller face on the cable bundle to avoid becoming stuck (see Figure 14.6). It is also advisable to tie on another pull string during the first pull so there is access for future pulls through the pathway.

When running a cable through a chase from the basement to the attic (or between other floors), it is advisable to run an additional pull string along with the cable. If multiple runs will be made through the chase, it is best to cut free

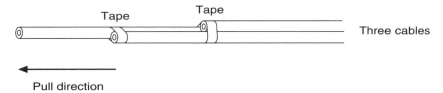

Figure 14.6   Reduce cross-section of cable bundle.

the pull string after each run in an attempt to limit the amount of twisting that may occur. Once subsequent cable pulls are tangled within each other, pulling tensions will exceed the recommended maximum pulling tension and the cables become stuck within the chase.

Specialty Tools

There are a number of specialty tools that can help fish walls and make a cable path through blind areas within walls and other building structures.

For vertical pathways between floors or from a ceiling down through a wall—if the wall cavity is free of insulation—a weighted metal chain can be attached to a pull string and dropped down the channel. An extendable, magnetic wand can be pushed through the hole at the far end of the channel to catch the weighted chain and pull through the pull string (see Figure 14.7). This can also help in routing pull strings from ceiling locations (for speakers, etc.) to the walls and down to the final location.

Some specialty tools include:

- Drill bits of various lengths and styles (e.g., flex, bell hangers, masonry).
- Metallic/nonmetallic fish tape.
- Grab poles.
- Push rods.
- Extendable magnetic rod with chain.
- Hemostats.
- Mirrors or fish eye scope to view wall cavities.
- Flexible light for illuminating wall cavities.
- Cable caster for long, open runs.
- Rotary cutting tool.
- Molding pry bar.
- Carpet installation tools.

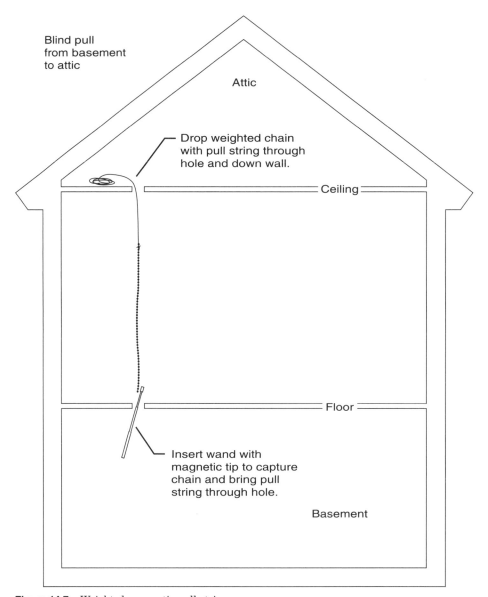

Figure 14.7 Weighted, magnetic pull string.

Other tools for pushing and pulling wires and cables within walls and other spaces include the:

- Telescoping retriever.
- Flexible retriever.
- Ball chain retriever.

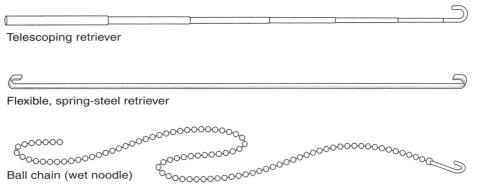

Telescoping retriever

Flexible, spring-steel retriever

Ball chain (wet noodle)

Figure 14.8 Telescoping, flexible, and ball-chain retrievers.

The telescoping retriever consists of a stiff, telescoping rod with a hook at the end. Its stiffness makes it less versatile than the flexible retriever, which is a spring-steel tape with a hook on each end. The ball chain, also known as the wet noodle, is extremely flexible, making it useful for fishing insulated walls or for hitting holes that are aligned vertically (see Figure 14.8).

In addition to the standard electrician's fish tape, there are thinner, more flexible wire fish tapes, as well as fiberglass fish tapes.

> **NOTE:** Be careful in using fiberglass fish tapes and rods; fiberglass does not bend easily, and may break if forced around corners.

Where a fish tape will not work, a stiff push or pull rod may be required. Some sectioned rods are made of fiberglass. A Grab-It™ is a telescoping, self-locking push/pull rod made of light metal.

Push/pull rods often have detachable tips that can be interchanged. A bull nose does not easily become caught on obstacles and includes an eye for the attachment of a pull cord. J-tip and Z-tip rods are useful for differing cable retrieval situations. An "eggbeater" end can be used where there is insulation or other obstacles to contend with (see Figure 14.9).

When pulling a cable, a pull sleeve can be used. With a ring on its tip for attachment of a pull cord, the pull sleeve slips easily onto a cable but it is difficult to dislodge when it is pulled from the ring end (see Figure 14.10).

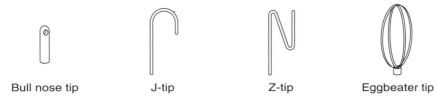

Bull nose tip   J-tip   Z-tip   Eggbeater tip

Figure 14.9 Bull nose tip, J-tip, Z-tip, and eggbeater tip.

Figure 14.10   Pull sleeve.

Long, thin drill bits with extension rods may be necessary to drill through from one floor to another or from an unfinished area within a basement to an outlet location in a finished section. Drill bits may be required to punch holes through a number of materials including wood, metal, and brick. Flexible drill bits can help make penetrations from the attic into the upper floor walls in locations where a low roof line limits accessibility (see Figure 14.11).

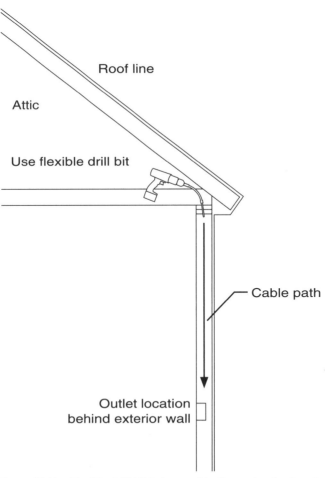

Figure 14.11   Flexible drill bit being used in the angle of a sloped roofline.

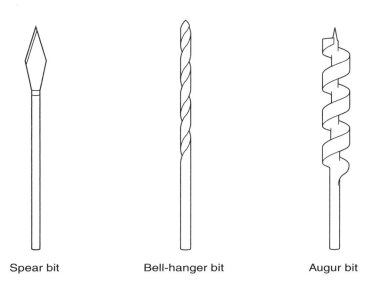

| Spear bit | Bell-hanger bit | Augur bit |

Figure 14.12 Spear, bell-hanger, and augur bits.

Spear-style drill bits will not "walk," and so are useful when aligned holes are needed. Standard bell-hanger drill bits are used for smaller holes, while augur-style bits can be used to drill large holes (see Figure 14.12).

A feed screw on the tip of a drill bit can keep it from walking, even when the bit is at the end of a long, flexible extension. Some drill bits have holes punched in them, so that a pull cord can be attached to the bit once it has penetrated to another floor or area. A shaft guide is helpful in guiding long, flexible drill bits around tight corners (see Figures 14.13, 14.14, and 14.15).

Figure 14.13 Feed screw on drill bit.

Figure 14.14 Augur bit with hole in tip for attachment of pull cord.

Since retrofit installations must work around existing walls, floors and ceilings, it is necessary to have tools to locate wires, pipes, and other invisible obstructions that may impede the cable pull. A standard stud finder is indispensable for this task, but it is by no means the only tool needed.

A mirror attached by a bent arm to an extendable rod is useful for looking through small holes (e.g., outlet cutouts). The mirror must be manipulated along with a flashlight, which can be clumsy. The Walleye™ solves the problem by combining a light source and mirror in the same device (see Figure 14.16).

In retrofit installations, it is important to cut holes in finished walls and ceilings with as little damage as possible. A cutting template used with a Drywall saw is effective for making outlet cutouts in Drywall. For older, lath-and-plaster construction, it is necessary to use a specialized cutout tool to avoid extensive damage to the wall or ceiling (see Figure 14.17).

Mount Retrofit Hardware

Once the cables are run, the residential cabling installer must mount the outlets and the enclosure. The outlets will require the use of a low-voltage mounting bracket or similar device.

The DD may be mounted on a finished wall, either recessed between studs or surface mounted. For surface-mounted installations, it is beneficial to mount a backboard first—screwed through the wall into the studs for additional strength—and then mount the DD to the backboard. Mounting a DD directly to the Drywall may not provide enough mechanical strength.

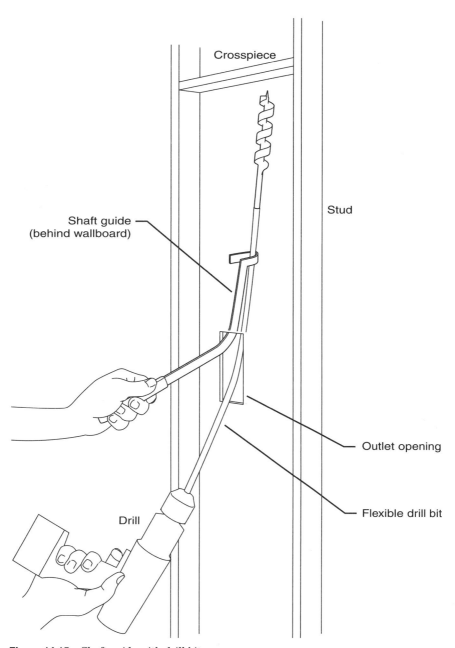

Figure 14.15   Shaft guide with drill bit.

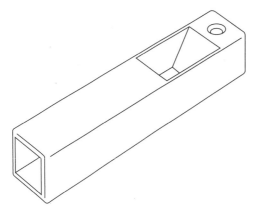

Figure 14.16 Walleye™.

The location for the DD must be determined prior to beginning the work to ensure that there is:

- Enough physical space to fit an enclosure capable of handling all of the required cabling and modules.
- Sufficient clearance to the front of the DD to allow the door to come off or swing open.
- Proximity to available power.
- Minimized cable run lengths.

Once the DD is mounted in position, the cables can be routed into the DD. Since the rough-in and trim-out in a retrofit application can be accomplished simultaneously, the cabling system can be terminated within the DD at this time. This can also include the ADO cables, unless the homeowner needs the existing telephone and CATV system to work prior to the completion of the total installation (a critical issue if the retrofit project will take more than a single day).

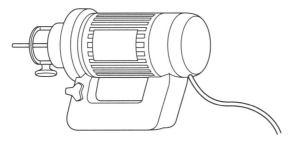

Figure 14.17 Cutout tool.

The outlet locations can be terminated as soon as the cable is put in place. The base of the low-voltage mounting bracket is pressed into the hole (making sure all the cables are routed through the opening of the low-voltage mounting bracket). Then, the feet are attached, pressing the Drywall between the

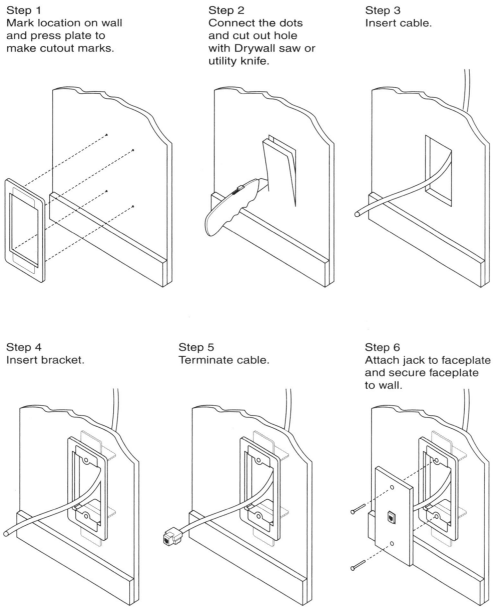

Step 1
Mark location on wall
and press plate to
make cutout marks.

Step 2
Connect the dots
and cut out hole
with Drywall saw or
utility knife.

Step 3
Insert cable.

Step 4
Insert bracket.

Step 5
Terminate cable.

Step 6
Attach jack to faceplate
and secure faceplate
to wall.

Figure 14.18   Installation of a retrofit ring.

retrofit ring base and the feet for the appropriate level of mechanical strength (see Figure 14.18).

The cables can then be terminated (see Chapter 15: Trim-Out Finish) and snapped into the outlet faceplate. The faceplate is then screwed into the low-voltage mounting bracket.

Brackets used for the mounting of audio speakers, intercoms, and other devices follow the same procedure. The cables can all be attached to the hardware when in position within the bracket cutout and the hardware can be screwed into the bracket.

Other Wall Materials

This chapter focuses on walls made of Drywall material, since this is the most common construction method to be experienced by the residential cabling installer. Retrofit applications encompass other wall materials.

Many older residences are constructed with plaster walls. In this case, the cutting of holes for outlets presents special challenges to the residential cabling installer. A rotary cutting tool is effective for making the fine cuts necessary to avoid damaging the plaster.

The other problem with plaster walls is that many are supported vertically, as with the studs behind Drywall, and also with horizontal pieces. This makes the routing of cable from above or below difficult. Again, long drills are needed, or the walls may have to be penetrated at the level of the horizontal obstruction, a situation the homeowner must be made aware of prior to starting the work.

For solid walls (e.g., exposed brick) in which the homeowner or residential cabling installer has chosen to use surface-mounted cable raceway, the outlet will likely be installed on an outlet box surface-mounted to the wall. This surface-mounted outlet box needs to be deep enough to allow the jacks to seat properly, while providing room for the cable to maintain the proper bend radius. The box is secured with screws or other fasteners (depending on wall material) and a knockout is removed from one side of the box in order to accept the incoming cable raceway. The cable raceway should be installed so the end is completely maintained within the box (see Figure 14.19).

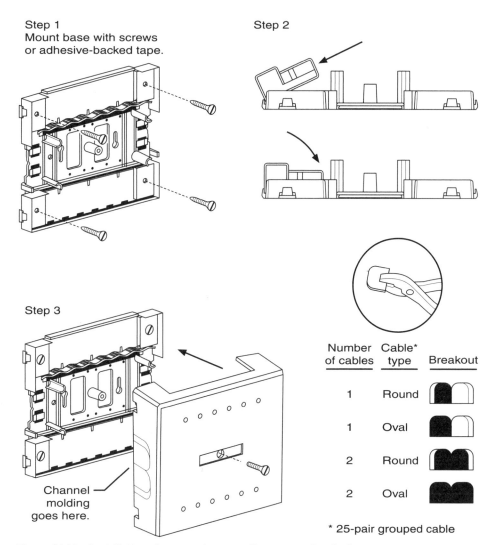

Figure 14.19   Installation of raceway into a surface-mounted outlet box.

Trim-Out Finish

Overview

Due to the wide variety of available connections in the residential structured cabling system (RSCS), it is beyond the scope of this manual to be all inclusive. Refer to the specific instructions provided by the manufacturer of each connector used to assure proper termination and performance.

Cable termination involves the organizing of cables by destination, forming and dressing cables, and proper labeling, as well as actually creating a connection with the appropriate media.

Proper cable termination practices are vital for the complete and accurate transfer of both analog and digital information signals. Insulation displacement connector (IDC) termination is the recommended method of twisted-pair copper termination that is recognized by ANSI/TIA/EIA-568-B.2, *Commercial Building Telecommunications Cabling Standard, Part 2: Balanced Twisted-Pair Cabling Components*. This method removes or displaces the conductor's insulation as it is seated in the connection. Specific tools designed for making IDCs are required. During termination, the cable is pressed between two edges of a metal clip, displacing the insulation and exposing the copper conductor. The copper conductor is held tightly within the metal clip, ensuring a solid connection.

Security and audio cables are usually connected with mechanical crimp or screw-type connectors, although some connections may require soldering.

The crimp connector for coaxial cable is dependent upon the diameter of the cable. Care should be taken to ensure that the proper crimp connector for the specific cable is being used.

Optical fiber cables are constructed differently from copper cables. The fiber core must be aligned with the fiber core of the connecting cable. This ensures that the maximum transfer of light pulse energy is obtained.

There are several applications relative to the termination of a cable. This section addresses the following:

- Termination blocks
- Modules
- Telecommunications outlet/connectors
- Coaxial cables
- Optical fiber
- Patch panels

The following types of cable currently recognized by the ANSI/TIA/EIA-570-A, *Residential Telecommunications Cabling Standard*, for use in the premises cabling are covered in this section:

- 100 Ω unshielded twisted-pair (UTP) copper cable
- 100 Ω screened twisted-pair (ScTP) copper cable
- Optical fiber cable
  - Singlemode
  - Multimode
- 75 Ω coaxial cable

In addition to the cables listed above, there are additional cable types used for audio, video, security, heating, ventilating, and air conditioning (HVAC), and other home automation applications.

Pretermination Functions

Overview

Proper preparation for cable termination improves the quality of the job, and also decreases the amount of time required for termination. Through proper preparation, the residential cabling installer can concentrate on doing the job right.

The performance of pretermination functions involves organizing the cable by destination. Cable to be terminated should be placed in close proximity to the point of termination and must be identified properly to ensure it is terminated in the correct position.

Forming and dressing the cable involves properly aligning and positioning the cables in a neat and orderly manner for termination. The length of cable needed to reach the termination location must be determined, taking into account enough slack to reterminate if necessary, and not placing undue pulling stress on the termination.

Proper cable management results in neat and orderly bundles of cables. This is not to say that cables should be neatly placed side by side.

> **NOTE:** Copper cabling placed neatly side by side may cause a condition called alien crosstalk loss between the cable and affect some applications.

Besides being aesthetically acceptable, proper cable management provides support and mechanical protection of the pairs.

Cable connection is not complete until all terminations are properly identified and labeled.

Pretermination—steps

Follow the steps below for pretermination functions.

> **NOTE:** If using a surface-mount, ensure grommets are installed in knock-out opening to protect cable from sharp edges.

Step	Pretermination Functions
1	Organize cable by destination.
	▪ Know the wiring scheme. If incompatible parts are used for terminating, drastic results occur. With twisted-pair copper IDC terminations, there are three predominant wiring schemes: T568A, T568B, and universal service order code (USOC). Only the T568A wiring scheme is compliant with the ANSI/TIA/EIA-570 standard. Each of the wiring schemes is shown in Figure 15.1.

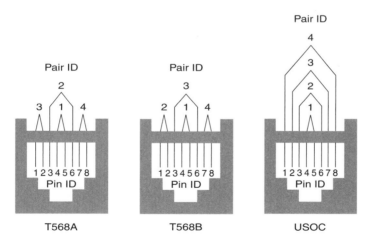

ID = Identification
USOC = Universal service order code

Figure 15.1 Wiring schemes.

Step	Pretermination Functions

- For telecommunications outlet/connectors, check that all cables are available and properly labeled at the wall outlet locations. In distribution devices (DDs) or multi-dwelling unit telecommunications rooms (MDU-TRs), ensure that blocks, panels, or modules are installed in accordance with the residential cabling designer's layout and manufacturer's recommendations.

- Verify that the right products are on hand for the application. For instance, modular furniture that may be placed in a home or MDU requires a different variety of outlets than Drywall. Care must be taken to ensure the proper product type and manufacturer is specified (see Figure 15.2).

2 Form and dress the cable.

- Prepare the cable for termination by bringing all the cables into a layout, which cascades into a sweeping curve to the destination.

- ANSI/TIA/EIA-570-A (Sections 8.2.1 and 8.4.1) sets out a specific bending radius for UTP cable and coaxial cable. This standard states that sharp or right angles should be avoided. The cable bend radius for UTP cable must not be less than four times the cable diameter, or about 25 mm (1 in) for 4-pair cables. The bend radius for Series 6 coaxial cable must not be less than 10 times the cable diameter. The bend radius for fiber cable must not be less than 10 times the cable diameter, or about 25 mm (1 in). These sweeps should be maintained both in cable runs and at termination (see Figure 15.3).

- If racks are used, dedicate a minimum of one rack space (45 mm [1.75 in]) of cable management for every two rack spaces (89 mm [3.5 in]) of patch panels.

- Dress the cable to form a neat, orderly bundle.

- Cables may be entering the DD or MDU-TR from multiple directions, which frequently result in cables of many different lengths. After determining the amount of slack necessary, the cables should be relabeled (see Figure 15.4).

- Each cable must be carefully remarked, using the same markings, prior to cutting off the excess cable containing the original markings. Attention should be given to this process to ensure the new labeling is correct.

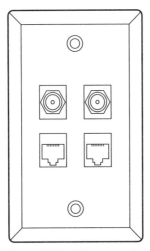

Figure 15.2 Telecommunications outlet/connector.

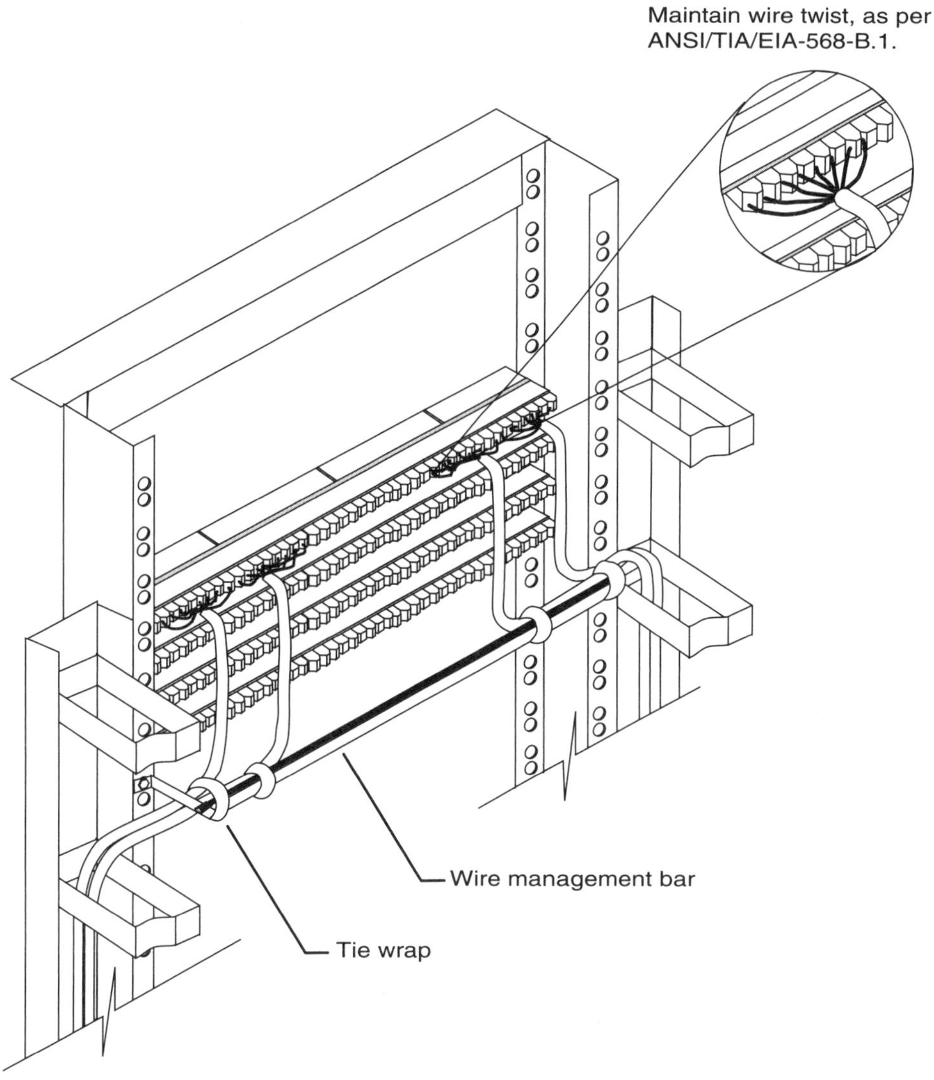

Maintain wire twist, as per
ANSI/TIA/EIA-568-B.1.

Wire management bar

Tie wrap

Figure 15.3   Example of form and dressing cables.

Step	Pretermination Functions
	■ Use plastic cable ties, or preferably hook and loop straps, to keep the cables secured. The plastic cable ties or hook and loop straps should be evenly spaced throughout the dressed length. Care should be taken to tighten plastic cable ties by hand only (see Figure 15.5).
3	Determine the length and slack required.
	■ Provide an adequate amount of cable to reach the destination point for termination.

Write on label and wrap label
around cable for pulling.

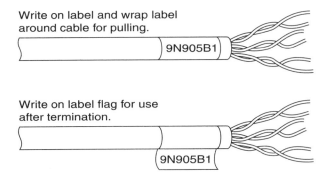

Write on label flag for use
after termination.

Labeling system

9	**N**	**905**	**B1**
Floor	Quadrant	Room	Jack position

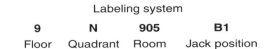

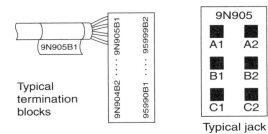

Typical
termination
blocks

Typical jack

Figure 15.4   Relabeling cables.

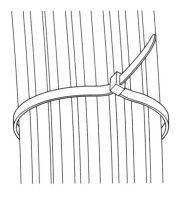

Cable tie

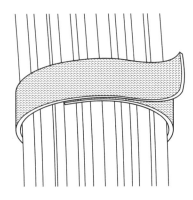

Hook and loop strap

Figure 15.5   Hook and loop straps.

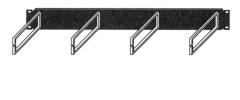

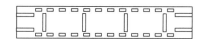

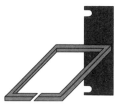

Figure 15.6   Cable management hardware.

Step	Pretermination Functions
	■ Leave enough cable to reach the farthest point in the distribution device/MDU-TR when running telecommunications outlet/connector cables.
	■ Backbone cables in MDU-TR applications running vertically should be laid with enough slack to reach the floor or ceiling, plus the distance across the space.
4	Use the proper cable management hardware.
	■ There are several different types and styles of cable management hardware. All cable management products are designed to properly support the in-place cables and relieve tension, as well as provide support for future cable, which may be installed as a result of moves, adds, or changes (MACs). Figure 15.6 shows examples of cable management hardware.

Copper Insulation Displacement Connector (IDC) Termination

Overview

There are five basic types of copper IDC terminations used in the termination of telecommunications outlet/connectors and backbone copper cabling. The most common IDC terminations are the:

- Crimp-style 8P8C (RJ-45) modular plugs.
- 110-type.
- 66-type.
- BIX™.
- Krone LSA.

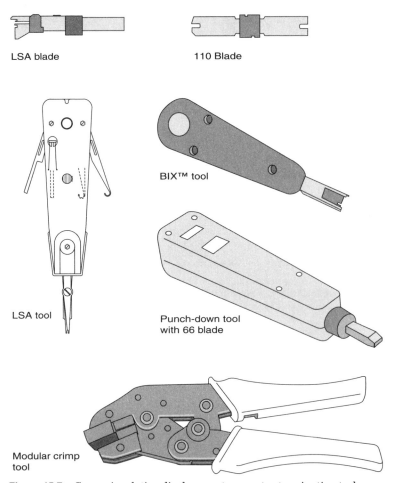

Figure 15.7   Copper insulation displacement connector termination tools.

These comprise the majority of the market. However, other devices are available.

Several manufacturers provide both rack-mountable and wall-mountable IDC termination hardware, which can house multiple termination blocks.

To ensure a good connection, care must be taken to closely follow the IDC connecting hardware manufacturer's specifications. Special attention should be given to complying with the proper procedures for:

- Determining the proper method and length of sheath removal.

- Length of pair untwisting permitted (ANSI/TIA/EIA-B.1, *Commercial Building Telecommunications Cabling Standard, Part 1: General Requirements*, recommends a maximum of 13 mm [0.5 in] untwist in pairs, measured from the last twist to the IDC).

Each type of IDC termination requires a specially designed terminating tool for performing the IDC termination correctly. There are several manufacturers that market termination tools that have the ability to interchange the blades for use on several styles of IDC termination blocks.

> **CAUTION:** Ensure that the brand of tool being used is compatible with the blade. Different blades sometimes look similar, but there are slight differences in design. Improper matching of termination tool handle and termination blade can lead to serious personal injury as well as poor IDC terminations.

The following procedures address the proper methods and tools required to terminate each type of IDC termination block.

Follow the manufacturer's and/or customer's specifications and guidelines regarding all terminations. Document all termination information properly. This identifies the origin, destination, and routing of all cables.

Copper Insulation Displacement Connector (IDC) Terminations—Steps

Follow the steps below when terminating copper IDC.

Step	Terminating Copper IDC (General)
1	Determine the method and length of sheath removal. Sheath removal is performed in three ways: ▪ Ring tool—The ring tool contains a razor blade, which is set to a depth that allows the sheath to be cut deep enough to sever the sheath but not nick the inner pairs. Ring tools are designed to save the residential cabling installer time by ringing the cable's outer jacket directly at the point where the jacket will stop. The excess jacket can then be easily removed by simply sliding it off the conductor ends. To ring a cable: – Determine where the cable jacket needs to stop and expose the conductors. – Slide the cable into the ring tool where the jacket needs to be cut.

Step	Terminating Copper IDC (General)

- Rotate the tool around the cable at least two complete revolutions.

- Slide the tool off the cable.

> **IMPORTANT:** Do not use the ring tool to pull the jacket off. This action can damage the tool's blade and skin the insulation off the conductors.

- Pull the severed jacket off the cable's conductors.

Cable jackets vary in thickness, therefore, it is important that a scrap piece of cable is tested to ensure that the tool's blade does not nick the conductor insulation. Use a ring tool to ring the cable 50 mm (2 in) from the end to expose the cable's rip cord. The rip cord is then pulled back against the cable sheathing until the proper length of sheath is split and is ready to be removed. A pair of electrician snips will then flush cut the slit jacket and rip cord without the danger of nicking the conductors or the insulation.

- Electrician snips—Use the electrician snips to carefully cut into the sheathing using a ringing motion similar to a ring tool. The sharp cutting edges of the inner blades score the jacket to a depth that exposes the rip cord. Use the rip cord to pull down the sheathing until the proper length of sheath is ready to be removed. Remove the split sheath and ripcord with the electrician snips.

- Slitter tool—The slitter tool is used on large pair-count cables. Carefully insert between the cable sheath and the pairs and slide the tool down the sheathing until the proper amount of sheathing is prepared for removal. Remove the severed sheath by using the electrician snips.

2   Use a proper sheath removal tool to remove the cable sheath in accordance with ANSI/TIA/EIA-B.1 and the termination equipment manufacturer's specifications (see Figure 15.8).

- Remove only as much of the sheath as is necessary to terminate the cable pairs and ensure that the twist of the pairs is maintained. A common fallacy is that only 50 to 75 mm (2 to 3 in) of sheath should be removed from the cable. This requirement can vary depending on the block, type, and size of the cable and type of IDC termination hardware manufacturer.

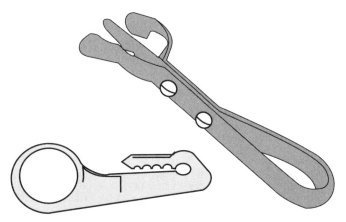

Figure 15.8   Sheath removal tool.

Step	Terminating Copper IDC (General)
3	Separate, identify, and tie off binder groups.

- Binder groups are associated with 50-pair and larger pair-count cables.

- A unique color code identifies each binder group. Cables are grouped in 25-pair increments with each 25-pair group (or subgroup) individually wrapped with a fabric or plastic tape to identify the groupings. See Table 15.1 color-code chart for clarification.

TABLE 15.1   Binder color-code chart

	Pair		Binder Group	
Number	Tip	Ring	Color	Pair Count
1	White	Blue	White-Blue	001–025
2	White	Orange	White-Orange	026–050
3	White	Green	White-Green	051–075
4	White	Brown	White-Brown	076–100
5	White	Slate	White-Slate	101–125
6	Red	Blue	Red-Blue	126–150
7	Red	Orange	Red-Orange	151–175
8	Red	Green	Red-Green	176–200
9	Red	Brown	Red-Brown	201–225
10	Red	Slate	Red-Slate	226–250
11	Black	Blue	Black-Blue	251–275
12	Black	Orange	Black-Orange	276–300
13	Black	Green	Black-Green	301–325
14	Black	Brown	Black-Brown	326–350
15	Black	Slate	Black-Slate	351–375
16	Yellow	Blue	Yellow-Blue	376–400
17	Yellow	Orange	Yellow-Orange	401–425
18	Yellow	Green	Yellow-Green	426–450
19	Yellow	Brown	Yellow-Brown	451–475
20	Yellow	Slate	Yellow-Slate	476–500
21	Violet	Blue	Violet-Blue	501–525
22	Violet	Orange	Violet-Orange	526–550
23	Violet	Green	Violet-Green	551–575
24	Violet	Brown	Violet-Brown	576–600
25	Violet	Slate	25th binder is not used.	

Step	Terminating Copper IDC (General)
	NOTE: The cable binders for cable up to 600 pairs have a white overall binder on each grouping. Cable exceeding 600 pairs will begin with a red overall binder on each grouping and continue the color sequence assigned to the tip conductor up to 2400 pairs.
	▪ Depending upon the manufacturer, 25-pair binder groups are combined into identifiable master groups.
	▪ Tie off binder groups to keep them identified until ready to terminate.
	NOTE: Use color-coded plastic cable ties of the same color as the binder group. Install them snugly, but not tightly, to the sheath end and the outside end of the unsheathed cable binder group. This helps to identify binder groups and keeps the binder groups together during termination, allowing for easier housekeeping.
4	Fan out and form cable pairs from each binder group.
	▪ Cable pairs should be uniformly placed so as to be not only aesthetically pleasing, but also, pairs should not cross or interfere with any other pairs.
	▪ Wire pairs should be parallel with no tension at the point of connection, and equal tension on all connections.

Crimp style—steps

Follow the steps below when terminating modular plug.

Step	Terminating Crimp-Style 8P8C (RJ-45) Modular Plug
1	Strip jacket to the appropriate length.
	▪ Make a clean 90-degree cut at the end of the cable.
	▪ Remove only enough jacket from the cable to reach the end of the plug. There should still be jacket under the cable-clamp portion of the modular plug.
2	Use the correct crimp tool.
	▪ Modular plugs are configured in 4-, 6-, and 8-pin combinations. The correct die for crimping all of the wires in one motion is required. Only 8-pin modular plugs are Category 5 compliant.
3	Verify the pin-wiring configuration.
	▪ Inspect the connection to make sure all the wires are seated properly and are in the correct position.

110-style hardware

When placing 110C-4 terminations inside the DD:

- Follow the color code on the 110C-4, matching up colored wires to their counterpart's locations on the 110C-4.

- A maximum of 12 mm (0.47 in) is allowed for untwisting of pairs from the termination point for each cable pair, if needed.

- To achieve maximum performance from the 110C-4 termination, install either the orange (#2) or green (#3) wire first. This will help center the cable.

  NOTE: If the blue pair is installed first and terminated close to the cable, the brown pair will extend further than 13 mm (0.5 in) from the jacket and be more exposed than the standard allows.

- Use a 110 termination tool for inserting the cable into the 110C-4 connector (see Figure 15.9). The termination tool may be used to trim the ends of the wires.

  NOTE: Make sure the tool is properly oriented before use.

110-style blocks

The 110-style IDC termination hardware is used in MDU-TR applications.

The majority of patch panels are wired in specific configurations (e.g., T568A, T568B, and USOC) and constructed with 110-style connectors. Telecommunications outlet/connector terminations are manufactured with 110-style hardware. Use a T568A wiring scheme on patch panels.

Cables are routed through the middle pathway of the 110-style blocks from either the top or bottom, and fanned into the wireway from alternate sides.

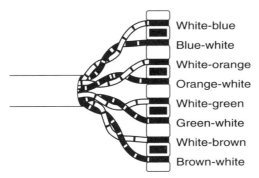

White-blue
Blue-white
White-orange
Orange-white
White-green
Green-white
White-brown
Brown-white

Figure 15.9   110C-4 termination.

The block wireway is designed such that one row is terminated on the wiring block by punching down from the bottom up. The next row is terminated from the top down. For the specific location of each termination, refer to the residential telecommunications designer's layout.

The 110-style frame does not contain an IDC. The back plane is used solely for fanning and positioning of cables. The IDC for this type of termination application is in the C-3 (3-pair), C-4 (4-pair), and C-5 (5-pair) connector block that is punched down on top of the 110-style frame to permit cross-connection. The connecting block pair count is determined by the application (e.g., 4-pair outlet cabling is terminated on a C-4 [4-pair] connecting block; backbone cables in 25-pair increments are terminated on C-5 [5-pair] connecting blocks).

> **NOTE:** If using six C-4 clips, one pair will not be covered by a clip at the row's end when terminating 4-pair cables. A C-5 clip is often used in the last position to fill the row.

Place the designation strips in the holder, which covers the terminated cables (see Figure 15.10).

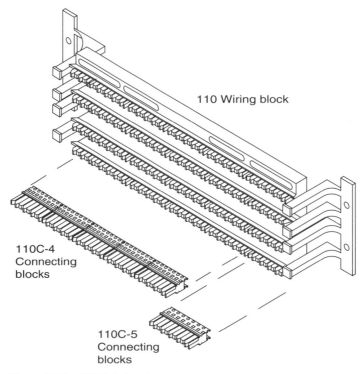

110 Wiring block

110C-4 Connecting blocks

110C-5 Connecting blocks

Figure 15.10   110-style hardware.

110-style block—steps

Follow the steps below when terminating a 110-style block.

Step	Terminating a 110-Style Block
1	Mount the 110-style base.
	• The 110-style base is available with or without stand-off legs, which are used to provide a space behind the wiring base for cable routing.
	• Mount the wiring base on an approved plywood surface, using two sets of legs and four wood screws.
	• Mount the wiring base without legs on a rack-mounted cross-connect frame using four machine screws.
2	Route the cables.
	• All 110-style bases are configured in multiples of two rows (e.g., an upper row and a lower row with a deep wiring channel in between the rows for cable routing). Cables are routed into the channel from the right and the left sides of the frame. Some wiring bases are designed with openings in the rear to allow cables to enter and route directly to their point of termination.
	• Determine the cable's termination location (e.g., left or right, upper or lower row on the wiring frame).
	• Route the 4-pair cables from behind the wiring base and out to the sides through the appropriate slots in the legs. The appropriate slot in the leg will line up with the wiring base's channel, located between the two termination rows.
3	Strip back the cable jacket.
	• Determine the 4-pair cable's termination location, and strip the jacket back so the jacket stops 13 mm (0.5 in) from where the shortest conductor pair will fan into the block. This allows a secure connection with equal stress on all pairs, while maintaining minimal sheath removal. Category 5e 25-pair cables will have the sheath stop before the conductors fan out from behind the wiring base. Category 4, 5, and 5e enhanced terminations require minimal sheath removal. The practice of removing the sheath behind the block so all the conductors fan out neatly within the wiring channel should not be used for data installations.
4	Fan the conductors into the block.
	• Category 5 and 5e enhanced terminations require the conductors to be fanned into their point of termination with a maximum of 13 mm (0.5 in) of pair untwist.
	• Route the conductor pairs to the point of termination and lace the conductors into the wiring base. Each row is designed to terminate 25 conductor pairs, using a series of fanning posts to hold the conductors in place. Every other fanning post has a pointed top designed to help separate the conductor pairs as the pairs are laced into position.
	• Cables terminating on the upper row will fan the conductors up and cut the excess conductor from the top. Cables terminating on the lower row will fan the conductors down and cut the excess conductor off the bottom.
5	Terminate conductors.
	• Terminate each conductor, using an impact termination tool equipped with a 110 blade installed with its cutting edge out. The impact setting can be set on either high or low for 110-block terminations. The preferred method of termination is to use a 5-pair termination tool to insert and cut up to five conductor pairs at once.

Step	Terminating a 110-Style Block
	■ The blade is orientated so the cutting edge is facing up or down toward the end of the conductor(s) to be removed.
	■ The blade is positioned directly in front of the conductor(s) and is pushed inward toward the block. As the blade moves forward, each conductor is pushed into the slots between the fanning posts, and the unwanted conductor ends are cut off when the blade impacts the rear of the block. This places the conductors into the wiring base, but no IDC connection has been made.
6	Install the 110-connecting block (C-clip).
	■ Insert the C-clip into the head of the 5-pair impact tool.
	■ Carefully align the C-clip over the wiring base, with the blue marking to the left side of the block. Slip the rear of the C-clip into position over the conductors, and push the tool's handle inward until the impact is felt and the C-clip is fully seated onto the plastic-wiring base. The cable conductors will make an IDC connection on the rear of the C-clip.
7	Cross-connect installation.
	■ Cross-connects make their IDC connection on the front of the C-clip and should never be routed to block the label that is inserted between the upper and lower termination rows. The upper row's cross-connects are routed up and to either side of the wiring base, and the lower row's cross-connects are routed down and out to either side of the wiring base.

66-block termination

The 66-termination block, an IDC, is the choice for connecting voice applications (e.g., private branch exchange [PBX], key telephone systems [KTS], and some local area networks [LANs]). Several manufacturers of 66-termination block designs have updated their termination blocks to handle high-speed data applications and to be compliant with Category 5 specifications.

> **CAUTION:** Ensure that the appropriate 66-termination block is installed in new installations.

There are several different 89-style brackets available for use in several applications (e.g., RJ-21X, side-mount and rear-mount 50-pin connector), but the most common are the 89B- and 89D-style (see Figures 15.11 and 15.12).

Typically, these blocks are mounted on backboards in vertical rows of four blocks each to accommodate up to a 100-pair termination. Several manufacturers provide a preassembled single- or double-sided distribution frame to provide increments up to 2700 pairs per side for large system installations.

Cable for a 66-termination block is routed through the 89-style bracket to allow the individually sheathed 4-pair cables, or a 25-pair cable's unsheathed pairs, to be fanned out from the rear into the guides in the side of the block. Each block has five-pair increments marked with a distinctive groove for ease of identification.

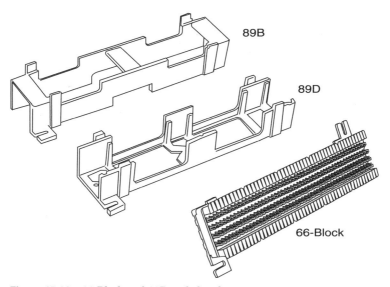

89B

89D

66-Block

Figure 15.11   66-Block and 89D-style brackets.

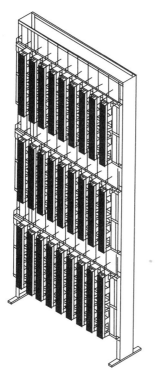

Figure 15.12   66-block distribution frame.

A 66M1-50 block provides the means of terminating two 25-pair or twelve 4-pair cables per block. These blocks have two rows of contacts, which are mechanically connected together to provide cross-connection capability. The 66M1-25 blocks have four rows of contacts connected together. Voice applications may use bridging clips to make a connection between the left and right set of contacts on a 66M1-50 block. By lifting the bridging clips, which opens the circuit, it is easy to test the voice circuit in both directions when troubleshooting.

> **NOTE:** Bridging clips are not Category 5 compliant. This description is for information purposes only.

A fine-tipped, indelible marker is generally used to designate cable-pair identifiers on the fanning strips of 66-type blocks. The preferred method is to label snap-on designation strips or to use color-coded hinged covers, which can be labeled on the inside of the cover.

66M1-50 block—steps

Follow the steps below when terminating a 66M1-50 block.

Step	Terminating a 66M1-50 Block
1	Mount the 89-style bracket.
	■ Mount the bracket on an approved plywood surface using two wood screws, or on a rack-mounted cross-connect frame, using two machine screws.
2	Route the cables.
	■ Determine the cable's termination location (e.g., left or right, upper or lower positions on the block).
	■ Route the 4-pair cables up or down the center of the bracket, and feed them out either the left or right side of the bracket through the upper or lower sections to their termination locations.
	NOTE: 25-pair cables will fill one side of the 66-block.
3	Install the 66-block.
	■ Snap the block onto the front of the 89-style bracket being careful not to crush any cables between the block and bracket.
4	Strip back cable jacket.
	■ Determine the 4-pair cable's termination location and strip the jacket back so the jacket will stop 13 mm (0.5 in) from where the shortest conductor pair will fan into the block. This allows a secure connection with equal stress on all pairs, while maintaining minimal sheath removal. Category 5e 25-pair cables will have their sheath stop just before the first slot on the side of the 89-style bracket. Thirteen conductor pairs will exit the first slot and the remaining twelve conductor pairs will exit through the second slot of the 89-style bracket.

Step	Terminating a 66M1-50 Block

- Category 4, 5, and 5e 4-pair cable terminations require minimal sheath removal. The practice of removing the sheath behind the block so all the conductors fan out neatly along the sides of the block should not be used for data installations.

5   Fan the conductors into the block.

- Category 4, 5, and 5e terminations require the conductors to be fanned into the block as pairs and not to be untwisted and split outside the fanning fingers.

- Fan the conductors as pairs through every other slot in the fanning fingers. The first pair of the block will fan into the first fanning slot beneath the top fanning finger. This aligns the conductor pairs so they enter the block between the two pin rows (Rows 1 and 2) where they will be terminated. The second fanning slot will be skipped. The second conductor pair will fan into the third fanning slot and enter the block between termination pin rows three and four.

- Tip conductors are fanned into the top or odd-numbered rows, while ring conductors are fanned into the bottom or even-numbered rows.

- Separate the conductors of each pair inside the fanning strip, and place in the appropriate quick clips (pins) closest to the fanning strip. Conductors may be routed over the top of each pin and inserted from the top down with the excess conductor removed from the bottom of the pins. The preferred method for data terminations is to feed the tip conductor up to its pin and cut off the excess conductor on top, while feeding the ring conductor down and removing the excess conductor from the bottom. This method maintains the pair twists much closer to the point of termination.

6   Terminate conductors.

- Terminate each conductor, using an impact termination tool equipped with a 66-blade installed with its cutting edge out. The impact setting can be set on either high or low for 66-block terminations.

- The blade is oriented so the cutting edge is facing up or down toward the end of the conductor to be removed.

- The blade is slipped over the pin and the tool is pushed inward towards the block. As the blade moves forward, the conductor is pushed into the base of the pin and the unwanted conductor end is cut off when the blade impacts the rear of the block.

7   Cross-connect installation.

- Voice cross-connects may be installed on the far right pin on the opposite side of the block and have a bridging clip inserted between the two sides to provide connectivity and ease of troubleshooting voice applications. Bridging clips are not Category 5 compliant. This allows 25 cable pairs to be terminated and cross-connected per block.

- Data cross-connects are fanned into the block as pairs and share the same fanning slot as the pair they are cross-connecting. They terminate on the inside pins on the same side of the block as the cable being cross-connected. 66M1-50 blocks have the pins in each row configured so the two left pins are electrically connected and the two right pins are connected. This allows 50 cable pairs to be terminated and cross-connected per block.

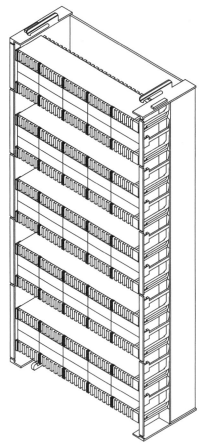

Figure 15.13   BIX™ mount with connectors.

BIX™ hardware

BIX-type termination hardware is similar to the 110-hardware previously described. Unlike 110-hardware, which places clips on top of the wiring block, BIX equipment is a one-piece "pass-through" unit that is reversed in its mount after termination of the cable to expose the opposite side to enable cross-connect capability.

BIX termination block assemblies are available in 50-, 250-, 300-, and 900-pair increments for wall mounting and with floor-frame assemblies for large-size installations.

BIX-type termination hardware is available in both patch panel and telecommunications outlet/connector configurations.

BIX termination block connectors are designed with four slots for inserting small plastic cable ties to support termination of cables. These connectors can be identified in 2-, 4-, and 5-pair increments for various applications.

All BIX mounts (frames) are configured in multiples of two rows (e.g., sets of an upper row and a lower row). Small installations may use the 50-pair mount that is actually one set of an upper and a lower block within the mount. Medium to large installations will use several of the 250-pair or 300-pair mounts that are actually five or six sets of upper and lower blocks respectively.

It is important to think of BIX blocks in sets of two because of the way each upper and lower block is terminated. All the residential cabling installation procedures for the upper blocks take place on the upper, or topside, of the block. All the installation procedures for the lower blocks take place on the lower, or bottom side, of the block.

Routing the cable into BIX mounts on the wall may be from the top down or from the bottom up through the inside of the mounts. There is a large hole in the rear of each mount that can be used if the cables are being routed in from behind a hollow cavity within the wall, or when mounted on a rack-mounted frame.

A properly installed BIX block will have a hinge or loop of conductors created during the residential cabling installation process that allows ample slack to troubleshoot or to make changes to the installation. The hinge may be on either the left or right side of the mount, but is always placed on the side of the mount closest to the cross-connect channel.

Cross-connects make their IDC connection on the front of the block, and should never be routed to block the label that is inserted between the upper and lower termination rows. The upper row's cross-connects are routed up and toward the hinge side of the mount, and the lower row's cross-connects are routed down and out toward the hinge side of the mount.

When terminating cables with larger than 25 pairs on BIX termination blocks, the fabric or plastic binder of each of the 25 pairs should be carried to the end of the connector. The designation strips are then placed in the holder that covers the terminated cable.

BIX™ block—steps

Follow the steps below when terminating a BIX block.

Step	Terminating a BIX Block
1	Mount the BIX mounts.
	▪ Select the proper size BIX mount based on the size of the installation.
	▪ Mount the BIX mount on an approved plywood surface, using two pan-head wood screws.
	▪ Mount the BIX mount on a rack-mounted cross-connect frame, using two machine screws.
2	Route the cables.
	▪ Cables are routed into the BIX mounts from either the top or bottom. BIX mounts also have a hole in the rear to allow cables to enter and route directly to their point of termination, without being seen on the backboard.

Step	Terminating a BIX Block

- Determine which side of the mount will be next to the cross-connect channel (the hinge side).

- Route the cables along the inside of the mount opposite the hinge side and secure them using a plastic cable tie. Plastic cable tie holders are located in the four corners of all mounts designed for 250 pair and larger. Leave a minimum of 410 mm (16 in) of cable slack extending beyond the end of the BIX mount.

3 Strip back cable jacket and identify each cable.

- Each pair of BIX blocks can terminate up to 50 pairs of conductors. This could be two groups of 25 pair from a multipair cable, or 12 individual 4-pair cables. Since the jackets are removed from all the cables at once for both blocks, it is imperative that the conductors be identified to ease in their identification later. For example, in the case of 4-pair cables, there will be 12 white/blue pairs hanging from the plastic cable tie and identification would be almost impossible.

- One method of identification is to ring each cable twice during the sheath-removal process. The first ringing takes place where the jacket is to stop, and the second ringing will be 50 mm (2 in) toward the conductor ends. The 356 mm (14 in) of sheath is removed and the 50 mm (2 in) piece is labeled with the termination location and slid down near the end of the conductors. Just prior to each cable's termination, the labeled piece of jacket is discarded.

- A second method is to keep each block's cables separated by twisting a scrap piece of conductor around each block's conductors to identify the group (e.g., a white/orange conductor indicates the second block's conductors). Each 4-pair cable's conductors would be separated within the group by twisting the end of its own pairs around the group. For example, the first cable would have its white/black wrapped around them, second cable—white/orange, third cable—white/green, fourth cable—white/brown, fifth cable—white/black and white/orange, and the sixth cable would have the white/green and white/brown conductors wrapped around the pairs. Using the conductors will not affect the quality of the termination because the conductors that are twisted are at the very end of the cable and will be cut off during termination.

- Determine which pair of blocks the 4-pair cables will terminate on, and strip the jacket back so the jacket will stop between the upper and lower blocks. Leave 410 mm (16 in) of unsheathed conductors extending beyond the end of the jacket.

NOTE: Each BIX block is almost 180 mm (7 in) long. An easy way to determine 410 mm (16 in) is to leave enough conductor to go approximately 2.5 times across the length of a BIX block.

4 Fan the conductors into the block.

- The QCBIXIA block (universal block marked every five pairs) can be inserted into the mount without concern for proper orientation. The QCBIX1A4 block (designed for 4-pair terminations and marked every four pairs) has six positions for 4-pair cables, and one pair (the orphan pair) that is not used on the far right-hand edge of the block. This block must always be inserted into the frame with the orphan pair to the right.

IMPORTANT: All instructions from this point forward apply to the installation of both the upper and lower blocks. All work on the upper blocks is done on the top and all work on the lower blocks is done on the bottom of the blocks.

Step	Terminating a BIX Block

- Insert a small plastic cable tie into the connector's plastic cable tie slot closest to the hinge side. The plastic cable tie is always inserted on the side of the block opposite from where the cables are routed within the frame. The block is always oriented so the orphan pair is to the right (e.g., if the cables are routed on the left side of the mount, the plastic cable tie is in the slot closest to the right side or the orphan pair for the QCBIX1A4). If the cables are routed on the right side of the mount, then the plastic cable tie is on the left side of the block, or away from the orphan pair. Top blocks have the tie on top; bottom blocks have the tie on the bottom.

- Place the wire bundle (the six 4-pair cables or 25-pair cable) along the top/bottom of the connector and loosely fasten the cable tie. The connector should be oriented so the end of the connector with the plastic cable tie is closest to the cable sheath.

- Slide the connector along the wire bundle to give approximately 180 mm (7 in [length of one BIX connector]) of slack from the end of the cable sheath to the cable tie. Tighten the cable tie and trim off the excess.

CAUTION: Overtightening of the plastic cable tie will degrade performance.

5   Insert the connector into the mount.

- Insert the connector (orphan always to the right) into the lower BIX connector position mount.

NOTE: Both the upper and lower connectors are placed in the lower position during termination and later rolled into their final locations.

Check that the:

– Hinge is opposite the side where the cable sheaths are routed.

– Plastic cable tie is opposite the side where the cable sheaths are routed.

– Plastic cable tie is closest to the side of the mount next to the cross-connect channel.

– Wire bundles should flow from the end of the sheath toward the far side of the mount. Gently make a 180-degree turn prior to the far side and route back through the plastic cable tie and across the top/bottom of the connector.

6   Fan the conductors into the block.

- Pick wire pairs and insert them into the connector. Conductors should run along the top/bottom of the connector and fan downward/upward, respectively, into the point of termination. Use the pair splitters on the front edge of the connector to minimize pair untwisting. Category 5 and 5e terminations require the conductors to be fanned into the point of termination with a maximum of 13 mm (0.5 in) of pair untwist.

- Cables terminating on an upper connector will fan their conductors down and cut the excess conductor from the bottom. Cables terminating on a lower connector will fan the conductors up and cut the excess conductor off the top.

7   Terminate the conductors.

- Terminate each conductor using a BIX wire insertion tool, or equivalent. The BIX blade uses a scissors action to cut the conductor and is not considered an impact tool that cuts the conductor by trapping it between the cutting blade and the rear of the block.

- The blade is oriented so the cutting edge is facing up or down toward the end of the conductor(s) to be removed.

Step	Terminating a BIX Block
	▪ The blade is positioned directly in front of the conductor(s) and is pushed inward toward the block. As the blade moves forward, each conductor is pushed into the slots between the fanning posts, and the unwanted conductor ends are cut off when the blade is fully inserted and starts to withdraw.
8	Rolling the connector.
	▪ The connector is removed from the mount and rolled like a log into its final position (e.g., bottom blocks are rolled 180 degrees toward the bottom and snapped back into the lower position. Upper blocks are rolled 180 degrees toward the top and are snapped into the upper position).
	NOTE: All conductors should be in-between the two rows and are covered by the labeling strip when installed.
9	Install the labeling strip.
	▪ Mark the labeling strip with the cable identifiers and snap it into place between the two connectors.
10	Cross-connect installation.
	▪ Cross-connects make their IDC connection on the front of the BIX connector and should never be routed to block the label that is inserted between the upper and lower termination rows. The upper row's cross-connects are routed up and toward the hinge side of the BIX mount, and the lower row's cross-connects are routed down and toward the hinge side of the BIX mount. The cross-connects exit the mount through the two small jumper wire slots on either side of the mount.

Krone LSA hardware

Krone LSA-type termination hardware provides silver-plated IDC contacts at a 45-degree angle, with the conductor being held in place by tension in the contacts. This hardware is available in patch panels, telecommunications outlet/connectors, and termination blocks. It also provides disconnect modules, connect modules, switching modules, and feed-through modules.

Disconnect modules are normally closed, two-piece contacts that can be disconnected by inserting a disconnect plug into the wire pair. This allows temporary or permanent disconnection of the circuit. A test cord can be inserted into a pair to test circuits both ways when testing is necessary. These modules are available in 8-pair or 10-pair increments (see Figure 15.14).

Connect modules use a one-piece contact that provides a continuous link between the cable and the cross-connect wiring.

Switching modules consist of a normally open, two-piece contact. Switching modules allow for high-density termination and patch cables.

Feed-through modules consist of a one-piece contact that passes the signal through the module, front to back, and provides a continuous link between feeder and jumper for high-density termination in small areas. These modules are available in 25-pair increments.

For the specific locations of each termination, refer to the designer's layout.

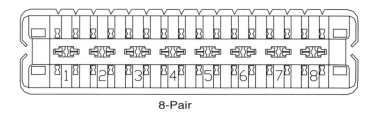

8-Pair

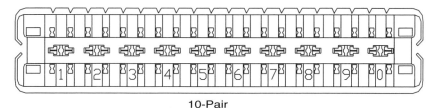

10-Pair

Figure 15.14   Krone LSA 8-pair and 10-pair block.

Krone LSA—steps

Follow the steps below when terminating a Krone LSA block.

Step	Terminating a Krone LSA Termination Block
1	Mount the mounting bracket.
	▪ Mount the bracket on an approved plywood surface, using four wood screws, or on a rack-mounted cross-connect frame, using plastic rivets provided with the frame.
2	Route the cables.
	▪ Determine the cable's termination location (e.g., left or right and row location of the block).
	▪ Route the 4-pair cables up or down the center of the bracket, or feed them through the back using the hole in the back of the mount. Allow 457 mm (18 in) of additional cable from where the cable reaches its block on the frame.
3	Install the Krone LSA blocks.
	▪ Snap the block onto the front of the mounting bracket, starting with the bottom block location and working upward.
4	Strip back the cable jacket.
	▪ Each block holds two 4-pair cables. Locate the two cables and route them so they hang over the top-front of the block to be terminated.
	▪ Strip the jacket back so the jacket will extend 300 mm (12 in) beyond the block, and 150 mm (6 in) of sheath will be removed from the conductors beyond the 300 mm (12 in) of sheath that is to remain on the conductors.

Step	Terminating a Krone LSA Termination Block
5	Fan the conductors into the block.
	▪ The conductors are routed from the rear of the block, through the small fanning fingers located across the top-rear of the block. The pairs are snapped into the fanning fingers following the color code from left to right. They are then fanned down into the appropriate termination locations across the front of the block. The cable sheath is positioned so it is centered between the white/orange and white/green pairs, and as close to the rear of the block as possible, without creating stress on the conductors.
	▪ Separate the conductors of each pair as they are fanned downward into their point of termination.
6	Terminate the conductors.
	▪ Terminate each conductor using a Krone wire insertion tool or equivalent. The Krone blade uses a scissors action to cut the conductor and is not considered an impact tool that cuts the conductor by trapping it between the cutting blade and the rear of the block.
	▪ The blade is oriented so the cutting edge is facing up or down toward the end of the conductor(s) to be removed.
	▪ The blade is positioned directly in front of the conductor(s) and is pushed inward toward the block. As the blade moves forward, each conductor is pushed into the slots between the fanning posts. The unwanted conductor ends are cut off as the blade is fully inserted.
7	Cross-connect installation.
	▪ Cross-connects are fanned into the block as pairs from the bottom of the block, and are routed to the cross-connect channel through the block's side wire management guides. They should not extend to the rear of the block or use the fanning tabs located on the bottom rear of the block.

Patch panels

Data and voice patch panels are available from various manufacturers and in many different styles and wiring configurations (see Figure 15.15).

Patch panels that feature 110-, BIX-, and Krone LSA-connections are available. Common configurations are 24-, 48-, and 96-port.

To properly terminate a 4-pair horizontal cable onto a patch panel, keep sheath removal to the minimum amount required. Category 5 cable pairs must remain twisted to within 13 mm (0.5 in) of the point of termination. When a cable is approaching the point of termination from the top or bottom, the sheathing should be centered on the termination block to allow for an equal tension on all conductors with respect to the sheath. When the cable is approaching the point of termination from the side, the insulation should stop approximately 13 mm (0.5 in) from the cable's closest terminated pair, in reference to the end of the cable's sheath. For example, the sheaths of cables approaching from the right will stop 13 mm (0.5 in) from where the w/brown pair will terminate, and the sheaths of cables approaching from the left will stop 13 mm (0.5 in) from where the

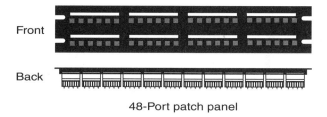

Front

Back

48-Port patch panel

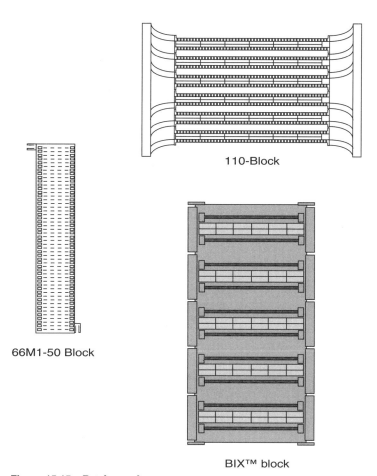

110-Block

66M1-50 Block

BIX™ block

Figure 15.15 Patch panels.

w/black pair will terminate. This allows each pair to flow gently into its point of termination with equal strain on each pair. Termination techniques are the same as described for the individual blocks, with the exception of the impact tools, which should never be set on high impact. Tools set on high impact can cause failure of printed circuit boards solder joints and compression connections.

Strain relief of the cables is accomplished by the use of plastic cable ties, or hook and loop straps, installed on a cable management bar that is installed at the rear of the patch panel (see Figure 15.16).

Care should be taken prior to actual termination to verify that the telecommunications outlet/connector wiring scheme (T568A or T568B) matches the patch panel wiring configuration (T568A or T568B) to ensure proper functioning. Identify patch-panel termination locations in the space allocated on the patch panel. The designer will provide the manner of labeling patch panels.

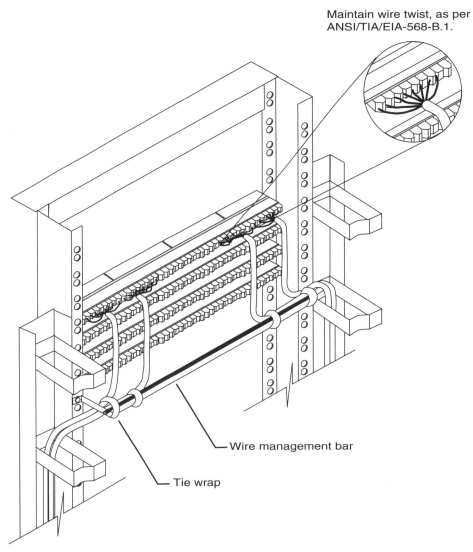

Figure 15.16 Cables terminated on rear of panel with cable management bar.

Telecommunications outlet/connectors

Many different styles of telecommunications outlet/connectors are provided for living areas. They can be terminated in single- or double-gang faceplates, with jack inserts ranging from low-density single-port to high-density eight-port devices. (See Figure 15.17 for an example of a 4-port faceplate.)

Termination of telecommunications outlet/connectors should be completed by removing the minimum required amount of the sheath, according to the manufacturer's specifications. Follow the manufacturer's specifications for proper termination techniques, being careful to ensure the 13 mm (0.5 in) maximum amount of untwisting of the cable pairs to maintain category compliance. Carefully coil the remaining slack (minimum of 300 mm [12 in]) into the termination box. Do not kink the cable or exceed the bend radius of four times the cable diameter.

It is always advisable to use the same manufacturer's patch panels and telecommunications outlet/connectors at a given project. This minimizes the possibility of component mismatch. Another aspect of component mismatch is the warranty issue for manufacturer-compliant residential cabling installation practices. Additionally, all patch cords should be obtained from the same source, pretested, and certified to category by the manufacturer.

Screened Twisted-Pair (ScTP)

Several manufacturers provide ScTP cable for use in areas of high electromagnetic interference (EMI) generation, and to protect against mechanical dam-

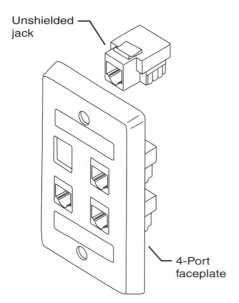

Unshielded
jack

4-Port
faceplate

Figure 15.17   4-port faceplate.

age. It is used for Category 5-compliant installations and carries the same impedance and electrical characteristics as UTP cable, except it has a metallic-coated screening around all the cable pairs and has a drain wire that is in contact with the metallic-coated screening.

An ScTP cable must be terminated in a screened modular jack. The screened jack is enclosed by a metallic-type, EMI-resistant housing. The dressing block for cable pairs fits within the wall of the connector. When terminated, the twisted wires and IDCs are totally enclosed by the metal, which forms an EMI shield.

The IDC terminations are made in the same manner as UTP, with the added requirement of terminating the screen shield. ScTP cable manufacturers provide detailed instructions on the proper termination of the shield. This procedure varies by manufacturer, and the specific method provided for the selected product must be followed for acceptable shield effectiveness over the full 100 megahertz (MHz) bandwidth.

The screen shield is effectively grounded at one end by attaching the drain wire securely to the screen shield of the modular jack, following the modular jack manufacturer's specifications. The other end is terminated to an ScTP patch panel. The IDC terminations are made to the ScTP patch panel in the same manner as UTP, with the added requirement of terminating the screen shield with the drain wire.

> **NOTE:** ScTP cable manufacturers provide detailed instructions on the proper termination of the shield. This procedure varies by manufacturer. The specific method provided for the selected product must be followed for acceptable shield effectiveness over the full 100 MHz bandwidth.

The EMI shielding capability of the ScTP cable is achieved by the internal contact with the drain wire to the metallic-coated screening. To ensure that the screening will effectively block EMI, care must be taken to ensure that the patch panel is grounded following the manufacturer's specifications.

Coaxial Cable Terminations

Overview

A coaxial cable consists of an inner conductor (solid or stranded wire), separated by a dielectric (core) from its outer conductor (single- or double-braided shield), and available in either plenum- or nonplenum versions (see Figure 15.18).

The predominant coaxial cables are Series 59, Series 6, and Series 11, which have a characteristic impedance of 75 Ω.

There are many types of coaxial cable connectors available in the industry. Termination techniques for different manufacturers may vary. Always refer to

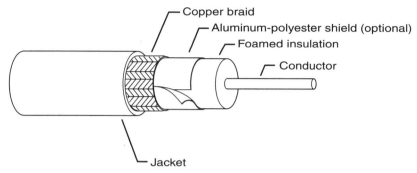

Figure 15.18 Typical coaxial cable construction.

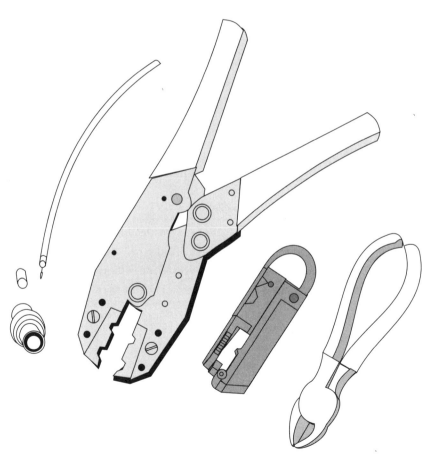

Figure 15.19 Coaxial cable termination tools.

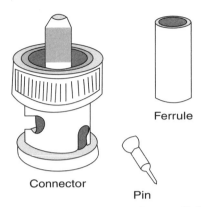

Figure 15.20   Captive-pin Bayonet Neil-Concelman connector.

manufacturers' guidelines for detailed information. Figure 15.19 illustrates typical termination procedures for Bayonet Neil-Concelman (BNC) and F connectors.

The tools required for proper stripping and terminating are typically a three-step, rotating stripper and a ratcheted crimping tool that is designed specifically for crimping of the BNC or the N connector.

Although other connector systems (screw-on style) are available, captive-pin connectors that assure positive retention of the center conductor of the coaxial cable are recommended to ensure proper data transfer (see Figure 15.20).

Coaxial cable BNC—steps

Follow the steps below when terminating coaxial cable BNC.

Step	Terminating Coaxial Cable BNC
1	Determine the proper method and length of sheath removal.

- Make a straight cut in the termination end of the coaxial cable.
- Place the connector ferrule over the end of the cable.
- Adjust the three-step stripping tool to meet the desired cable diameter and stripping requirements. The stripper should be adjusted to expose 4.6 mm (0.18 in) of the center conductor, 1.5 mm (0.06 in) of dielectric without foil screen, and 8 mm (0.3 in) of braid.
- Insert cable into the stripper.
- Rotate the three-step stripper three to five full turns.
- Remove the stripper from the cable.
- Remove the severed sheathing, shielding, and dielectric material.
- Inspect the cable for stripping quality and ensure that the center conductor and the insulation are not nicked or scored, and that any stray strands of the braided shield are pushed away from the center conductor.
- Pull all of the outer braid away from the dielectric.

Step	Terminating Coaxial Cable BNC
2	Terminate the cable.
	▪ Seat the connector's center pin on the center conductor.
	▪ Use the small diameter die of the crimping tool to crimp the pin to the center conductor.
3	Install the connector onto the cable.
	▪ Place the connector body onto the cable by aligning it so that its shaft fits over the center conductor pin, and between the dielectric and the braided shield. Slide the connector until the center pin locks into place and its tip is flush with the face of the connector.
	▪ Pull the braid forward over the rear shaft of the connector.
	▪ Slide the connector ferrule up to cover the exposed braided shield.
	▪ Make sure the crimping tool has the correct die for the cable.
	▪ Place the crimp tool's larger opening over the connector ferrule and squeeze the tool until the die is completely closed.
4	Final inspection.
	▪ Inspect the connection for neatness (no exposed braiding strands). The connector has to be tight.

Coaxial cable F connector—steps

Follow the steps below when terminating coaxial cable F connectors.

Step	Terminating Coaxial Cable F Connectors
1	Determine the proper method and length of sheath removal.
	▪ Make a straight cut in the termination end of the coaxial cable.
	▪ Adjust the two-step stripping tool to meet the desired cable diameter and stripping requirements. The stripper should be adjusted to expose 8 mm (0.375 in) of the center conductor and 3 mm (0.125 in) of braid.
	▪ Insert cable into the stripper.
	▪ Rotate the two-step stripper three to five full turns.
	▪ Remove the stripper from the cable.
	▪ Remove the severed sheathing, shielding, and dielectric material.
	▪ Inspect the cable for stripping quality and ensure that the center conductor and the insulation are not nicked or scored, and that any stray strands of the braided shield are pushed away from the center conductor.
	▪ Pull all of the outer braid away from the dielectric.
2	Place the connector on the cable.
	▪ Slide the F connector onto the cable so the inner sleeve goes between the braid and the foil around the dielectric and the outer sleeve goes over the cable jacket.

Step	Terminating Coaxial Cable F Connectors
	■ The connector is fully seated when the dielectric is flush within the base of the connector when viewed by looking into the connector's open end.
3	Crimp the connector.
	■ Make sure the crimping tool has the correct die for the radio grade (RG)-6 or plenum-rated sheath.
	■ Place the crimp tool's opening over the connector ferrule and squeeze the tool until the die is completely closed.
4	Final inspection.
	■ Inspect the connection for neatness (no exposed braiding strands). The connector has to be tight.
	■ Use an ohm-meter to test for unwanted shorts between the center pin and outer housing.

Fiber Terminations

Overview

Optical fiber cable has a greater bandwidth than copper cable and can transmit more information through a smaller, lighter cable. It is popular in new residential cabling installations and retrofits. As bandwidths increase, there is an increasing use of optical fiber as outlet cable, ultimately ending at the outlet.

Optical fiber cables use light generated by a laser or a light-emitting diode (LED) to carry signals. Laser light can be very intense and may be invisible to the human eye.

> **WARNING:** Never look into a terminated fiber as the light may damage the eyes, even though the light will probably be invisible. Viewing it directly does not cause pain, but the iris of the eye will not close involuntarily as when viewing a bright light. Serious damage to the retina of the eye is possible.

Optical fiber cable may also be combined with copper cable in a telecommunications system with the optical fiber carrying information from building to building in a campus installation.

Optical fiber cable is becoming popular for both backbone and horizontal cable, since it offers a much lower signal loss than copper cable. It is also immune to EMI and lightning.

In telecommunications, there are two specific types of optical fiber cable used: multimode and singlemode. Each specific type of fiber has its own characteristics.

Standards-compliant multimode optical fiber has an outside diameter of 125 μm. The glass core of the fiber, which carries the optical signal, has a diameter of 50 or 62.5 μm. A micron is equal to one millionth of a meter. A protective acrylate coating is typically added to the glass fiber for protection, raising the diameter to 250 μm.

Singlemode optical fiber has an outside diameter of 125 μm. The glass core, which carries the optical signal, has a diameter of 8-9 μm. A protective acrylate coating is typically added to the glass fiber for protection, raising the diameter to 250 μm (see Figure 15.21).

An optical fiber, whether singlemode or multimode, has two distinct areas known as the core and the cladding. Although depicted below as being separate, the two areas are a single solid piece of glass. The difference between the two areas is the amount of additional materials, called dopants, which are added to the glass during manufacture to change the index of refraction. The index of refraction changes between the core and cladding, enabling the optical signal to remain in the fiber core.

Optical fiber cables may be cleaved or scored prior to termination. Proper preparation ensures maximum transmission of light from the optical fiber and reduces diffusion and reflection of the light at the termination. After termination of the optical fiber by connectorization, the fiber end is typically polished. This step helps to ensure good light transmission through the end of the optical fiber, and provides the proper alignment of the connection to maintain the integrity of the signal.

Optical fibers are immune to EMI and do not require shielding, although some optical fiber cables (normally outdoor and aerial cables) may have a metallic armored shield. Optical fiber cables usually do not contain metallic components; therefore, they do not require the same grounding and bonding considerations as copper cable. However, if there is a metallic component as part of the cable, it should be properly grounded and bonded, as with any ground or shield member.

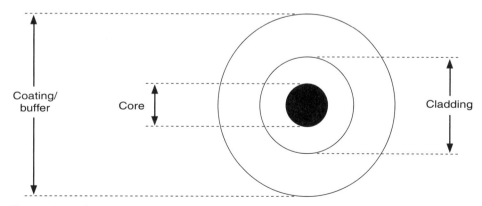

Figure 15.21   Cross-section of an optical fiber cable.

Specific termination kits are required for optical fiber connectors. Selection of the proper kit is dependent on the cable system, type of cable, and the specific connector requested by the customer or selected by the designer.

The most common connectors for multimode fiber cable are the straight tip (ST) and subscriber connector (SC) connectors, which are normally field terminated (see Figure 15.22).

Termination procedures for optical connectors in the field vary by manufacturer. The various methods of termination are:

- Epoxy (typically heat cured).
- Preloaded adhesive (hot melt).
- Ultraviolet (UV) light cured.
- Mechanical (without epoxy).
- Anaerobic adhesive (nonheat cured).

Epoxy terminations can be either ambient-cured (at room temperature normally for 24 hours) or heat cured by inserting the assembled connector into a curing oven for as long as 30 minutes. This can be a time-consuming process. Typical curing ovens hold as many as 24 connectors at a time.

Hot-melt connectors are preloaded with an adhesive that is normally hard and becomes soft when exposed to high temperatures. Connectors are placed in

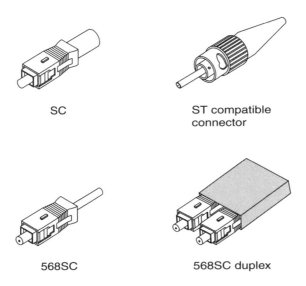

SC

ST compatible connector

568SC

568SC duplex

SC = Subscriber connector
ST = Straight tip

Figure 15.22   Fiber connectors.

an oven from two to five minutes until the adhesive becomes soft enough to insert the fiber. The connectors are then cooled for five minutes, scored, and polished. This is a time-consuming process if only one connector is needed. Typical ovens can hold four or more connectors at a time to enable an assembly-line process to speed up the per connector time.

UV-cured terminations are cured under an ultraviolet light and the process takes less than a minute.

Enhancement of optical connectors has resulted in the development of connectors without epoxy, which require no curing process.

Anaerobic termination connectors typically use a two-part chemical application—the first is the epoxy and the second part is the primer or catalyst. When the catalyst makes contact with the epoxy, a hardening of the epoxy occurs. Typically, the hardening takes 10 to 20 seconds, but it can take substantially longer (see Figure 15.23).

The 62.5/125 μm multimode fiber is covered by a protective coating with an overall diameter of 250 μm and possibly a tight buffer with an overall diameter of 900 μm. This buffer coating must be removed prior to connectorization. Removal of the buffer and protective coating is accomplished by use of an appropriate stripper (see Figure 15.24).

The following termination procedures are designed to be generic in nature for each of the four methods of termination. Refer to vendor termination instructions, which vary significantly for specific connectors. Termination procedures for both ST and SC connectors are interchangeable in most cases, but

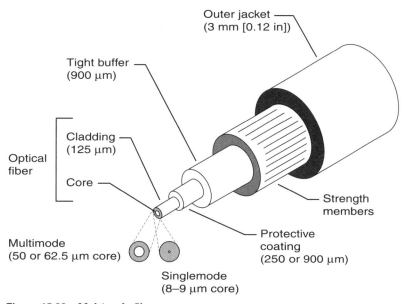

Figure 15.23 Multimode fiber.

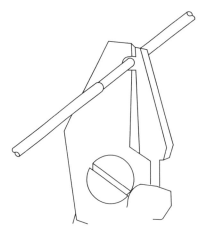

Figure 15.24 Fiber stripping tool.

the manufacturer specifications have to be followed to ensure a properly aligned low-loss termination.

Heat-cured fiber—steps

Follow the steps below when terminating heat-cured fiber.

Step	Terminating Heat-Cured Fiber
1	Slide the boot of the connector over the end of the cable end to be connectorized. There are typically two sizes of connector boots—3 mm (0.12 in) and 900 μm. The 3 mm (0.12 in) boot is used with patch cables, the 900 μm boot is used with 900 μm buffered fiber and pigtails. This procedure is for 900 μm fiber connectorization with ST connectors (see Figure 15.25).

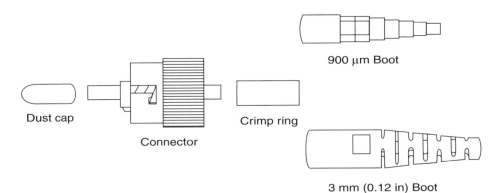

Figure 15.25 Exploded view of ST connector.

Step	Terminating Heat-Cured Fiber
2	Carefully mark the length of buffer coating to be removed per the manufacturer's specifications (see Figure 15.26). Remove the buffer in 6.0 mm (0.236 in) to 8 mm (0.3 in) increments, employing the buffer removal tool (see Figure 15.27). Strip the buffer to the mark to obtain the specified amount of bare fiber. Make sure the buffer coating has no jagged edges and it is cut cleanly. Refer to the instructions packaged with the removal tool for further details.
3	Use an alcohol wipe or bifurcated swab to clean the stripped fiber. Fold the wipe over the fiber and squeeze gently on the fiber inside the wipe, while pulling the fiber through the wipe.

> **NOTE:** If using a bifurcated swab, carefully insert the fiber into the swab, gently squeeze the foam sides of the swab, and pull the fiber straight through the swab.

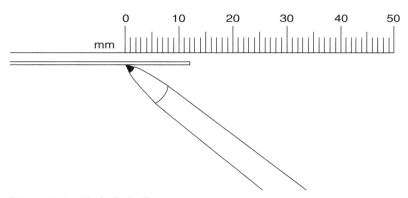

Figure 15.26   Mark the buffer.

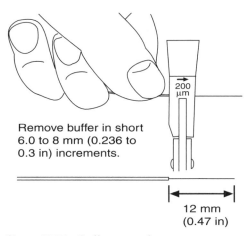

Figure 15.27   Buffer removal.

Step	Terminating Heat-Cured Fiber
4	Prepare the epoxy by mixing it in accordance with the manufacturer's instructions.
5	Remove the cap from the syringe. Load the syringe with the mixed epoxy. Reinstall the cap into the syringe.
6	Select the connector to be terminated. Remove and discard the small black cap on the rear of the connector. Remove and retain the dust cap from the front of the connector.
7	Insert the epoxy syringe tip into the rear of the connector until it bottoms on the ceramic ferrule. Place a mark, with a black marking pen, on the syringe tip just below the connector (see Figure 15.28). Remove the connector. Place a second black mark on the syringe tip 2 mm (0.08 in) to 3 mm (0.12 in [or whatever distance the manufacturer specifies]) from the first mark and toward the tip of the syringe. These marks enable the residential cabling installer to repeatedly and accurately fill all connectors.

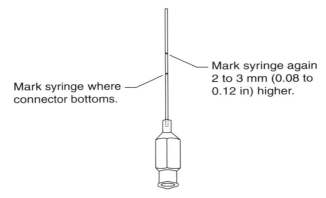

Mark syringe again
2 to 3 mm (0.08 to
0.12 in) higher.

Mark syringe where
connector bottoms.

Figure 15.28   Mark syringe.

8	With the syringe pointed up, place the connector back on the syringe tip and while holding the connector, slowly inject epoxy. The moment that epoxy is visible at the tip of the ferrule, let go of the connector. Continue injecting epoxy and allow the connector to rise on the syringe tip. When the rear of the connector rises past the second black mark, stop injecting epoxy. Remove the connector from the syringe tip.
9	Insert the ferrule into the tube of the load adapter of the curing oven.
10	Place a thin layer of epoxy along the length of the bare fiber (see Figure 15.29). Be sure to coat the junction where the bare fiber enters the buffer. This keeps the epoxy bead intact during fiber insertion into the connector (see Figure 15.30).
11	Slide the fiber into the connector until the buffer seats fully on the rear of the ferrule. Pull the boot back up the fiber cable. Slide the spring clip on the load adapter down, and position the boot back in the jaw of the clip. The load adapter clip now rests on the boot and not directly on the fiber.
12	Remove any excess epoxy at the rear of the connector, being careful not to unseat the ferrule.
13	Place the connector and the load adapter into the oven (see Figure 15.31). Temperature of the oven should be within the 105 to 115 °C (220 to 240 °F) range. Depending on the type epoxy used, cure time is anywhere from 6 to 20 minutes. Refer to the manufacturer's specifications for details.

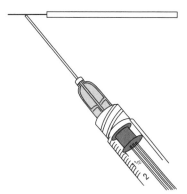

Figure 15.29   Apply epoxy.

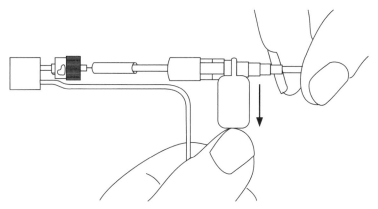

Figure 15.30   Install connector.

Step	Terminating Heat-Cured Fiber
14	Allow the load adapter to cool after the prescribed cure time. The epoxy should change color when cured. Remove the connector from the load adapter. Be careful to not break the fiber protruding from the tip of the ferrule.
15	Use a precision scribing tool (cleaving tool) to carefully scribe or nick the excess fiber at a point approximately two times the diameter of the fiber from where it exits the bead of epoxy (see Figure 15.32). Pull the scored fiber straight up to complete the separation. Safely dispose of the detached fiber (see Figure 15.33).
16	Slide the 900 μm boot onto the back of the connector. The connector is now ready to be polished.
17	Use a 2 μm lapping film formed into a U-shape with the abrasive face inward to remove the fiber nub by making 25 mm (1 in) circles inside the U-shaped disk (see Figure 15.34). Stop when the fiber nub no longer scratches the lapping film (see Figure 15.35).

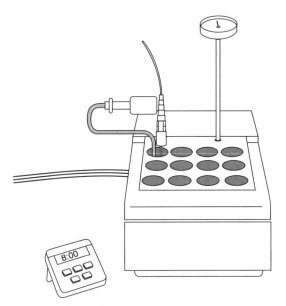

Figure 15.31   Curing oven.

Step	Terminating Heat-Cured Fiber
18	Select the first connector to be polished and check for excess epoxy on the side of the ferrule. Remove any stray epoxy with a razor blade. Test fit the ferrule in a polishing jig. It should move freely and seat on the jig top. Polishing jigs can be purchased that will polish six connectors at once.
19	Place the silicone rubber pad from the terminating kit on the glass plate provided. Clean the pad with alcohol followed by a dry wipe, and blow with dry air. Place a 3 μm disk and a 0.3 μm disk on the plate.

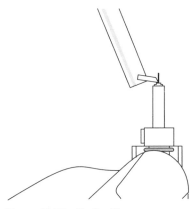

Figure 15.32   Scribe fiber.

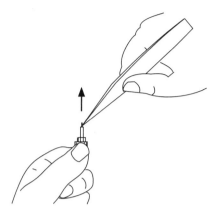

Figure 15.33 Remove excess fiber.

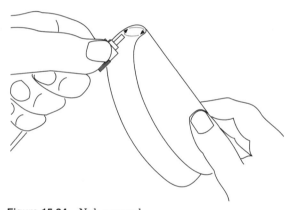

Figure 15.34 Nub removal.

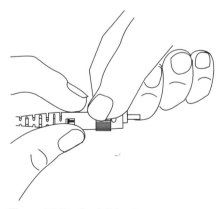

Figure 15.35 Install boot.

Step	Terminating Heat-Cured Fiber
20	Place a separate polishing jig on each abrasive disk and use each jig only on its corresponding abrasive disk (see Figure 15.36). This prevents contamination of the white adhesive disk. **NOTE:** If only one polishing jig is used, clean the jig with alcohol after each step.
21	Place a puddle of water 25 mm (1 in) in diameter on each disk, and use the polishing jig to uniformly wet the surface.
22	Check the end of the connector. A small amount of epoxy should be present. In cases where no epoxy is visible, go directly to the white abrasive disk.
23	Place the connector into the first jig and onto the 3 μm disk, making one or two initial figure-eight patterns (about 75 mm [3 in] high), using a gentle downward pressure. This step ensures that the fiber nub is flush with the epoxy bead.
24	Once the fiber nub is flush, make single figure-eights using medium pressure. Check the epoxy bead after each figure-eight to confirm the epoxy bead is gone. Once the epoxy is almost gone, remove and clean the connector with a clean, dry wipe.
25	Place the connector in the second polishing jig and onto the wet 0.3 μm abrasive disk. Make eight to ten figure-eight patterns with only enough pressure to keep the connector steady.
26	Remove the connectors and clean the end face with a clean wipe moistened in isopropyl alcohol (99 percent pure); finish cleaning with a blast of compressed air. Ensure that the air is manufacturer-approved as suitable for cleaning fiber terminations (e.g., without Freon®).
27	Insert the connector into the adapter end of a 100x microscope to inspect the end face (see Figure 15.37). Replace any connector that contains a crack. Small pits are acceptable and typically occur when the connector is ground too long during the dark gray abrasive disk polishing step (see Figure 15.38).
28	After passing visual inspection, place the original dust cap over the end of the connector ferrule.

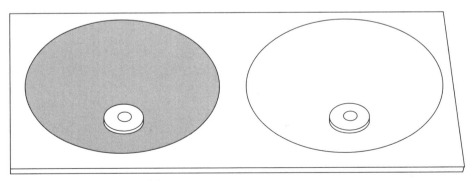

Figure 15.36   Polish jigs on disks.

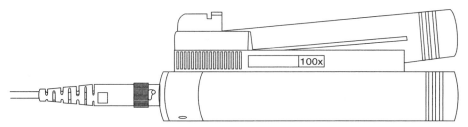

Figure 15.37   100x microscope.

Acceptable

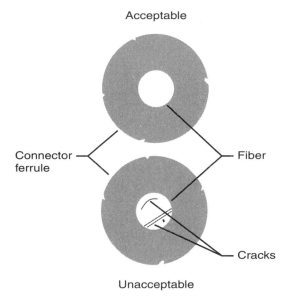

Connector
ferrule

Fiber

Cracks

Unacceptable

Figure 15.38   Fiber end examples.

Hot-melt connector—steps

Follow the steps below when terminating hot-melt connector fiber.

Step	Terminating Hot-Melt Connector Fiber
1	Preheat oven for five minutes.
2	Remove the connector's dust cap and place the connector into the holder. Place in the oven for two to five minutes.
3	Inspect the cable to ensure that the aramid yarn and fiber are all visible at the end of the jacket. Cut back, as necessary, to have all three components flush together.
4	Slide the boot of the connector over the end of the cable end to be connectorized. There are typically two sizes of connector boots—3 mm (0.12 in) and 900 μm. The 3 mm (0.12 in) boot is used with patch cords; the 900 μm boot is used with 900 μm buffered fiber or 250 μm fibers built up with a breakout kit. This procedure is for 3 mm (0.12 in) jacketed fiber connectorization with ST connectors.

Step	Terminating Hot-Melt Connector Fiber
5	Measure and remove 25 mm (1 in) of outer jacket. **NOTE:** The boot is 25 mm (1 in).
6	Hold the aramid yarn away from the buffer and carefully mark the buffer coating 3 mm (0.12 in) from the jacket. Remove the buffer in 6.0 mm (0.236 in) to 8 mm (0.3 in) increments, employing the buffer removal tool (see Figure 15.39). Strip the buffer to the mark to obtain the specified amount of bare fiber. Make sure the buffer coating has no jagged edges and it is cut cleanly. Refer to the instructions packaged with the removal tool for further details.
7	Use an alcohol wipe or bifurcated swab to clean the stripped fiber. Fold the wipe over the fiber and squeeze gently on the fiber inside the wipe, while pulling the fiber through the wipe (see Figure 15.40).

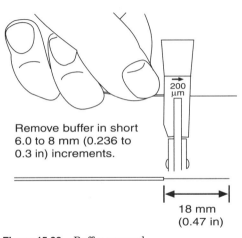

Remove buffer in short
6.0 to 8 mm (0.236 to
0.3 in) increments.

200
μm

18 mm
(0.47 in)

Figure 15.39 Buffer removal.

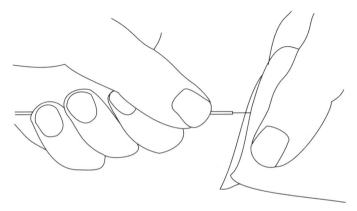

Figure 15.40 Clean fiber.

Step	Terminating Hot-Melt Connector Fiber

> **NOTE:** If using a bifurcated swab, carefully insert the fiber into the swab, gently squeeze the foam sides of the swab, and pull the fiber straight through the swab.

8   Trim the aramid yarn 4.6 mm (0.18 in) from the jacket; massage jacket to distribute the aramid yarn around the buffer.

9   Remove the connector from the oven and insert the fiber, aramid yarn, and jacket fully into the rear of the connector. Secure the outer jacket in the holder's strain relief clip.

A. Place the hot connector in cooling rack for five minutes or until cool to the touch.

B. Remove the cooled connector from the holder. Be careful to not break the fiber protruding from the tip of the ferrule.

C. Use a precision scribing tool to carefully scribe or nick the excess fiber at a point approximately two times the diameter of the fiber from where it exits the bead of epoxy (see Figure 15.41). Pull the scored fiber straight up to complete the separation. Safely dispose of the detached fiber (see Figure 15.42).

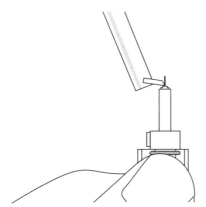

Figure 15.41   Scribe fiber.

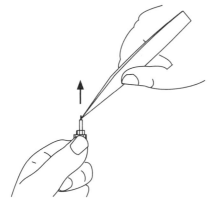

Figure 15.42   Remove excess fiber.

Step	Terminating Hot-Melt Connector Fiber
	D. Slide the 3 mm (0.12 in) boot onto the back of the connector (see Figure 15.43). The connector is now ready to be polished.
10	Use a 0.3 µm lapping film formed into a U-shape with the abrasive face inward to remove the fiber nub by making 25 mm (1 in) circles inside the U-shaped disk (see Figure 15.44). Stop when the fiber nub no longer scratches the lapping film. Be careful to not over air polish. If the blue adhesive starts to lighten in color, stop polishing.
11	Clean the rubber pad from the terminating kit. Clean the pad with alcohol followed by a dry wipe, and blow with dry air. Place a 0.3 µm disk on the rubber pad shiny side down.
12	Check the end of the connector. A small amount of epoxy should be present. In cases where no epoxy is visible, be careful to not apply too much pressure during the first few figure-eights.
13	Place the connector into the polishing puck, or jig, and onto the 0.3 µm disk, making one or two initial figure-eight patterns (about 75 mm [3 in] high), using a gentle downward pressure. This step ensures that the fiber nub is flush with the epoxy bead.

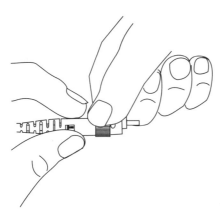

Figure 15.43 Install boot.

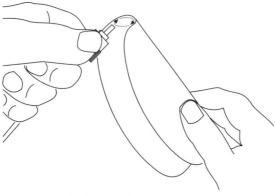

Figure 15.44 Nub removal.

Step	Terminating Hot-Melt Connector Fiber
14	Once the fiber nub is flush, make figure-eights using medium pressure. Slowly reduce the amount of downward pressure as the connector nears its final polishing. If the resistance suddenly becomes smooth or after 10–15 figure-eights have been completed, check the epoxy bead.
15	Inspect the fiber tip to ensure all of the blue adhesive has been removed. If it still remains, continue additional polishing.
16	Remove the connector and clean the end face with a clean wipe moistened in isopropyl alcohol (99 percent pure); finish cleaning with a blast of compressed air. Ensure that the air is manufacturer-approved as suitable for cleaning fiber terminations (e.g., without Freon®).
17	Insert the connector into the adapter end of a 100x microscope to inspect the end face (see Figures 15.45 and 15.46).
18	After passing visual inspection, place the original dust cap over the end of the connector ferrule.

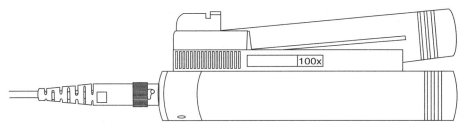

Figure 15.45   100x microscope.

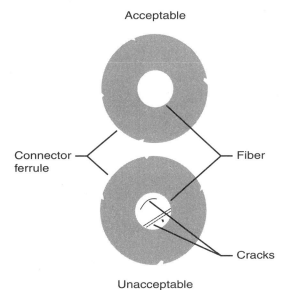

Acceptable

Connector ferrule

Fiber

Cracks

Unacceptable

Figure 15.46   Fiber end examples.

UV-cured cable—steps

Follow the steps below when terminating UV-cured cable.

Step	Terminating UV-Cured Cable
1	Slide the boot of the connector along with the strain relief collar over the end of the cable to be terminated. This procedure is for a 900 μm fiber.
2	The orientation of the strain relief collar is extremely important. Ensure that the hexagonal end is toward the boot.
3	Carefully mark the length of buffer coating to be removed per the manufacturer specifications (see Figure 15.47). Remove the buffer in 6.0 mm (0.236 in) to 8 mm (0.3 in) increments, employing the buffer removal tool. Strip the buffer to the mark to obtain the specified amount of bare fiber. Make sure the buffer coating has no jagged edges and it is cut cleanly. Refer to the instructions packaged with the removal tool for further details. Clean the stripped fiber with a lint-free tissue soaked in alcohol. Gently squeeze the fiber inside the tissue while pulling the fiber through the tissue (see Figure 15.48).

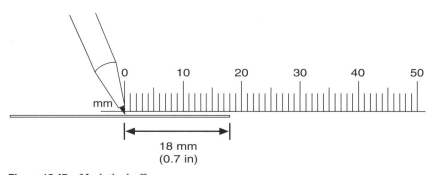

Figure 15.47   Mark the buffer.

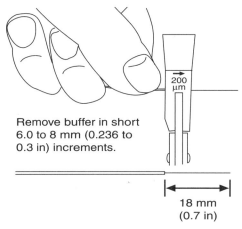

Figure 15.48   Buffer removal.

Step	Terminating UV-Cured Cable

4   Clean the stripped fiber using an alcohol wipe or bifurcated swab. Fold the wipe over the fiber and squeeze gently on the fiber inside the wipe, while pulling the fiber through the wipe. If using a bifurcated swab, carefully insert the fiber into the swab, gently squeeze the foam sides of the swab, and pull the fiber straight through the swab (see Figure 15.49).

5   Remove and discard the cap on the rear of the UV connector assembly.

6   Insert the syringe tip into the rear of the connector assembly until it bottoms on the ferrule. Place a mark on the syringe tip just below the connector (see Figure 15.50). Remove the connector. Place a second mark on the syringe tip 2 mm (0.08 in) to 3 mm (0.12 in) away (or a distance specified by the manufacturer) from the first mark toward the tip of the syringe. These marks enable the technician to repeatedly and accurately fill all connectors with adhesive.

7   Place the connector back on the syringe tip and inject the UV adhesive slowly until it begins to come out of the tip of the ferrule. Let go of the connector. Continue to inject adhesive, allowing the connector to rise up on the syringe tip. When the rear of the connector rises to the second mark, stop injecting adhesive (see Figure 15.51).

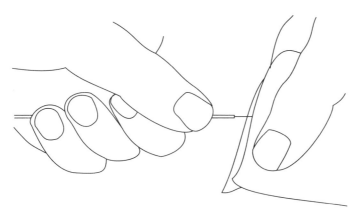

Figure 15.49   Clean fiber.

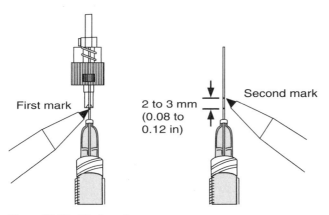

First mark

2 to 3 mm
(0.08 to
0.12 in)

Second mark

Figure 15.50   Mark syringe.

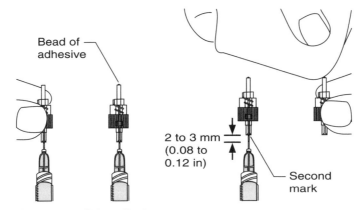

Figure 15.51   Inject adhesive.

Step	Terminating UV-Cured Cable
8	Remove the connector assembly from the syringe tip. Wipe off the adhesive bead on the end face, employing a lint-free tissue.
9	Hold the fiber vertically, slide the connector onto the fiber, and then release it. Gently wiggle the fiber until the connector slides down the fiber and the rear of the ferrule is fully seated on the buffer. When properly seated, approximately 5 mm (0.2 in) of bare fiber should be sticking out of the end of the ferrule. The back of the connector assembly should be 3 mm (0.12 in) or less from the end of the buffer. A small amount of adhesive should now be present around the fiber (see Figure 15.52).
10	Refer to the specific instructions for placing the connector in the UV curing lamp. Some curing lamps have timers and LED indicators, which change color to indicate the curing cycle is complete. The normal cure time under UV lamps is less than one minute (see Figure 15.53).
11	Carefully remove the fiber from the curing lamp. Be careful to not break the fiber that is protruding from the ferrule tip.

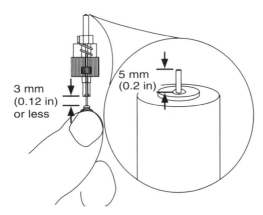

Figure 15.52   Place connector on fiber.

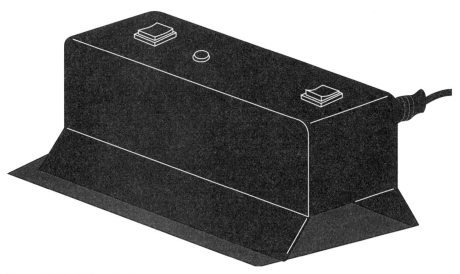

Figure 15.53   UV curing lamp.

Step	Terminating UV-Cured Cable
12	Use a fiber scribe to score the excess fiber where it emerges from the adhesive bead on the ferrule tip. Gently and carefully pull the fiber straight up to remove excess fiber (see Figure 15.54). Safely dispose of the detached fiber (see Figure 15.55).
13	Slide the strain-relief collar against the back of the connector (see Figure 15.56). Carefully thread the collar onto the threaded portion of the lead-in tube. Tighten the strain relief until it no longer turns. Do not apply excessive force.
14	Slide the 900 µm boot onto the back of the ST-compatible connector (see Figure 15.57).
15	Clean the surface of the glass plate with alcohol followed by a clean dry tissue. Blow the surface with compressed air.

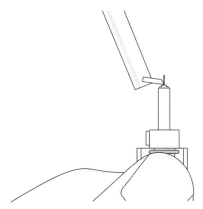

Figure 15.54   Scribe fiber.

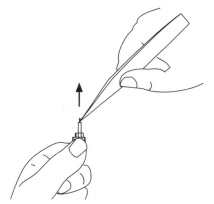

Figure 15.55   Remove excess fiber.

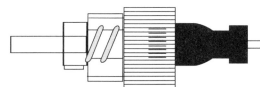

Figure 15.56   Install strain-relief collar.

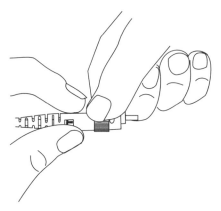

Figure 15.57   Install boot.

Step	Terminating UV-Cured Cable
16	Remove the backing from the 0.3 μm and the 5 μm abrasive disks. Carefully apply the disks to the glass plate, forcing any air bubbles from underneath the lapping film.
17	Place a separate polishing jig on each abrasive disk. Use each jig only on its respective disk to prevent contamination of the fine disk.
18	Wet the surface of both disks with water (25 mm [1 in] diameter). Use the polishing jig to spread the water evenly on the surface (see Figure 15.58).
19	Take a loose 5 μm disk and form it into a U-shape with the abrasive face inward. While making 25 mm (1 in) circles, remove the nub of adhesive from the end of the ferrule. Stop when the fiber nub no longer scratches the disk.
20	Test fit the connector ferrule into the polishing jig. It should move freely up and down and be seated on the jig top. Check the condition of the connector's end face. A small amount of the adhesive should be present.
21	Place the connector into the jig on the gray abrasive disk, and make one or two figure-eight patterns about 75 mm (3 in) high (see Figure 15.59). This step is to make sure that the fiber nub is flush with the adhesive bead.
22	Polish the connector with approximately 20 strokes, or until the UV adhesive bead is removed.
23	Gently wipe the end face and ferrule surfaces of the connector with a lint-free tissue soaked in alcohol. Wait at least five seconds for the surfaces to dry.
24	Complete the polishing by placing the connector into the second jig on the 0.3 μm disk and make about 15 figure-eight patterns.
25	Clean the connector again in the same manner as in Step 23.
26	Keep the air nozzle approximately 13 mm (0.5 in) away from the connector (see Figure 15.60). Ensure that the air is manufacturer approved as suitable for cleaning fiber terminations (e.g., without Freon®).
27	Secure the connector into the adapter end of the 100x microscope supplied with the tool kit, and visually inspect the connector. The surface should have a mirror finish with the edge of the fiber face perfectly circular.

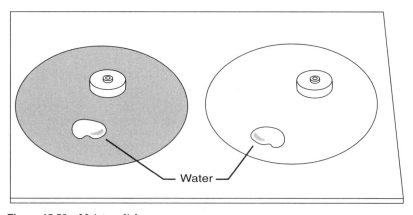

Water

Figure 15.58   Moisten disks.

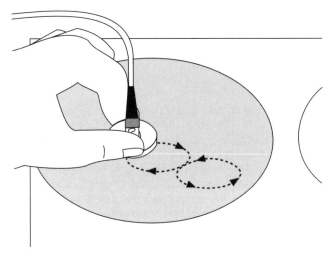

Figure 15.59 Polish fiber.

13 mm (0.5 in)

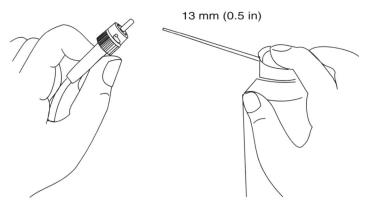

Figure 15.60 Compressed air cleaning.

Step	Terminating UV-Cured Cable
28	If scratches or other imperfections are visible, repolish the connector with the 0.3 μm lapping film.
29	After passing visual inspection, place the dust cover over the end of the connector ferrule.

Crimp fiber—steps

Follow the steps below when terminating crimp fiber.

Step	Terminating Crimp-Style Fiber
1	Slide the 900 μm boot (small end first) down the fiber until out of the way. This procedure is for termination of either ST or SC crimp connectors onto 900 μm fiber (see Figures 15.61 and 15.62).
	Different manufacturers provide distinct tools for crimping connectors to fiber. Read and follow the specific instructions for the particular manufacturer's connector being used (see Figure 15.63).
2	Carefully mark the length of buffer coating to be removed from the manufacturer's specifications (see Figure 15.64). Remove the buffer in 6.0 mm (0.236 in) to 8 mm (0.3 in) increments, employing the buffer removal tool. Strip the buffer to the mark to obtain the specified amount of bare fiber. Make sure the buffer coating has no jagged edges and is cut cleanly. Refer to the instructions packaged with the removal tool for further details. Mark the buffer an additional 11 mm (0.43 in) back from the strip point. Clean the bare fiber with two passes of an alcohol wipe or a lint-free tissue that has been soaked in isopropyl alcohol (99 percent pure). Do not touch the bare fiber after cleaning and do not remove the 11 mm (0.43 in) mark.

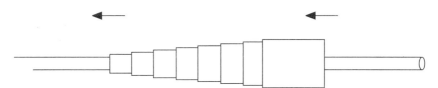

Figure 15.61   Preinstall boot.

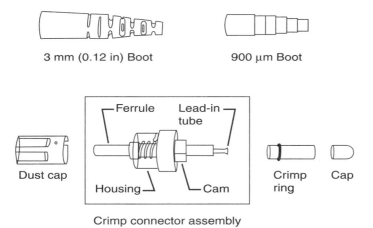

3 mm (0.12 in) Boot   900 μm Boot

Crimp connector assembly

Figure 15.62   Straight tip components.

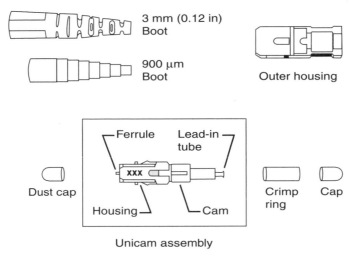

Figure 15.63   Subscriber connector components.

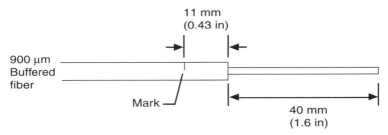

Figure 15.64   Mark the buffer.

Step	Terminating Crimp-Style Fiber
3	Cleave the fiber as described in the instructions provided with the cleaving tool. Cleave the fiber to 8.5 mm (0.34 in) or to the specified length (see Figure 15.65). Safely dispose of the fiber scraps.
4	Insert the crimp connector into the crimping tool following the manufacturer's instructions (see Figure 15.66).
5	Carefully insert the cleaved fiber into the lead-in tube until feeling it firmly stop against the connector's fiber stub. Be sure to guide the fiber in straight. Do not bend or angle the fiber (see Figure 15.67). Observe the 11 mm (0.43 in) reference mark.
6	If resistance is felt at the entry tunnel, rotate the fiber back and forth while applying a gentle inward pressure.
7	Carefully, push the 900 μm buffer fiber into the fiber clamp on the tool. Maintaining pressure on the fiber, form a slight bow between the connector and the clamp. This bow is very important, as it helps the fibers make contact in the connector during the next step.

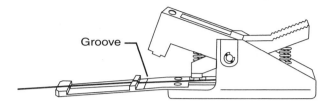

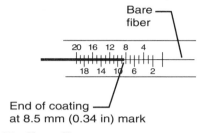

Figure 15.65   Cleave fiber.

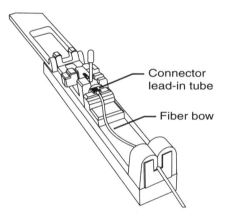

Figure 15.66   Cam tool.

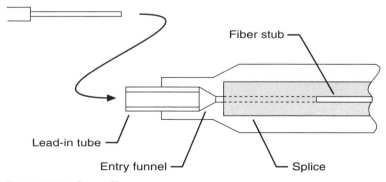

Figure 15.67   Insert fiber.

Step	Terminating Crimp-Style Fiber
8	Rotate the wrench of the crimp tool past a 90-degree angle to seat the connector (see Figure 15.68).
9	The splice now holds the fiber within the connector.
10	Carefully flip the crimp handle 180 degrees until it contacts the crimp tube. Push down firmly to crimp. A flat impression can be seen in the crimp tube. The tool cannot over crimp the connector (see Figure 15.69).
11	Flip the crimp handle back. Leave the wrench handle down. Remove the connector by lifting it straight up and out of the tool. Do not pull on the fiber.

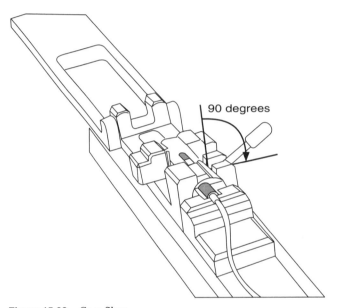

Figure 15.68   Cam fiber.

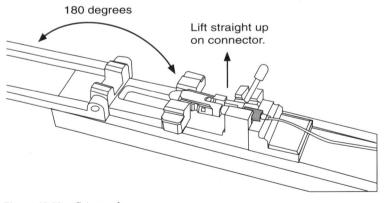

Figure 15.69   Crimp tube.

Step	Terminating Crimp-Style Fiber
12	Slide the boot up to the back of the connector until it reaches the cam (see Figure 15.70).
13	The outer housing of a multimode connector is beige. To install the connector assembly into the SC outer housing, line up the bevel edges on the inner housing with the key on the outer housing. Using the boot, push the assembly into place. It may be necessary to wiggle the parts to make them snap together (see Figure 15.71).
14	The connector is now ready to use. Leave the front dust cover on until ready to insert the connector into a sleeve.

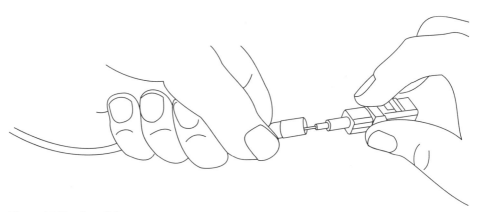

Figure 15.70   Install boot.

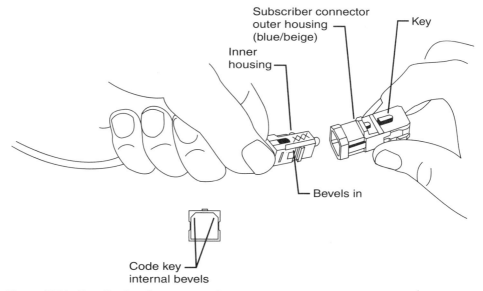

Figure 15.71   Install subscriber connector housing.

Mechanical fiber terminations—steps

Mechanical termination methods are widespread and have been used for years in the optical fiber community. The method described here is given as an example of a compression-style connection.

Step	Mechanical Fiber Termination
1	Select the connector to be terminated. Remove and retain the dust cap from the front of the connector.
2	Slide the boot of the connector over the cable end to be connectorized. The following procedure is for 3 mm (0.12 in) jacketed fiber connectorization with ST connectors.
3	Remove fiber jacket to length following the connector manufacturer's instructions.
4	Gently insert the fiber into the connector's buildup sleeve until the fiber jacket is butted up against the build-up sleeve and the aramid strands are flared around the build-up sleeve.
5	Carefully mark the length of buffer coating to be removed per the manufacturer's specifications. Remove the buffer in 6.0 (0.236 in) to 8 mm (0.3 in) increments, employing the buffer removal tool (see Figure 15.72). Strip the buffer to the mark to obtain the specified amount of bare fiber. Make sure the buffer coating has no jagged edges and is cut cleanly. Refer to the instructions packaged with the removal tool for further details.
6	Trim the aramid strands even with the end of the build-up sleeve.
7	Clean the stripped fiber using an alcohol wipe or bifurcated swab. Fold the wipe over the fiber and squeeze gently on the fiber inside the wipe while pulling the fiber through the wipe.
	NOTE: It is important that all traces of gel or any other contamination are removed and that the fiber is absolutely clean.
8	Distribute aramid strands evenly over the outer surface of the buildup sleeve.

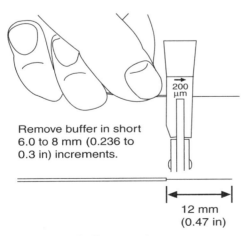

Remove buffer in short 6.0 to 8 mm (0.236 to 0.3 in) increments.

200 µm

12 mm (0.47 in)

Figure 15.72 Buffer removal.

Step	Mechanical Fiber Termination
9	Fold the retention sleeve over the buildup sleeve and the aramid strands aligning the end of the buildup sleeve with the threaded end of the retention sleeve leaving 3 mm (0.12 in) to 6.3 mm (0.25 in) of the buffered fiber exposed.
10	While firmly holding the threaded retention sleeve, insert the exposed fiber into the connector housing.
11	Rotate the connector housing over the threaded retention sleeve. Do not rotate the fiber or retention sleeve during this process. Most manufacturers offer a tool to grip the connector housing to facilitate this step.
12	Slide the 900 μm boot onto the back of the connector (see Figure 15.73).
13	Use a precision scribing tool (or cleaving tool to carefully scribe or nick the excess fiber at a point approximately two times the diameter of the fiber from where it exits the ferrule (see Figure 15.74). Pull the scored fiber away from the connector in a direction parallel to the fiber to complete the separation. Safely dispose of the detached fiber (see Figure 15.75).

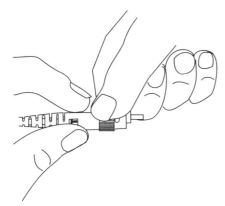

Figure 15.73   Installing the boot.

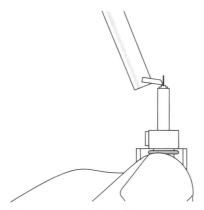

Figure 15.74   Scribing fiber.

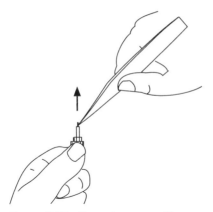

Figure 15.75   Removing excess fiber.

Step	Mechanical Fiber Termination
14	Hold a 12 μm lapping film (dull side up) loosely in one hand. Gently touch the tip of the connector to the top of the film. Once contact is made, move the connector in a circular motion gently while keeping contact with the lapping film. Slightly increase the pressure and air polish the excess fiber away.
15	Listen to the sound the fiber makes as the connector is air polished. A scratching sound will be heard. This sound decreases as the excess fiber is removed. When the scratching sound has diminished, the first polishing step is complete. Fifteen to eighteen 25 mm (1 in) circles should be adequate for the air polish.
16	Place the polishing pad from the terminating kit on a firm, smooth surface. Clean the pad with alcohol followed by a dry wipe, and blow with dry air. Place a 3 μm polishing film on the pad.
17	Use an alcohol wipe to clean the polishing jig and the end of the connector.
18	Insert the connector firmly into the polishing jig.
19	Use a figure-eight pattern and light pressure to polish the connector for 18 to 20 rotations (see Figure 15.76).

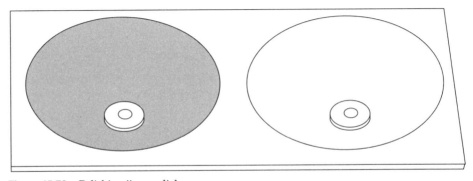

Figure 15.76   Polishing jigs on disks.

Step	Mechanical Fiber Termination
20	Remove the connector from the jig and clean both with alcohol.
21	Replace the 3 μm lapping film with a 0.3 μm lapping film and repeat the above polishing process in Steps 19–21.
22	Clean the connector with a lint-free wipe.
23	Insert the connector into the adapter end of a 200x microscope to inspect the end face (see Figure 15.77). Replace any connector that shows a crack under magnification. Small nicks are acceptable and typically occur when the connector is polished too long (see Figure 15.78). **NOTE:** Some manufacturer's connectors can be removed and reterminated if the polishing process is unsuccessful.
24	After passing visual inspection, place the original dust cap over the end of the connector ferrule.

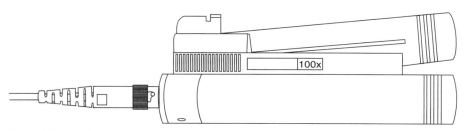

Figure 15.77   Microscope.

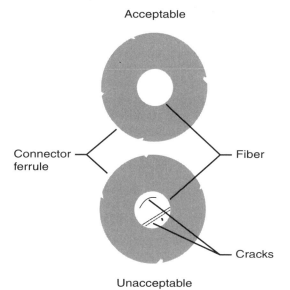

Figure 15.78   Fiber-end examples.

Anaerobic-style fiber—steps

Follow the steps below when terminating anaerobic-style fiber.

Step	Anaerobic-Style Fiber Termination
1	Slide the 3 mm (0.12 in) boot (small end first) down the fiber until out of the way. Then, slide the crimp collar down the fiber. This procedure is for termination of either ST or SC connectors onto 3 mm (0.12 in) jacketed fiber.
2	Follow the manufacturer's specifications to measure and mark the jacket at the specified distance from the end of the fiber. Use the proper tool to sever the jacket (or sheath) and remove it from the cable.
3	Follow the manufacturer's specifications to measure and mark the aramid yarn the specified distance from the end of the fiber.
4	Carefully mark the length of buffer coating to be removed per the manufacturer's specifications. Remove the buffer in 6.0 (0.236 in) to 8 mm (0.3 in) increments, employing the buffer removal tool. Strip the buffer to the mark to obtain the specified amount of bare fiber. Make sure the buffer coating has no jagged edges and is cut cleanly. Refer to the instructions packaged with the removal tool for further details. Clean the bare fiber with two passes of an alcohol wipe or a lint-free tissue that has been soaked in isopropyl alcohol (99 percent pure). Do not touch the bare fiber after cleaning.
5	Insert the syringe tip into the rear of the connector assembly until it bottoms on the ferrule. Slowly squeeze the adhesive into the connector until the adhesive appears on the tip of the ferrule. Withdraw the syringe slowly, while continuing to lightly squeeze adhesive into the barrel of the connector.
6	Set aside the connector and paint the bare fiber with the primer (catalyst). Ensure that the bare fiber, as well as the exposed buffer coating, is primed for insertion. **CAUTION:** Do not dip the fiber into the bottle of primer, as a slight bump of the bare fiber could cause it to break off and fall into the primer, creating a safety hazard.
7	Carefully and steadily, insert the fiber into the connector until the connector reaches the buffer tube. **NOTE:** Depending on the manufacturer and type of adhesive and primer used, there may be only 15 seconds for the insertion to take place before the catalytic action hardens the chemicals and causes a solid bond.
8	Carefully slide the crimp ring up the buffer tube and then fan the aramid yarn around the connector base. Slide the crimp connector onto the base, and with the specified crimping tool, crimp the crimp ring onto the body of the connector. Depending on the manufacturer, a second crimp may be required around the jacket and aramid yarn.
9	Slide the boot up the jacket to the back of the connector and over the crimp ring.
10	Using a fiber scribe, score the excess fiber where it emerges from the adhesive bead on the ferrule tip. Gently and carefully pull the fiber straight up to remove excess fiber. Safely dispose of the detached fiber.
11	Clean the surface of the silicone pad with alcohol followed by a clean, dry tissue. Blow the surface with compressed air.

Step	Anaerobic-Style Fiber Termination
12	Use a 5 μm lapping film formed into a U-shape with the abrasive face inward to remove the fiber nub by making 25 mm (1 in) circles inside the U-shaped disk. Stop when the fiber nub no longer scratches the lapping film.
13	Place the connector into the polishing jig on the 3 μm abrasive disk, and make figure-eight patterns about 75 mm (3 in) in diameter. This step is to ensure that the adhesive bead has been removed. Clean the end of the ferrule with alcohol and inspect the connection with a microscope to ensure no shattering of the fiber's end.
14	Place the connector into the polishing jig on the 0.3 μm abrasive disk and make figure-eight patterns about 75 mm (3 in) in diameter. This step is to ensure that the end of the fiber is polished to eliminate power loss due to irregular polishing. Clean the end of the ferrule with alcohol. Inspect the connection with a microscope to ensure that the connection is clear and without blemishes. The surface should have a mirror finish with the edge of the fiber face perfectly circular.
15	If scratches or other imperfections are visible, repolish the connector with the 0.3 μm lapping film.
16	After passing visual inspection, place the dust cover over the end of the connector ferrule.

Testing the Cabling System

The RSCS must be tested in accordance with ANSI/TIA/EIA-B.1 requirements. Procedures for performing these tests, as well as descriptions of the test equipment needed and considerations for its purchase and maintenance, are covered in detail in Chapter 16: Testing and Troubleshooting.

Attaching Cables to Consumer Component Devices

The residential cabling installer's work usually ends with the installation, testing, and documentation of the RSCS. This is because the eventual homeowner may not be known and specific consumer components will not be available. However, in other cases, once the RSCS has been installed and tested, the residential cabling installer may be required to attach various consumer components to the system.

These consumer electronics will include devices that support communications, entertainment, security, and control applications in the home. The components are inserted into, or mounted near, the DD, or reside near the telecommunications outlet/connectors.

In terms of telecommunications solutions run over the twisted-pair cabling system, patch cords in the DD attach the cabling system to a networking hub, router, or telephone block. For voice services, the patch cords may include six- or eight-position modular plugs. For data applications, the patch cords will have eight-position modular plugs. Pre-terminated patch cords are recommended.

There are three primary methods for terminating twisted-pair outlet cables within the DD:

- Outlet cables terminated directly onto 110 or other IDC termination blocks
- Outlet cables terminated to 110 or other patch panel devices and then use 8P8C (RJ-45) patch cords
- Outlet cables terminated directly with 8P8C (RJ-45) modular plugs

For voice applications, the critical factor is to determine which telephone line is needed at each outlet. The patch cords can then connect the particular outlet cables within the DD to the appropriate incoming telephone-line service. Many manufacturers offer solutions that provide two or more telephone lines running over a single cable to the outlet to support multi-line phones.

It is important to properly connect these phones to the telephony block in the DD per manufacturer instructions. Many vendors provide an RJ-31X jack on the telephony board for connection to the home's security system. Ensure that only the security system plugs into this device.

Most data-networking solutions do not require a cable to be installed in a particular port within a networking hub or router. However, to expand the hub to additional hubs, a patch cord may have to be inserted between specific ports. Follow manufacturer's instructions when this is necessary.

At the outlet, patch cords are routed between the 8P8C (RJ-45) eight-position jacks to the computers, fax machines, telephones, and other consumer products. The residential cabling installer needs to determine whether the patch cords must include six-position or eight-position plugs. Patch cords should be sized long enough to reach the consumer electronic product, even when rearranged within the work space or living area, but kept short to reduce clutter. Once installed, the slack cable, if any, may be looped and kept orderly by using a cable tie or other cable management device, making sure that bend-radius control is maintained and the individual cables are not cinched too tightly.

Many of the other consumer components (e.g., entertainment systems) are discussed in Chapter 6: Consumer Component System Integration. In general, the residential cabling installer must determine the proper type of cable, type of connector, and cable length for each connection. The cables are then routed, and if necessary dressed, between the outlet and consumer electronics.

If a number of electronics are to be maintained in the same area (e.g., a stack of audio electronics), a cable management rack or enclosure may be preferred.

Power cords are an essential part of a properly designed and maintained system. The system may include various devices that provide electrical protection for the consumer electronics.

Run Cross-Connects—Multi-Dwelling Units (MDUs) and Single-Family

General

Jumpers and patch cords are used to configure MACs to the residential telecommunications system. Cables are typically terminated on connecting blocks that a designer systematically designed into an RSCS. This systematically designed termination field should provide for adequate routing of patch cords and jumpers.

After cable is terminated on connecting blocks, the media on the termination fields needs to be cross-connected or interconnected to extend services. For example, telephone service that is delivered on a cable pair within a backbone cable that is terminated on one field needs to be cross-connected to an auxiliary disconnect outlet (ADO) cable terminated on another field that extends the service into an MDU. An interconnect may be used to connect an Ethernet switch directly to outlet cabling. Most important, when either cross-connecting or interconnecting, like media is used for jumpers or patch cords as that of the cabling that it is being connected to. For example, a Category 5 jumper or patch cord should connect Category 5 cables, also a 62.5/125 μm should not be used to connect 50/125 μm fibers. See Figure 15.79 for an illustration of a cross-connect and an interconnect.

Twisted-pair

Cross-connects are typically installed with twisted-pair wires, called jumpers, that are available in 1-pair, 2-pair, 3-pair, and 4-pair. In some cases, the pairs of wires are encased in an overall sheath. The process of terminating a 1-pair jumper is the same as that for terminating 4-pair jumpers with the exception that there are three more pairs to terminate.

Step	Terminating Jumpers
1	Insert the jumper wire into the connecting block slots containing the assigned cable pair; apply light finger pressure to push it in as far as possible.
2	Use the impact tool to seat the wires.
3	Route the wire to the edge of the connector block; form the wire into a loop to create some slack.
4	Route the jumper wire to a corresponding connector block in the other termination field. Form a loop, then place the wires in the connecting block slots to secure the wires to the assigned cable pair on that field.
5	Terminate the wires with the impact tool.

Interconnects are typically made with patch cords, although some cross-connects are also made with patch cords. Patch cords are typically made with the same plugs on both ends of the cordage.

Step	Making Interconnects
1	Select the length of patch cord needed. Ensure it has the appropriate number of pairs, and that the position of the pairs in the plug delivers the signal over the correct cable pairs.

> **NOTES:** Using a 4-pair straight-through patch cord generally ensures that signals are delivered over the correct cable pairs. Special cords known as "cross-over" patch cords are used when connecting two switches such as two 100Base-T switches.
>
> Per ANSI/TIA/EIA-570-A, to prevent damage of the eight-position (8P8J [RJ-45]) outlet/connector when mating to a six-position plug, the tab width for a six-position plug shall be 6.0 mm (0.236 in) to 6.2 mm (0.243 in).

> **CAUTION:** All 6-position plugs do not meet the requirements of ANSI/TIA/EIA-570-A standard, especially old cords imported from outside the United States. Inserting a noncompliant plug into an 8-position modular jack can spread the outer pins of the jack, causing it to become ineffective for some systems.

Step	Making Interconnects
2	Plug one end of the cord into the connector on one connector block.
3	Route the cordage to the corresponding connector block or equipment and plug it into the connector.
4	Take the excess cordage and route it into the jumper/patch cord pathway to avoid congestion of the pathway.

Coaxial

Cross-connects and interconnects are made with coaxial patch cords. In most cases, the patch cord is made with the same connectors on each end of the cord (e.g., F connectors). However, there are instances when a coaxial patch cord may have different connectors on each end of the cord.

Step	Coaxial Patch Cords with Cross-Connects and Interconnects
1	Select the length of patch cord with the appropriate connectors on each end.
2	Plug or screw one end of the cord into the connector on one connector block.
3	Route the cordage to the corresponding connector block or equipment and plug it into or screw it on the connector.
4	Take the excess cordage and route it into the jumper/patch cord pathway to avoid congestion of the pathway.

Optical fiber

Cross-connects and interconnects are made with optical fiber patch cords. In most cases, the patch cord is made with the same connectors on each end of the cord (e.g., SC connectors). However, there are instances when a patch cord may have different connectors on each end of the cord.

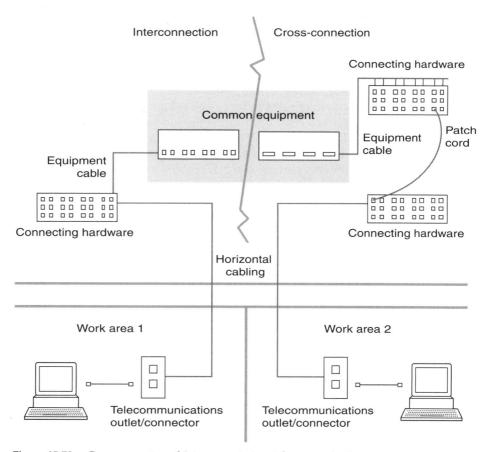

Figure 15.79   Cross-connects and interconnects in a telecommunications room.

NOTES: The horizontal cabling to work area 1 is interconnected to the common equipment.

The horizontal cabling to work area 2 is cross-connected to the common equipment.

Step	Optical Fiber Patch Cords with Cross-Connects and Interconnects
1	Select the length of patch cord with the appropriate connectors on each end.
2	Plug one end of the cord into the connector on one connector block.
3	Route the cordage to the corresponding connector block or equipment and plug it into the connector.
4	Take the excess cordage and route it into the jumper/patch cord pathway to avoid congestion of the pathway.

Powering Up and Performing an Operational Test

If the residential cabling installer has installed consumer products, these applications may need to be tested prior to the completion of the job. Prior to any operational test, the cabling should be tested as described in Chapter 16: Testing and Troubleshooting. In some cases, this testing can mitigate any disputes over the cable being suitable for the applications being deployed.

For telephone service, the NID cables are connected to the incoming service lines and a handheld telephone field test instrument can be plugged into every outlet location that is connected within the DD to the telephone services. If dial tone is present, the cable run is working properly. If multiple telephone lines are run into the home, the residential cabling installer should call the number that is to be made available at each port from the corresponding telecommunications outlet/connector to ensure that the correct number is available.

For data networks, each personal computer and networked peripheral device should be plugged into the port on the telecommunications outlet/connector and powered up. The residential cabling installer should be able to log into the home's LAN. The residential cabling installer may also try to log on to a dial-up or broadband Internet connection to ensure proper functionality.

Security systems, if monitored, are tested when establishing the outside connection to the monitoring service. The residential cabling installer may also manually trip each device to ensure the monitoring component is working properly.

Sound systems can be tested by attaching the electronic equipment (e.g., amplifiers, preamplifiers, receivers, etc.) and driving music through each room in the home, varying the volume if volume controls are present and controlling the source via infrared controls, if installed.

Control systems for lighting and other household functions can be tested after programming is completed. Since many functions are time sequenced, temporary settings may be put in place to test the system's communication and control success.

Completing Documentation

Labeling telecommunications outlet/connectors depends on the preference of the homeowner or builder. Every customer does not want writing to be visible on the face of the outlet.

Labeling at the DD is usually attached to the inside cover of the enclosure.

A separate listing of cables, locations, and other system features will be provided to the homeowner or builder, along with company contact information if maintenance or additions are required. This documentation includes a listing of cables by number, or a floor plan showing the exact locations and numbering sequence for the cabling.

Documenting the Distribution Device (DD)

It is important to document the DD used in residential cabling.

At a minimum, use a simple form to document the faceplates and their associated contents (see Example 15.1).

The preferred documentation should include a computer-aided design (CAD) drawing of the residence with an accompanying legend, designating the cabling (see Figure 15.80).

The CAD approach is preferred; however, the use of the spreadsheet-based documentation is more cost-effective for most applications.

Note that the completed documentation should be either laminated to the inside cover of the DD on a 216 mm (8.5 in) × 279 mm (11 in) adhesive-backed label, or placed within a transparent pouch inside the DD. Other documentation related to the DD may be included in the pouch.

When labeling telecommunications cables, identify cables at each end with a permanent tab or label. Use the same alphanumeric identifiers at both ends of the cable. In systems that are simple, the numbering scheme can be a simple number sequence. In complicated systems, the labeling may indicate the type, function, and terminating position.

It is helpful to color code the cables by function (e.g., local area network/voice/fire alarm/environmental control) with a high-quality colored vinyl tape at each termination and whenever the cables are accessible. An alternative is the use of different color-jacketed cable for voice/data, etc.

Connecting hardware

Connecting hardware items (e.g., cross-connect fields and workstation outlets) require a unique, alphanumeric identification (e.g., three-level scheme). The first level is the termination field or patch panel. Use color coding or other labeling to uniquely identify each termination field (e.g., voice and data) on a common mechanical assembly. The second level is the terminal block within a given field or patch panel, which could be a row of insulation displacement connectors, optical fiber connectors, or modular jacks. The third level defines the individual position within a given terminal block or patch panel.

For further information on this subject, consult ANSI/EIA/TIA-606-A, *Administration Standard for the Telecommunications Infrastructure of Commercial Buildings*.

Residential Structured Cabling System Documentation Form

Development: <u>Highlands Estates</u>   Builder: <u>ABC Builders</u>   Date: <u>10/24/01</u>

Division: <u>3</u>   Building: <u> </u>   Lot/Unit: <u>14</u>   Address: <u>12345 Sixth Avenue</u>

Feeder Locations:

Voice NID location: <u>East side garage</u>   ☐ CAT 3   ☒ CAT 5   ☐ CAT 5e

Coax NID location: <u>East side garage</u>   ☒ Series 6   ☐ Series 11

Fiber NID location: <u>N/A</u>   ☐ Vault   ☐ Splice case

Distribution Device (DD) Location:

☐ Guest closet ☐ Mstr closet ☒ BR 2 closet ☐ BR 3 closet ☐ Garage ☐ <u>   </u>

AC power in DD? ☒ Yes   ☐ No

Cable Labels (per Faceplate):

Faceplate Location	Voice 4x C3	Voice 4x C5	Voice 4x C5e	Data 4x C3	Data 4x C5	Data 4x C5e	Coax Ser 6	Coax Ser 11	Other (Type and quantity)
Living-A		1			1		2		
Living-B									
Kitchen-A		1					2		
Kitchen-B									
Mstr BR-A		1			1		2		
Mstr BR-B		1			1		2		
BR 2-A		1			1		2		
BR 2-B									
BR 3-A		1			1		2		
BR 3-B									
BR 4-A									
BR 4-B									
Den-A		1			1		2		(1) 2c/18 for future electronics
Den-B									
Family-A		1			1		4		
Family-B									
Exterior									
Video Cam-A									
Video Cam-B									
Antenna									
DSS-A		1			1		2		
DSS-B									
Demarc		1			1		1		(1) 2-strand multimode optical fiber
Security									

Notes and comments: <u>   </u>

Installed by: <u>AAA Cabling Company</u>   Date: <u>10/24/01</u>   Phone: <u>(123) 456-7890</u>

Note to Installer: Leave a copy of this documentation inside the DD.

Example 15.1 Completed documentation form.

Performing Cleanup

All scrap materials should be removed from the residence upon completing the trim out. Since this is often one of the final steps in new construction prior to closing, this cleanup is critical. A small, battery-operated vacuum may be needed to remove the Drywall dust that often settles to the floor during the

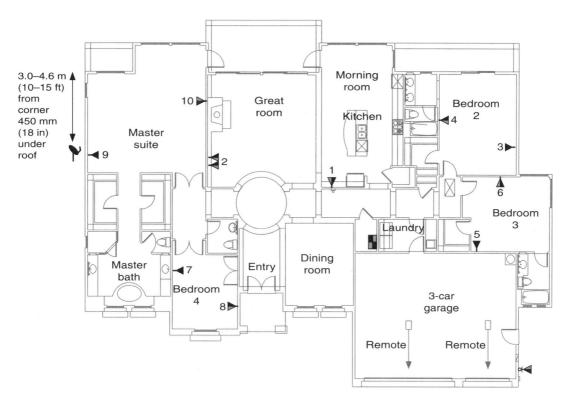

3.0–4.6 m (10–15 ft) from corner
450 mm (18 in) under roof

Figure 15.80 Structured cabling plan.

termination of jacks and connectors and the attachment of the telecommunications outlet/connector.

The cover of the DD should be attached or closed. All exposed cable runs should be visually inspected to ensure cleanliness and safety.

Providing Orientation

If the customer is a homeowner (and particularly in the case of a retrofit application), it is advisable to review the functions of the system upon completion of the trim out.

Provide the customer with as-built system documentation and certified test results. If the system is covered under an extended vendor-sponsored performance warranty, review the warranty in detail with the customer, including the products covered and the proper steps to be taken in case of a system failure. Review basic troubleshooting techniques with the customer. These may include, but are not limited to, those steps outlined in Chapter 16: Testing and Troubleshooting.

Walk through the home with the homeowner to ensure the residential cabling installation is aesthetically acceptable. Turn on speakers, security panels, and other functional subsystems, if available. Review any special issues the homeowner or builder may face in running these systems.

Chapter 6

Testing and Troubleshooting

Testing Residential Structured Cabling Systems (RSCS)

Overview

In this chapter, requirements, equipment, and procedures for testing RSCSs are described. The second part of the chapter explains troubleshooting and procedures needed to identify, isolate, and analyze the kinds of faults and problems commonly encountered in residential telecommunications cabling.

Testing the RSCS is a must due to residential cabling installation irregularities that may go unseen. A Category 5e cable may be unknowingly terminated with a Category 5 outlet/connector or the connection of the wire to the insulation displacement connector (IDC) may not be complete due to a worn tool. Additionally, cables may be kinked and twisted to the extent that the performance is degraded or the conductors nicked while the walls are being enclosed with sheetrock.

All cabling should be thoroughly tested by the residential cabling installer. It is recommended that all cables be tested after they are placed in the structure, prior to enclosing the walls and ceilings. When performing a residential cabling installation or even a rearrangement of equipment in an RSCS, test the system for compliance with ANSI/TIA/EIA-570-A, *Residential Telecommunications Cabling Standard.*

If testing identifies any problems or deficiencies, locate and correct them.

Additional information on field testing can be found in the BICSI *Telecommunications Distribution Methods Manual (TDMM)*, 9th edition, Chapter 10: Field Testing, and the *Telecommunications Cabling Installation Manual (TCIM)*, 3rd edition, Chapter 8: Testing Cable.

Twisted-pair circuits must be verified for proper wire mapping and certification to the installed category. Test results should include all test data results and a statement of compliance—"pass" or "fail."

All coaxial cable must be certified for continuity and shorts. The residential cabling installer, signifying that no tested circuits have shorts, and that every cable has continuity, should submit a statement of compliance, indicating pass or fail. Reference Society of Cable Telecommunications Engineers (SCTE)-ISP-100, *Specification for Trunk, Feeder, and Distribution Coax Cable* and SCTE-ISP-404, *"F" Connector (Male Indoor) Installation and Performance* for testing requirements.

All optical fiber cabling must be certified for length, polarity, and attenuation. The residential cabling installer should include all test data results and a statement of compliance.

Testing requirements

The requirements for testing RSCSs are outlined in Annex B, "Field Test Requirements," of ANSI/TIA/EIA-570-A. Although included in an annex, the requirements are considered to be part of the standard.

The standard specifies testing procedures for three cabling media, as follows:

- 100 Ω, twisted-pair.
- 75 Ω coaxial.
- Optical fiber.

In addition, the standard establishes two administrative requirements for testing, namely, that all:

- Field test instruments be calibrated and maintained in accordance with the manufacturer's specifications.
- Test data be recorded and provided to the building owner or agent.

Twisted-pair cabling testing

ANSI/TIA/EIA-570-A describes the following aspects of twisted-pair cabling testing in homes:

- Properties of field test instruments
- Performance parameters to be measured for twisted-pair cabling
- Basic test configurations for twisted-pair cabling
- Performance parameters for backbone and other twisted-pair testing

Properties of field test instruments. A wide variety of copper cable field test instruments are on the market, but only a fraction of these perform all of the tests described below at the required level of category certification and accuracy. These devices are called certification field test instruments.

The two most important features of these devices are that they:

- Perform the required tests needed to qualify twisted-pair cabling to the specified category.

- Operate to the level of accuracy required by the residential cabling standard.

Performance parameters for twisted-pair cabling.   The cabling that runs from the distribution device (DD) to the telecommunications outlet/connectors in living spaces must be field tested for the following performance parameters:

- Wire map
- Length
- Insertion loss (formerly called attenuation)
- Near-end crosstalk (NEXT) loss

Additional field tests are required for Category 5e twisted-pair cabling, as follows:

- Propagation delay
- Propagation delay skew
- Equal level far-end crosstalk (ELFEXT) loss
- Power sum near-end crosstalk (PSNEXT) loss
- Power sum equal level far-end crosstalk (PSELFEXT) loss
- Return loss

Continuity tests are required for audio, security, and home automation cabling.

> **NOTE:** Detailed instructions for performing these tests are provided later in the chapter.

Test configurations for twisted-pair outlet cabling.   As is the case for horizontal copper cable testing in commercial environments, twisted-pair outlet cabling in residential applications is tested as both a link and as a channel. The link includes the components and cable between the DD and the telecommunications outlet/connector, consolidation point (CP), or transition point (TP).

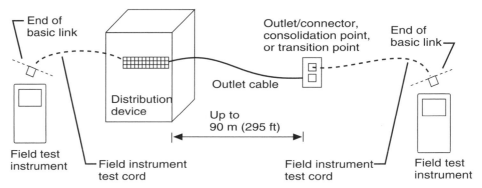

Figure 16.1   Basic link test configuration.

NOTE: The test cords may be up to 2 m (6.6 ft) in length for the basic link test.

The channel includes all cable and components in the link, with the addition of the patch cord in the DD and the equipment cord at the telecommunications outlet/connector to the customer premises equipment (CPE).

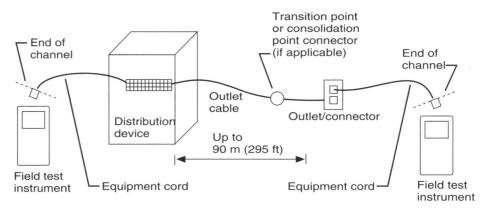

Figure 16.2   Channel test configuration.

Performance parameters for backbone and other twisted-pair cabling.   A wire map test must be performed for all wire pairs in twisted-pair backbone cabling, as well as for all other 4-pair twisted-pair cabling. The wire map test to be performed and recorded for all cable segments includes:

- Continuity to the remote end of the pair.
- Crossed pairs.

- Reversed pairs.
- Split pairs.
- Any other miswiring.

Coaxial cabling testing

The minimum test requirements for 75 Ω coaxial cable include:

- Continuity for the center conductor and shield.
- Attenuation.
- Length.
- Other tests, as required by local, state, and federal authorities.

Optical fiber cabling testing

Optical fiber cabling may be used in single-family residential cabling, but is prevalent in multi-tenant buildings.

Properties of optical fiber field test instruments.   Field test instruments for multimode optical fiber must meet the requirements specified in ANSI/TIA/EIA-526-14-A, *Optical Power Loss Measurements of Installed Multimode Fiber Cable Plant*. Field test instruments for singlemode optical fiber must meet the requirements specified in ANSI/EIA/TIA-526-7, *Measurement of Optical Power Loss of Installed Singlemode Fiber Cable Plant*.

Performance parameters.   The field tests required for residential optical fiber cabling include:

- Length.
- Polarity.
- Attenuation.

Field test instruments

Each field test instrument is designed to perform a specific function or a range of functions required for the testing and certification of a specific cable type. Field test instruments vary greatly in capability and price. It is recommended that the most cost-effective unit capable of performing the specific tests required be selected.

Multimeter.   The multimeter is one of the most basic and most widely used field test instruments available for metallic cabling. Both analog models and digital models are available. A multimeter can be used to measure voltage, current, and resistance in metallic conductors. With the use of a shorting device on one end of a pair of conductors, continuity can be tested. When testing cable, it is possible to calculate the length of the cable by determining the actual loop resistance. In normal practice, tests are performed on cabling that is disconnected from any active equipment, so voltage on the cable would be an unexpected condition, and need to be corrected before proper testing can commence.

Inductive amplifier/tone generator. Also known as a toner or cable tracer, this field test device provides the ability to identify a specific cable pair by generating a tone on one end of a pair of conductors, with an inductive amplifier (wand) identifying it at the opposite end. Some induction amplifiers also provide the ability to trace the pair throughout its entire length by this same method. This field test device is primarily used for cable identification and troubleshooting. Most units are now a combination of tone generator and continuity field test instrument.

> **NOTE:** A cable passing a tone does not guarantee that there is continuity from end-to-end. Residential cabling installers must use the continuity test to ensure end-to-end continuity.

Wire map field test instrument. Wire map field test instruments are low-cost cable field test instruments that usually test for opens, shorts, crossed pairs, and miswiring in twisted-pair cabling. Some field test instruments in this category will also test for split pairs. These units are good for quick, basic tests, but usually lack the sophisticated diagnostic capabilities of more expensive field test instruments. Most field test instruments in this category are designed exclusively for unshielded twisted-pair (UTP) and screened twisted-pair (ScTP) cables.

There are two basic types of wire map field test instruments. Instruments with battery-powered active remote modules are generally capable of identifying complex wire map problems and isolating single conductor faults. Instruments with passive remote modules use loop-back techniques and displays a single conductor break as a pair break.

Cable-end locator kit. Sometimes called an office locator kit, this is a set of numbered 8-pin modular plugs that can be identified by the cable test equipment. The standard practice is to insert the plugs into a telecommunications outlet/connector and then search with the field test instrument at the patch panel or termination block until it finds the plug at the opposite end of the cable.

Certification field test instrument. Certification field test instruments are used to verify that an RSCS meets the transmission performance requirements specified in the applicable standard. These units will test a cabling system up to at least 100 megahertz (MHz), and in the autotest mode, include length, attenuation, wire map, and NEXT loss.

When the field test instrument is operated in autotest mode, it compares the actual measured values with required values for the appropriate category of cabling and displays pass or fail for the entire series of tests. The field test instrument displays the actual tested values for each parameter. These field test instruments are capable of other measurements, which may include impedance, capacitance, resistance, delay, delay skew, ELFEXT loss, and attenuation-to-crosstalk ratio (ACR) loss calculations.

Many certification field test instruments are configured to test continuity and length of coaxial cable. Several have the ability, with an add-on module, to perform an optical power meter test for attenuation of optical fiber cable. Certification field test instruments can store test data and export it to a database or output it to a printer. Most certification field test instruments also have the ability to be controlled by a PC and downloaded and stored directly to the PC hard drive or a floppy disk.

Certification field test instruments should test a twisted-pair cabling system to at least 100 MHz and may measure and record the following parameters:

- Delay skew
- Equal level far-end crosstalk (ELFEXT) loss
- Insertion loss (attenuation)
- Length
- Near-end crosstalk (NEXT) loss
- Power sum equal level far-end crosstalk (PSELFEXT) loss
- Power sum near-end crosstalk (PSNEXT) loss
- Propagation delay
- Return loss
- Wire map

The autotest feature compares the actual measured values with required values for the appropriate category of cabling performance and displays a pass or fail message for the entire battery of tests. The equipment will also display pass or fail messages and the actual measured values for each test, individually. These units may be capable of other tests, including time domain diagnostics, tone generation, telecommunications outlet/connector identification, impedance, capacitance, and loop resistance. Certification field test instruments can store data and export it to a database or output it to a printer.

Cable tracer. A cable tracer (or toner) is an accessory for certain cable field test instruments and certification field test instruments. It consists of a signal generator and a signal receiver. The generator is connected to one end of the cable, and the receiver is then used to follow the path of the cable through the wall, floor, or ceiling.

Time Domain Reflectometer (TDR).   The TDR has the ability to locate cable defects, breaks, splices, and connectors and offer loss values for each occurrence. In addition, the TDR is used to measure the electrical length of a cable. Developed originally for use with coaxial networks, it is also an excellent troubleshooting tool for use with twisted-pair cabling. The measurement of the cabling is accomplished by transmitting a high-rise-time pulse into the cabling and then analyzing the shape and delay time of the reflections caused by impedance mismatches.

Mismatches caused by kinks or splices are displayed as well as a large mismatch at the end of the cabling. The reflections are displayed either on a screen or in a printout. Certification tools and some wire map field test instruments typically measure length via TDR methods.

Optical fiber flashlight.   This is also called an optical fiber light-emitting diode (LED) or visible light source. It is used to test and troubleshoot continuity in optical fibers. By connecting the optical fiber flashlight to one end of a fiber and then viewing the other end, the residential cabling installer can determine continuity if the light is seen at the other end.

> **NOTE:** Never look into optical fiber strands unless personally confirmed that it is disconnected from potential laser light sources. Laser light sources may cause eye damage.

Infrared (IR) conversion card.   An IR conversion card allows the residential cabling installer to visually detect an IR signal on optical fiber cabling when that signal is directed at the card's phosphorus material.

Low-intensity laser.   A low-intensity laser (hot red light) operates in the visible light range. It is used to identify individual fibers within a cable by sending a red light down the fiber. When used as a troubleshooting tool, the fiber will glow red at the point of a break. Test instruments that pulse on/off are easier to visualize when looking for a break. These tools can also be used to detect a damaged fiber ferrule. Normally red laser light should appear as a sharp bright dot in the center face of the end of the white ferrule. If the ferrule is cracked or damaged, the entire ferrule will glow red during this test.

Light source and optical power meter.   A light source and optical power meter are required to perform loss (attenuation) tests on optical fiber. Some instruments also include the ability to measure the length of a fiber.

Optical Time Domain Reflectometer (OTDR).   The OTDR has the ability to locate and test all cable defects, splices, and connectors and give loss values for each occurrence. The OTDR is also used to measure the optical length of a fiber. These instruments vary in complexity from the complete stand-alone unit to small units that may require the use of a laptop computer.

Telephone field test instrument.   A telephone field test instrument (butt set) is used to test voice circuits and to perform the following functions:

- Simulate the user's telephone equipment
- Identify circuits
- Circuit diagnostics and troubleshooting

Test adapters, leads, and cables.   The residential cabling installer must have proper adapters available to connect test equipment to the cable under test.

Field test equipment selection

There are several factors that must be considered when choosing field test instruments.

Basic test instruments include such items as the multimeter, inductive amplifier, wire map field test instruments, and fiber flashlight. Each of these represents a comparatively small investment. The purchase of field test equipment (e.g., a TDR, certification field test instrument, light source and optical power meter, OTDR) involves higher investment items and requires more research prior to purchasing.

The first question to be answered is what types of media will be tested. Will it be copper or fiber? If copper, which of the several types will be primary? If fiber, is it multimode or singlemode fiber? Is the primary usage for new installations or for troubleshooting? These are some of the questions to ask before making a major purchase.

Copper media.   In addition to the basic field test instruments listed above, probably the most versatile is the certification field test instrument. Certification field test instruments have the ability to accomplish all required tests at one time by using an autotest function. Most certification field test instruments have the ability to store results; the most sophisticated can store data on removable memory cards.

In addition to the certification tests, the instrument may also be used for troubleshooting. Tests such as time domain analysis (NEXT loss or return loss vs. length), impedance, capacitance, loop resistance, and noise are available on some units. When NEXT or return loss faults are detected, time domain tools will show the location and magnitude of such faults. These tools are invaluable in troubleshooting.

If active circuits are to be tested, the field test instrument will require the ability to monitor a data circuit and provide real-time analysis. Other specialized products (e.g., handheld protocol analysis tools) are available for this purpose.

Make sure the certification field test instrument chosen can do all of the required tests easily, quickly, and accurately. Controls should be easy to operate and easily accessible without going through layers of menus. Remember that each added function has a cost.

Optical fiber media.   The four primary types of field test equipment for use with optical fiber media include:

- Optical fiber flashlight.
- Visual fault locator (VFL).
- Light source and optical power meter.
- Optical time domain reflectometer (OTDR).

These test instruments are available in many different configurations and price ranges. Advantages and disadvantages of some test equipment are described in the following list.

- Optical fiber flashlight—A small pocket flashlight equipped with a fitting that will mate with a fiber connector. The flashlight injects light into the core of the fiber and can be seen at the other end on a good circuit. It also works well for fiber identification.

  Flashlight advantages:
  - Inexpensive
  - Easy to use
  - Minimum training required
  - No calibration or referencing is required

  Flashlight disadvantages:
  - Short range
  - No quantitative measurement
  - Not useful in locating broken fibers
  - Cannot be used on working circuits

- Visual fault locator (VFL)—A device containing a laser diode as the light source. The wavelength of the light from the laser diode is usually 660 nanometers (nm), which is in the visible light range. The connection method is the same as for the flashlight and visible LED sources.

  VFL advantages:
  - Inexpensive
  - Easy to use
  - Minimum training required
  - Useful in locating broken fibers or those points along a fiber having high loss

- No calibration or referencing is required
- Medium to long range

VFL disadvantages:

- No quantitative measurement
- Cannot be used on working circuits

- Light source and optical power meter—A multimode light source has a LED that provides an output at either 850 nm or 1300 nm, or at both wavelengths. A singlemode light source has a laser diode providing an output at either 1310 nm or 1550 nm, or at both wavelengths. Optical power meters contain a photo detector sensitive to a wide range of wavelengths; a switch selects the chosen wavelength. An optical power meter is usually capable of measuring both multimode and singlemode wavelengths.

Light source and optical power meter advantages:

- Easy to use
- Some training required
- Quantitative measurements (the most accurate of any method)
- Some meters offer internal data storage and download capabilities

Light source and optical power meter disadvantages:

- May require two residential cabling technicians, depending on the number of tests to be completed
- Does not give the location of high loss locations
- Requires access to both ends of the fiber, and both ends must have connectors installed
- Cannot be used on a working circuit

- Optical time domain reflectometer (OTDR)—An optical fiber test instrument that uses a device to send out a light pulse on a fiber and then measures the time and amplitude of the reflected signal. The results are displayed as a trace on a screen. It is used as a troubleshooting and certification testing measurement tool.

OTDR advantages:

- Will provide approximate link attenuation measurements, but with less accuracy than a light source and optical power meter
- Provides a visual trace of loss (e.g., faults, splices, connectors, segments of fiber)
- Provides visual trace of the distance measurements to events (e.g., faults, splices, connectors) and to the end of the fiber
- Has the capability to print and store all displays
- May be accessed from one end of a fiber

OTDR disadvantages:

– Expensive

– Requires more training to properly operate and correctly interpret results over that of a source and optical power meter

– Should not be used to make accurate link attenuation measurements

Testing cabling

Characterizing cabling provides useful data to identify the ability of the cabling system to support existing and future applications. The certification of the transmission performance of the cabling system through field measurement results in a more professional handoff from the residential cabling installation contractor to the customer. This should improve the overall effectiveness of providing premises networking solutions, as well as establishing a baseline for future troubleshooting, resulting in higher customer satisfaction.

This section addresses the field testing of 100 Ω twisted-pair, coaxial, optical fiber, audio, security, and home automation cabling.

Testing the cabling infrastructure is most effectively performed using an organized, systematic method to verify that the installation has been completed in accordance with all of the terms and conditions of the contract and industry standards. This method is made up of three distinct phases. They are:

- Phase 1—Visual inspection.
- Phase 2—Testing with the various test equipment as required for each type of cable.
- Phase 3—Documentation.

The visual inspection (Phase 1) should include an inspection of all cabling, pathways and spaces (where possible), DDs, and multi-dwelling unit telecommunications rooms (MDU-TRs). Items to be inspected include but are not limited to the following:

- Infrastructure
  – Cable placement
  – Cable termination
- Grounding and bonding
- Equipment and patch cords
- Proper labeling of all components

Once the visual verification has been completed and all discrepancies corrected, it is now possible to move to Phase 2, testing with the various test equipment as required for each cabling type.

Phase 3, documentation, is probably the most prone to neglect. Documentation should be in the form requested by the customer or, if not specified, in accordance with ANSI/TIA/EIA-606, *Administrative Standard for the Telecommunications Infrastructure of Commercial Buildings.*

Testing copper cable

This section covers the testing of the copper media used for outlet cabling.

Continuity testing.   This is the most basic test to establish proper verification of the installed cabling. It is also referred to as a wire map test when utilizing handheld field test instruments. At a minimum, it searches for:

- Open circuits—Indicates incomplete terminations, faulty connectors, or broken cables.
- Short circuits—Indicates faulty connectors or crushed cables.
- Improper termination.
- Drain wire or shield continuity when testing ScTP cable.

A multimeter can be used to measure direct current (dc) loop resistance of twisted-pair and ScTP; however, this is impractical because of the time involved in checking combinations. A multimeter may be better used for testing audio and security cabling. Wire map verification field test instruments, also known as wire mappers, are more practical because they quickly identify open or short circuits and verify wiring positions for multiple pairs terminated in a connector. Wire mappers do not test cabling performance, but provide the functions of verifying continuity. Some units additionally indicate cable length, and may provide office identification, toner, and active equipment tests.

Step	Using a Wire Map Verification Field Test Instrument
1	Ensure active equipment is disconnected from both ends of the cable being tested.
2	Attach wire map verification field test instrument to one end of cabling. (This can also include patch cords.) Attach the wire map verification field test instrument's remote device to the opposite end.
3	If the wire map verification field test instrument is unable to locate the distant end, determine the problem.
4	Diagnose and repair faulty cabling.
	NOTE: In the case of a break, use TDR (if available) to isolate the location of the break to speed repair time. In most cases, this will occur at a connection point.
5	Record the successful test results for the customer's records.

When using a certification field test instrument for twisted-pair cable, follow the manufacturer's guidelines for operation of the specific test equipment. In general, the steps include:

Step	Using a Certification Field Test Instrument
1	Power the field test instrument.
2	Calibrate the field test instrument.
3	Configure the field test instrument to test the circuit.
4	Attach the field test instrument to the circuit being tested.
5	Run autotest function.
6	If the field test equipment indicates "pass," save the results.
7	Record the successful test results for the customer's records.
8	If the field test equipment indicates "fail," troubleshoot, diagnose the problem, and repair as needed.
9	Retest until passing results are achieved.
10	Repeat starting at Step 3 for all other circuits to be tested.

Troubleshooting a failure.   In the event that a permanent link or channel fails to meet specifications during testing, it is important to immediately stop and understand what may be the cause of the failure.

> **NOTE:** A failure is a measurement failure and could be caused by the cabling, components, test equipment, or improper installation practices.

Follow the procedure below to investigate a failure:

- Follow the troubleshooting guidelines in Table 16.1 to identify various problems.
- Identify the trouble and take corrective action to repair the cabling.
- Retest to ensure that the corrective action has worked effectively.

TABLE 16.1   Troubleshooting

Problem	Solution
Cannot turn on field test instrument	▪ Recharge or change batteries.
Field test instrument cannot perform or fails remote calibration	▪ Be sure both units are turned on and batteries are charged. ▪ Replace field test instrument cord.

continued on next page

TABLE 16.1   Troubleshooting (continued)

Problem	Solution
Field test instrument set for incorrect cable type	▪ Reset field test instrument parameters for proper cable type and calibrate nominal velocity of propagation (NVP).
Field test instrument set for incorrect link configuration.	▪ Reset field test instrument to basic link or channel, as required.
Link channel fails autotest	Check area(s) of failure:
a. Fails wire map	▪ Check that field test instrument is set to match existing pin wiring. Examine both connections for open, split, or crossed pairs.
b. Fails length	▪ Check and recalibrate NVP with known cable. ▪ Check total length of patch/equipment cords.
c. Fails attenuation	▪ Check all cables in link channel for suitable category rating.
d. Fails NEXT loss	▪ Check all components for suitable category rating. ▪ Check quality of all wire terminations.
Field test instrument will not run an autotest.	▪ Check settings on control knob or menu. ▪ Check to see if unit has failed calibration. ▪ Check field test instrument connections at both ends of link.
Field test instrument will not store an autotest result.	▪ Check that a unique name is selected for test results. ▪ Check amount of free memory available.
Field test instrument will not print stored results.	▪ Check that serial interface to autotest printer and field test instrument are set to same parameters/emulation. ▪ Check that results are selected for printout.

More information on troubleshooting is provided in the second section of this chapter.

Testing coaxial cable

Unlike the testing of balanced cabling, when testing unbalanced cabling (e.g., coaxial cable), the residential cabling installer is primarily concerned with two parameters: continuity and length. Continuity and length may be tested using a multimeter and a shorting plug. The certification field test instruments also have this capability, which will simplify the process. Some verification test instruments also include this functionality.

If performing the length test with a TDR or other field test instrument, perform the test in accordance with the field test instrument manufacturer's instructions. For the reading to be accurate, the residential cabling installer must either refer to the cable manufacturer's specifications for NVP, or determine the NVP of the cable using the field test instrument and the field test instrument manufacturer's instructions.

The continuity test for 75 Ω coaxial cable requires the use of a multimeter and a shorting plug. This procedure also requires the removal of amplifiers and passive devices, so the system may have to be tested one section at a time. Observe the following procedures if using a multimeter to test the cabling.

Step	Using a Multimeter to Test Cabling
1	Set the multimeter for alternating current (ac) and read, then for direct current (dc) volts and read, while checking the cabling for unwanted signals.
2	Remove any equipment that may have been generating unwanted signals.
3	Set the multimeter to the ohm position.
4	Calibrate (zero out) the meter.
5	With the multimeter, ensure that there is no continuity with the far end open.
6	Install a shorting plug on one end of the cabling.
7	Using the multimeter, read the dc loop resistance of the cable segment.
8	Calculate the dc loop resistance from the manufacturer's data for the cabling.
9	Compare the calculated value with the reading from the multimeter.
10	Diagnose and repair faulty cabling.
11	Maintain a record of the test result.

Using the dc loop resistance reading previously taken, calculate the length of the segment as follows:

Step	Calculating the Length of the Segment
1	Since the resistance values of the core and the shield are not the same, using the cable manufacturer's data, add the nominal dc resistance of the core to the nominal dc resistance of the shield (resistors in series).
2	Since this specification is given in either resistance per 1 km or 1000 ft segment, divide this number by 1000 to provide the loop resistance per meter (foot).
3	Divide the reading received during the continuity test by the dc resistance per meter (foot) derived from the manufacturer's data. The result is the length of the cable in meters (feet).

Testing optical fiber cable

Testing of an optical fiber link is conducted for the following reasons:

- Verification testing
- Certification testing
- Troubleshooting

The following standards are applicable to the field testing of premises optical fiber cabling:

- ANSI/TIA/EIA-455-171-A, *Attenuation by Substitution Measurement for Short Length Multimode and Graded Index and Singlemode Optical Fiber Cable Assemblies*
- ANSI/TIA/EIA-455-61, *Measurement of Fiber or Cable Attenuation*
- ANSI/TIA/EIA-526-7, *Measurement of Optical Power Loss of Installed Singlemode Fiber Cable Plant*
- ANSI/TIA/EIA-526-14-A, *Optical Power Loss Measurements of Installed Multimode Fiber Cable Plant*

An optical fiber link is a path consisting of one fiber, which has a connector on each end. Optical fiber links normally begin and end in administration housings called fiber distribution units (FDUs). Links are connected to electronic devices (e.g., hubs, multiplexers, and routers with jumpers). Multiple links can be connected in the FDUs, using jumpers to create circuits.

An optical fiber link or circuit should be tested before it is put into service and at other times, such as during troubleshooting. A light source and optical power meter or an OTDR can be used to measure the attenuation of the optical fiber link or circuit.

The following are the benefits of verification and certification testing:

- Establishes that the total attenuation of all passive components in the link is within the design parameters (loss budget)
- Establishes that the passive components were installed properly
- Minimizes downtime because of maintenance on improperly installed passive components
- Establishes accountability when circuits are configured with multiple links connected together and installed by more than one vendor
- Provides a benchmark for comparing future measurements

Verification and certification testing of intrabuilding fiber is most often performed with a light source and optical power meter. Two people may be required to perform this test. They may need some form of communications (e.g., a fiber talk set, portable two-way radios, or telephones). The light source and optical power meter are capable of more accurate insertion loss attenuation measurement than an OTDR.

Acceptance testing of a new installation, using a light source and optical power meter, should be performed on individual links and not on a circuit that has one or more jumpers. This is because the circuit may be reconfigured in the future and the attenuation value for each link will be needed by the customer or technician to properly design the new configuration. The results of each measurement should be recorded on the optical fiber link attenuation record.

Light source and optical power meter

Performing optical fiber cable testing is the process of certifying that the optical fiber cable installation functions as expected. Testing includes a physical inspection of each cabling link, verification of the cabling labeling, and actual testing of the optical fiber's physical condition and attenuation to determine whether it meets specifications. Field testing of optical fiber cabling covers the following activities:

- Calibrating the test equipment to be used
- Setting the test equipment to measure a selected wavelength
- Disconnecting active electronic equipment
- Recording the test result values
- Calculating the attenuation
- Documenting the results

If a test fails, the cable link should be corrected so that it can pass.

NOTE: Many field test instruments calculate the attenuation and document results.

Test configuration methods.   The residential cabling installer must be aware of both the expected test results and the manner in which the test should be conducted. ANSI/TIA/EIA-526-14-A outlines three methods to be used to configure the test equipment and the cable under test. The primary difference in the method is how the test jumpers are used in conjunction with the cable and testing equipment. Each method provides a slightly different result. It is important that the method to be used is identified prior to the start of testing and that the same method is used consistently throughout the installation.

These three methods differ in how the reference (the starting point) is established rather than the actual physical connection of the cable and test equipment. All three methods physically use a light source, optical power meter, and two jumpers: one to attach each test component to the cable under test. The testing methods differ in the number of jumpers that are connected together to establish the reference reading for the test equipment. Method A uses two jumpers, Method B uses one, and Method C uses three. The various configurations represent the following:

- Method A (the more traditional method of testing insertion loss)—Measures the actual cable loss plus the loss of one connection.
- Method B—Measures the actual cable loss plus the loss of two connections.
- Method C—Measures only the loss in the cable under test without including any connection losses.

Depending on the test method used and the manner in which the cable is physically joined, the final loss figures may be over or under the anticipated levels. For this reason, Method B is typically used for testing residential optical fiber cable.

> **NOTE:** Because optical fiber can be used for distances greater than most building applications (up to 60 km [37 mi] for singlemode), cable length is not an issue for most indoor installations.

The following steps are used for Method B.

Step	Testing Optical Fiber Cable—Light Source and Optical Power Meter
1	Disconnect active equipment. For retrofits, contact the equipment user prior to completing the retrofit to ensure that no customer operations are interrupted.
2	Obtain appropriate light source for singlemode (1310 nm or 1550 nm), multimode (850 nm or 1300 nm), and optical power meter.
3	Set source and meter to the proper wavelength. Calibrate the equipment before each test.
4	Obtain proper test jumpers and couplers. These should be part of the light source and optical power meter kit.
5	Connect a jumper (containing the same fiber size as the system fiber) to the optical source and the optical power meter.

- Turn the unit on.

- Record the reference power reading (P_{ref}), displayed in decibel milliwatt (dBm). This power level reference is the output power of the optical source coupled with the jumper.

- When using the single reference jumper method, zero the meter.

> **CAUTION:** Never disconnect or adjust the jumper connection at the optical source after recording the reference value. This may change the value and cause final test readings to be inaccurate.

| 6 | Using an adapter, insert a second jumper (Test Jumper 2) between the jumper used in Step 5 and the optical power meter. This jumper must have the same size fiber as the first jumper. |

- Use the following equation to ensure the attenuation added by the second jumper is not greater than 0.75 dB:

$$P_{ref} - P_{check} \pm 0.75 \text{ dB}$$

0.75 dB is commonly used. It may be replaced with the guaranteed maximum mated pair loss for the specific connector used.

- If the equation above is satisfied, continue to Step 7. If it is not satisfied, clean all connectors except the optical source connection point and repeat the procedure beginning with Step 6. If the result is still greater than 0.75 dB, replace Test Jumper 2 and repeat the procedure again, beginning with Step 6.

To ensure the connector and adapters are properly tested at each end of the system, always test the system with two jumpers.

Step	Testing Optical Fiber Cable—Light Source and Optical Power Meter
7	Leave the jumpers attached to the optical power source and optical power meter.

- Disconnect the two jumpers at the adapter.
- Attach the optical source/Test Jumper 1 to one end of the system fiber to be tested.
- Attach the optical power meter/Test Jumper 2 to the other end of the system fiber.
- Record the test power (P_{test}).
- Subtract the test power (P_{test}) from the reference power (P_{ref}), recorded in Step 5, to determine the end-to-end attenuation:

$$\text{Attenuation (dB)} = P_{ref} - P_{test}$$

| 8 | Document results. At the end of the process, record the test result information. |

NOTE: The information may need to be keystroke captured on a computer or written on some other permanent record.

Troubleshooting Residential Structured Cabling Systems (RSCS)

This section describes the methods, tools and equipment, and knowledge required to troubleshoot telecommunications cabling systems. It also includes specific steps for locating, diagnosing, and repairing problems in twisted-pair, coaxial, and optical fiber systems.

Most tasks in the troubleshooting process apply to all types of cabling systems. The residential cabling installer must identify, locate, and repair the problem and then test the repaired system to make sure it works correctly. The following sections describe the equipment and skills required to provide the foundation for efficient and effective troubleshooting.

Troubleshooting methods

Effective troubleshooting requires experience, knowledge, patience, and skill. The effective troubleshooter:

- Has good communications skills.
- Is able to gain useful information by asking questions in a manner that does not alienate the person being questioned.
- Has learned how to observe objects and events in order to extract information about the possible causes of problems.

Additionally, an effective troubleshooter knows how and where to locate information: how to get to the right people on vendor technical support lines, how to use online resources, and how to use manuals and technical documents. A good troubleshooter also understands the application of each of the available tools and keeps them in proper working order.

Troubleshooting is an investigative process. A troubleshooter looks for clues and follows leads. A troubleshooter must avoid making assumptions about the cause of problems—things are not always as they first appear, and false assumptions may cause hours of wasted time. Finally, a troubleshooter must know when to take a different approach or get outside help. Sometimes, the most obvious things are difficult to identify because the residential cabling installer is close to the problem.

Although there is no set of rules to cover all situations, the residential cabling installer can use some of the following general guidelines (steps) as a starting point.

Step	Troubleshooting
1	Gather relevant information. This includes locating cabling diagrams and documentation, asking questions of users, and observing the situation. Specifics may include:

- Type of system (e.g., local area network).

- Device experiencing problems (e.g., PC, telephone, etc.).

- Type of cable (e.g., twisted-pair, coaxial, optical fiber).

- Location of the MDU-TRs and ERs that house the problem circuit connections.

- Location of the telecommunications equipment.

- Location of the user equipment.

- Details of the circuit.

- Cabling diagrams.

- Certification test data.

- Previous service records.

- Other information (components, etc.).

2	Contact the end user before beginning work and ask the following questions:

- What is the specific problem? This initiates the troubleshooting process and helps to narrow the possible problem.

- When did the problem begin?

- How often does this problem occur?

- What else was happening in the area at that time? The problem could be related to the installation of other equipment or to another event that could help determine the cause of the problem.

- What other actions can be taken that will help in narrowing the cause? For example, if the user's PC is working properly, but cannot connect to the network, have the user check the connection. If the connections are good, ask the user to check the faceplate. If the user has a "Data 1" and "Data 2" connection, have the user change the connection and try to access the network. If the trouble is a telephone set, again have the user check the connections.

3	Perform a visual inspection. Check out the overall site, as well as the specific problem area. Look for evidence of equipment movement, construction, previous sloppy work, incorrect cable or components, and anything else that can aid in determining the cause of the problem.

Step	Troubleshooting
4	Complete all actions that the troubleshooter would normally have the user do when talking over the telephone.
5	If the problem has not been resolved, make sure all equipment is removed from the cable in question and perform the appropriate diagnostic tests.

If the problem has not been identified using the above procedures, break the problem into smaller parts and test those parts separately.

It may be necessary to obtain assistance in locating a problem. Modern networks are complex, and one person cannot be expected to know everything. An expert in a specific area may be needed. In other cases, someone with a fresh viewpoint looking at the situation may identify the problem.

Troubleshooting tools and equipment

A variety of tools are available to help the troubleshooter find the source of cabling problems. The troubleshooter should have appropriate test equipment available and know how to use it. Depending on the type of cabling system in question, this could include the following:

- Cable tracer
- Cable-end locator kit
- Certification field test instrument
- Infrared (IR) conversion card
- Light source and optical power meter
- Multimeter
- Optical fiber flashlight
- Optical time domain reflectometer (OTDR)
- Telephone field test instrument (butt set)
- Appropriate test adapters, leads, and cables
- Tone generator/wand
- Visual fault locator (VFL)
- Wire map verification field test instrument

Other troubleshooting equipment

Other equipment that may be required to effectively perform troubleshooting tasks includes:

- Alcohol and wipes.
- Compressed air.

- Documentation.
- Inspection microscope.
- Ladder.
- Notebook
- Two-way radio.

Alcohol and wipes.   Isopropyl alcohol (99-percent pure) and lint-free wipes, including cotton swabs, can be used to clean dirty electrical and optical connections.

Compressed air.   Compressed air is used for removing loose dust and other debris from fiber adapters.

Documentation.   The process of troubleshooting can be greatly eased when appropriate documentation is available. This should include:

- Cabling diagrams.
- Description and functioning of the equipment attached to the cabling system.
- Certification test data for the network.

Inspection microscope.   An inspection microscope is used to examine in detail the condition of the end of a fiber cable or a fiber connection.

Ladder.   When troubleshooting cabling problems, a ladder may be necessary to access difficult-to-reach areas, such as the space above a dropped ceiling.

Notebook.   Data should be documented in a notebook if the test equipment does not store these results.

Two-way radio.   Many troubleshooting tasks require two residential cabling installers—one positioned at each end of a cable link or data channel. A two-way radio can be helpful when residential cabling installers at opposite ends of a cable need to communicate with each other.

> **NOTE:** Interference from such radios has been known to affect the operation and accuracy of some handheld cable field test instruments.

The use of two-way radios may not be allowed in some facilities for safety or security reasons. Check with the job supervisor before using this equipment. Additionally, be aware that two-way radios do not function properly in all environments.

Some light-source and optical power meter equipment include a communications function. This is only useful, however, when both parties are connected to opposite ends of the same cable.

Troubleshooting knowledge

Troubleshooting involves a broad range of skills. In addition to knowing how to select and use the appropriate test equipment, the residential cabling installer must have a solid understanding of communication skills, industry standards and practices, and safety procedures.

Communication skills. Communication skills are an important element when troubleshooting a cabling system. A troubleshooter must be able to:

- Read technical manuals, instructions, catalogs, and other documents.
- Verbally communicate with customers, coworkers, contractors, and other support personnel.
- Understand blueprints and drawings that relate to residential cabling installation.

Ideally, a residential cabling installer will have computer skills to add, look up, update, and interpret cabling database information, view electronically stored documents, and obtain support information via online resources.

Industry and standards knowledge. It is recommended the troubleshooter be familiar with:

- The *National Electrical Code® (NEC®)*.
- ANSI/TIA/EIA standards.
- Basic electrical and electronic principles.
- Basic optical fiber and light theory.
- Basic troubleshooting techniques.
- Test equipment operation.
- Voice and data network basics.

Troubleshooting tips for isolation

Becoming an effective troubleshooter takes practice and experience. Modern cable field test instruments can make the job easier. Here are a few troubleshooting tips:

- Determine the nature of the problem and make sure it is cabling related. Many network problems can appear to be cabling related. Someone without the proper diagnostic skills and experience can easily misdiagnose a defective NIC as a cabling problem. Conversely, many cabling problems can appear to have other causes. For example, cabling that does not meet specifications can result in transmission errors, which will often be manifested as poor performance. The first thing to do is talk to the user. Determine the following:

- Has the permanent link or channel in question ever worked?
- Has the permanent link or channel been tested previously?
- When and how did the problem begin?
- Is there some specific event that correlates with the beginning of the problem? Correlation does not necessarily mean causation. Do not make assumptions. Sometimes, equipment movement, construction work, or other events can be related and may indicate possible causes of problems.
- Why is the problem believed to be cabling related? Again, assume nothing. Just because someone believes a problem is cabling related does not mean that it is. In either case, the users' response can often help locate the source of the problem.
- Does the problem affect all activity or specific applications? If only specific applications are affected on a network link, chances are the problem is not cabling related. However, some applications are more sensitive to cabling variations than others; some cabling problems may only be associated with specific applications.
- What is the scope of the problem? Is it one connection or many? If the problem is affecting a single connection, the chances are it is within that specific channel or the equipment directly connected to it. If the problem affects multiple links, it could be related to backbone cabling, a link to a shared service (such as a network file server), or with such equipment as a hub or a server.

- Look around for anything unusual. Check the work area for anything that suggests intervention. Examine the network interface device (NID), DD, or MDU-TRs for clues, such as sloppy wiring or evidence of recent changes. Be alert for signs of recent construction or electrical work.

- Use a certification field test instrument to check suspected cabling segments. The results will often provide a good indication of possible causes, as well as the location of a problem. Be sure to look at the results of the individual tests.

  Break the problem into smaller parts. If performing a channel test, examine each component (e.g., the connections, outlet cable) separately. Look for defective terminations or improperly installed connectors. Loose, improperly installed, or damaged connectors are a major cause of cabling problems. Check for intermittent problems by moving or bending the cable under test near the point of termination.

- Ensure that the proper outlets and other connections are employed. For example, outlets designed for Category 3 use will probably not meet Category 5 specifications.

- Patch cables and plugs are another potential trouble source. Although the standards call for the use of stranded wire for twisted-pair patch cords, many people construct their own using solid nonstranded wire. This practice is not recommended.

- Check to make sure the appropriate type of cable for the application has been installed.

- Direct attention to the cabling installation practices. Residential cabling installers who do not follow proper installation practices can create many problems—most commonly, not maintaining twists close enough to the point of termination and splitting pairs.

  Equipment and patch cords should be examined carefully, especially Category 5e or higher installations. Patch cords are easily damaged, especially at or near the point of termination. Patch cords also may not be properly made. It is difficult, for example, to maintain the pair twists in close proximity to the modular connector when making Category 5e or higher patch cords; therefore, it is recommended that they be purchased from a reputable manufacturer rather than being made by the residential cabling installer.

- Check to determine whether the cable been damaged. If an equipment cable is laid across the floor, it can easily be damaged. In addition, improper installation practices, building construction, or remodeling can damage installed cable. Furniture pushed against a wall plate may also create a bend radius smaller than the standards allow.

  Look for user intervention. Inspect for sources of interference (e.g., cable running over fluorescent lights, down elevator shafts, or in parallel with and in close proximity to ac power lines may cause problems).

- Inspect for sources of interference (e.g., cable running over fluorescent lights, down elevator shafts, or in parallel with and in close proximity to ac power lines may cause problems).

  Check for excessive cabling distances.

Testing and troubleshooting audio systems.   Testing of a residential audio system must be performed before connecting the amplifier, but after all speaker and volume connections are complete. Perform this test using the following steps.

Step	Testing a Residential Audio System
1	Turn all the volume controls to their highest setting in all the rooms.
2	Measure the impedance of the load. The signal that the speaker wiring and associated equipment work with is a low-level ac voltage. Impedance is the resistance to an ac load and is often slightly higher than the dc resistance that most multimeters measure. It is, therefore, important that a meter is used that is capable of measuring speaker impedance.
3	Connect the meter leads to the same jack to which the amplifier is connected. Test both left and right connections. The measurement must fall into the range of the capabilities of the intended amplifier. This range will be printed on the back of the amplifier near the speaker terminals or in the specifications section of the amplifier's owner's manual.

Step	Testing a Residential Audio System

CAUTION: Do not connect the speaker system if the measured load is below the range of the amplifier.

4   If the measurement is more than twice the highest number in the amplifier's impedance range, double-check the settings on any analog volume controls to be sure that all the volume controls are at their maximum settings. It is acceptable to have the system impedance slightly higher than the range of the amplifier, but if it is too high, the homeowner will experience diminished volume output.

5   If analog volume controls are installed in the home, changing the system impedance is usually a simple procedure. Refer to the volume control manufacturer for instructions on how this is accomplished. Generally, to increase the total system impedance, turn the tap settings in the smaller rooms to the next higher setting (see the volume control instruction sheet for details on this operation), then recheck the system load at the amplifier location. To decrease the total system impedance, turn the tap settings in the larger rooms to the next lower setting (see the volume control instruction sheet for details on this operation), then recheck the system load at the amplifier location. This should correct any issues. If the impedance does not change after changing the impedance settings on the volume controls, damaged speaker cabling should be suspected.

6   Isolate suspected shorts or opens by disconnecting each volume control one at a time from the audio distribution module in the DD until the problem is found.

7   When testing shows that the amplifier will see a load in the acceptable range as described above, document the final reading and connect the amplifier according to the following steps.

- Before powering on the amplifier, be sure to turn the volume in all the rooms down about one-third.

- Set the volume on the amplifier to zero and power it on. Select a source for testing (using a known good source such as a compact disc [CD] player and familiar CD makes testing easier).

- Slowly turn the volume up on the amplifier until it is at about one-half the maximum. You should now hear the source playing softly.

NOTE: Do not attempt to play things loudly now, as the amplifier could be pushed to distortion with the room volume controls at one-third.

8   Turn up the volume control in each room. Momentarily turning the volume to maximum in each room will show any flaws in the room (e.g., loose screws on the speaker mounts or debris in the speaker). Now turn the volume down and proceed to each room, repeating the test.

Troubleshooting video—Community Antenna Television (CATV) and Direct Broadcast Satellite (DBS).   Typical problems with video networks can be determined by symptoms displayed on video monitors (televisions [TVs]) on the network. In addition, many CATV- and DBS-related issues are not due to the wiring system, but to improper settings on the TV. Be sure to set the TV signal source (e.g., antenna, cable, line, etc.) and/or TV station based on the recommendation of the service provider (SP).

Prior to attempting the troubleshooting suggestions in the following text, be sure that the manufacturer's directions for installation of the required receivers and hardware for the system were followed.

NOTE: If the problem cannot be identified or solved using the information shown below, contact the manufacturer of the equipment or the SP for assistance.

- Symptom: Herringbone pattern on TV
  - Problem: Cross modulation, which is caused by overdriving small signals within a 6 MHz bandwidth (the bandwidth allowed for individual signals).
  - Correction: If the amplifier has an attenuator, reduce the level of signal amplification, or pad the input signal going into the amp.
- Symptom: Multiple images on TV
  - Problem: Ghosting, which is caused when two signals from the same source are on the system at the same time. Usually caused by a leak of signal into (ingress) the system.
  - Correction: Be sure that all components in the system are properly shielded and that any unused connections are properly terminated.
- Symptom: Dark scrolling band at bottom of TV
  - Problem: Hum, which is caused by problems in the power system.
  - Correction: Make sure that ground loop is properly installed. Isolate individual components to determine source.
- Symptom: Venetian blind effect, distorted sound
  - Problem: Cochannel interference, which is caused when two pictures are received at the same time from two different TV transmitters in different locations, but which share the same TV channel.
  - Correction: If the amplifier has an attenuator, reduce the amount of signal amplification or pad the input signal going into the amp.
- Symptom: Snowy image on TV
  - Problem: Weak signal at device.
  - Correction: If the amplifier has an attenuator, increase the level of signal amplification.

Testing and troubleshooting security and home automation.   Security and home automation wiring should be tested for continuity. A simulated functional test is also recommended to make sure that sensors and other devices will be functional after they have been installed and activated. In a prewiring scenario, the wire testing and functional testing will usually be performed at different times. In a rewiring scenario (or wiring of existing residences), both wire and functional tests will be performed at the same time.

Functional tests may be grouped into the following tasks by verifying that:

- The panel is properly powered and the back-up battery is properly installed.
- All sensors (window, door, proximity, smoke, and heat) are properly installed.
- All keypads and wireless controls are properly powered and wired.
- The alarm system can be armed and disarmed properly.
- The RJ-31X connection is properly wired, seizes the line, and dials the monitoring stations.
- All sounders (horns and strobes) are properly wired and powered.
- The installer default password is programmed. Provide this to the resident, along with instructions for changing the password.
- The entire alarm and security system has been installed and programmed in accordance with manufacturer instructions and that arm/disarm functions operate properly.

Follow the steps below for preliminary wire testing.

Step	Preliminary Wire Testing
1	Test alarm and security wiring for continuity.
2	Verify that 20 AWG (0.81 mm [0.032 in]) does not exceed 15.2 m (50 ft), 18 AWG (1.0 mm [(0.039 in]) does not exceed 30 m (100 ft), and 16 AWG (1.3 mm [0.051 in]) does not exceed 76 m (250 ft).
3	Verify that the plug from panel to RJ-31X device is properly wired as noted: ■ Handset: Brown (tip) ■ Handset: Grey (ring) ■ Incoming telecommunications: Green (tip) ■ Incoming telecommunications: Red (ring)
4	Perform a preliminary functional test as follows: ■ Temporarily connect a 2000 Ω end-of-line resistor to each of the wired zones. ■ Power-up system temporarily by plugging in the transformer. ■ Wait for "READY" to be displayed on keypads. ■ Power-down the system.

Follow the steps below for functional system testing.

Step	Functional System Testing
1	Test a functional system as follows:
	▪ Power-up the system.
	▪ Make sure that no sensors are in a faulted state.
	NOTE: Passive infrared (PIR) devices may need to be temporarily covered.
	▪ Keypads should read "Disarmed—Ready to Arm."
	▪ The system should display faulted zones, if any.
	▪ Correct faulted zones and repeat the test.
	▪ Power-down system.
	NOTE: Manufacturer's instructions should be followed.
2	Test sounders as follows:
	▪ Connect the power source.
	▪ Enter the installer code.
	▪ Press "Test."
	▪ The sounder should sound for approximately one second.
	▪ Press "Off."
	NOTE: Manufacturer's instructions should be followed.
3	Test the monitoring function as follows:
	▪ Disarm the system and check that all zones are intact.
	▪ Enter the security code.
	▪ Notify the monitoring station (central station) that a test is to be performed.
	▪ Fully test the system following the manufacturer's instructions.
	▪ Notify the monitoring station at conclusion of testing.
	▪ Verify identification (ID) codes of event status reporting.

Follow the steps below to troubleshoot alarm and security systems.

Step	Alarm and Security Troubleshooting
1	Low battery
	▪ Check if the battery is missing or discharged in the panel.
	▪ Check the power supplied to remote wired keypads.
	▪ Check the batteries in wireless devices.

Step	Alarm and Security Troubleshooting
2	Periodic beeps
	▪ Check if the system is in test mode.
	▪ Check for low battery indication.
	▪ Perform a supervision check.
3	Nuisance alarm
	▪ Check the sensors, which may not be properly installed, wired, or monitored.
	▪ Check the wireless transmitter, which may be programmed incorrectly.
4	Intrusion alarm
	▪ Check if doors or windows are open while the system is armed.
	▪ Check for improper settings or operation of door exit/entry delays.
	▪ Check if sensor magnets are located too far from switches. This is usually caused by door or window misalignment.

Troubleshooting copper cable

Troubleshooting a customer's problem in a copper cabling system involves four steps:

1. Verify the problem.
2. Isolate the source of the problem.
3. Repair the problem.
4. Test the repaired system to ensure it functions correctly.

Verify the problem. Using the troubleshooting methods described in this chapter, discuss the problem with the user to define the nature and extent of the problem. Important information includes:

▪ The specific difficulties experienced by the user.

▪ When the problem began.

▪ The frequency of the problem (e.g., intermittent or constant).

▪ Other conditions that may relate to or cause the problem (e.g., an electrical outage or installation of other equipment).

Isolate and repair the problem. After defining the specific nature of the problem, the residential cabling installer can select the appropriate test equipment to isolate the difficulty. A good cable field test instrument should help locate the sources of most cabling problems.

The results of the individual tests can provide a wealth of information. When reading results, be aware that there may be more than one potential cause for a particular test result; in some cases, there may be multiple contributing causes. The following sections list some test results and possible associated causes.

High insertion loss. Insertion loss is the loss in power of a transmitted signal as it travels along a cable. If the loss is too high, transmission reliability will be affected. Loss increases with temperature, frequency, and cable length. High loss can be caused by:

- High temperature—Cables installed in areas that may be subjected to high temperature may exhibit an increase in attenuation.

- Wrong (lower) grade or category of cable—The cable under test may be unsuitable for the data rate at which it is being used. In this case, it may be necessary to replace the cable.

- Cable is too long—Make sure the cable length is within the specified maximum length. Some residential cabling installers, afraid of waste, will leave excess cable coiled in the ceiling or MDU-TR. Excessive lengths of patch and equipment cables also can be a problem.

- Incorrect equipment patch cable—Make sure equipment and patch cables meet requirements for the desired installation (e.g., Category 3, Category 4, Category 5, and Category 5e).

- Improper terminations—Ensure all connections are terminated properly. Make sure that connecting hardware, including telecommunications outlet/connectors, DDs, and connecting blocks, meets the transmission performance requirements of the desired category.

Excessive Near-End Crosstalk (NEXT) loss. After wire map failures, the most common fault encountered is NEXT loss faults. Excessive NEXT loss may be attributed to:

- Wrong grade/category of cable—Make sure the grade of cable being used is correct for the application.

- Improper termination practices—If proper twisting is not maintained when terminating twisted-pair cable, NEXT loss may increase. Ensure the twisting is maintained as specified for the category of cable being used.

- Split pairs—Proper pairing must be maintained. The twisting of the pairs tends to minimize interference radiating from the cable pair and that received by the cable pair. By splitting pairs (using wires from two different pairs as a signal circuit), the beneficial effects of pair twisting are lost.

- Incorrect or substandard components—The quality of the permanent link or channel is determined by the weakest component. If a component is poorly made or does not meet the requirements for the desired installation category, the component should be replaced.

- Incorrect or defective cables and test adapters—Make sure that proper cables are used.

▪ Unmatched components—This is generally only a significant issue for Category 6 links and channels. Components manufactured before 2002 are not universally electrically compatible. It is recommended that Category 6 systems be constructed only out of supplier-approved components.

▪ Reflected far-end crosstalk (FEXT) loss—This is generally only a significant issue for Category 6 channels where return loss and FEXT loss can combine to produce up to 2 dB of unexpected NEXT loss. It can be reduced or eliminated by replacing patch cords with ones of more uniform impedance (better return loss).

Troubleshooting NEXT loss is greatly facilitated by time domain analysis tools that are capable of showing NEXT loss magnitude vs. length. This allows quick identification of the source and magnitude of NEXT loss faults.

Incorrect cable length.   Because of the twisting of the pairs, the actual physical length of a twisted-pair cable will appear slightly shorter than the electrically measured length.

▪ Length too short—A length test that displays a significantly shorter than actual length ($> 15\%$) may indicate an open or short in the cable. It can also indicate a poor termination at an intermediate connection, such as a punch-down block, patch panel, or wall outlet. If part of a certification test, this problem will be caught by the wire map test.

▪ Length too long:

 – Check for excess cable coiled in wall, ceiling, MDU-TR, or DD.

 – Verify cable length runs from blueprints.

 – Replace long patch cords with shorter cords where possible.

 – Notify job supervisor or customer as required.

▪ Inaccurate length measurement—Make sure that the NVP setting of the cable field test instrument matches the cable under test.

Excessive loop resistance.   Loop resistance may be available as a measurement in a certification field test instrument. Alternatively, it can be determined by shorting both conductors of a copper pair at one end of a cable and measuring the total resistance of both conductors from the opposite end. Excessive resistance can be caused by:

▪ Excessive cable length—Make sure the length of the cable is within specifications for the application.

▪ Poor connection at twisted-pair—Improperly punched down terminations or damaged wall outlets, connecting blocks, and patch panels could be to blame.

▪ Defective shorting plug, test cable, test adapter, equipment cable, or patch cable—Test each component separately.

▪ Oxidation within a connection—Wipe contacts clean with a pencil eraser and retest.

Wire map failure. This problem includes crossed pairs, miswires, split pairs, opens, and shorts. When testing a channel, investigate each cabling component separately. Possible causes include:

- Incorrectly installed or damaged cable or termination.
- Wrong patch cables—Patch cables wired to older universal service order code (USOC) specifications will create split pairs for T568A applications.
- Mismatched terminations—The wiring scheme for the T568A and T568B connectors is different. Do not use a T568A connector on one end of a cable and a T568B connector on the other.
- ScTP or International Organization for Standardization (ISO) tests may require shield continuity.
- Use TDR to measure the length of a shorted or open pair in order to determine the location of the fault.

Incorrect impedance. Impedance mismatches are associated with:

- Poor installation techniques.
- Incorrect or defective cable.
- Unmatched components.
- Patch cords (often $< 100 \, \Omega$).

Return loss. Return loss is a measure of impedance uniformity in a link. Thus, troubleshooting return loss reduces looking for sources of impedance changes. Return loss faults are often associated with:

- Substandard components (both cable and connectors).
- Generic patch cords (not provided or approved by link manufacturer). Such cords can be either low impedance or unstable impedance with handling.
- Poor installation practices (including rough cable handling).
- Slack loop coils above the closet. Do not coil excessive cable into a neat tied loop, as this can add 2–3 dB of unexpected return loss to the link measurement. If necessary, leave such loops loosely coiled in a nonoverlapping manner.

Troubleshooting return loss is greatly facilitated by time domain analysis tools that are capable of showing impedance or return loss magnitude vs. length. This allows quick identification of the source and magnitude of return loss faults.

Equal Level Far-End Crosstalk (ELFEXT) loss. Troubleshooting ELFEXT loss problems reduces to troubleshooting insertion loss and FEXT loss problems. Since insertion loss rarely fails, this usually means excessive FEXT loss is the cause. The coupling mechanisms for FEXT loss are the same as for NEXT loss. This means that there will be NEXT loss failures if there are ELFEXT loss failures, so troubleshooting ELFEXT loss is best accomplished by first solving any NEXT loss failures. Once this is done, it is likely that ELFEXT loss results will pass.

Power Sum Near-End Crosstalk (PSNEXT) loss and Power Sum Equal Level Far-End Crosstalk (PSELFEXT) loss.   These calculations are derived from their component measurements. Troubleshooting PSNEXT loss and PSELFEXT loss faults is done by first eliminating NEXT loss and ELFEXT loss faults. Once NEXT loss and ELFEXT loss measurements pass, PSNEXT loss and PSELFEXT loss results should pass.

Noise.   Noise is unwanted electrical activity that interferes with the desired signal transmission. Noise can be induced directly by a connected signal source, (e.g., a computer with a defective, noise-generating component). It can also be radiated from an external source (e.g., an electric motor, copier, or electrical power cable) in close proximity to a data cable. There are two types of noise to be addressed—impulse noise and continuous noise.

Impulse noise refers to discrete noise spikes that occur on a regular or irregular basis. These spikes can be as high as 350 volts (V). In some instances, impulse noise can damage equipment, but more often it causes machine lock-up (network file servers, network hubs, etc.) or corrupt data. Impulse noise can be caused by an electric motor, copier, laser printer, elevator, air conditioner, or any device that produces a large power disruption upon startup or shut down.

Continuous noise is usually below three V. While it can affect transmission on a cable, it generally does not result in hardware damage. Continuous noise can be caused by a fluorescent light, computer, ac power line in close proximity to a data cable, and many other sources.

If access to a spectrum analyzer is available, the frequency of continuous noise can provide a clue as to its source:

- Under 150 kilohertz (kHz)—Noise in this range is usually due to sources such as ac power lines, fluorescent lighting, and machinery.

- 150 kHz–20 MHz—In this range, noise is often attributed to light dimmers, medical equipment, computers, copiers, and laser printers.

- Above 20 MHz—Such noise may result from radios, cellular telephones, wireless telephones, television sets, microwave ovens, and broadcast equipment.

Most cable troubleshooting equipment will measure for continuous noise. Some can monitor for impulse noise. A field-strength meter can also be used to monitor for impulse noise.

There are no cabling standards for noise requirements. There is no value for noise that is correct, but it should be as low as possible.

Here are a few things the residential cabling installer can do about noise:

- Move cables away from possible sources of interference. ANSI/TIA/EIA-569-A, *Commercial Building Standard for Telecommunications Pathways and Spaces*, provides guidelines for separating twisted-pair cabling from common sources of noise. Ensure that these guidelines are followed.

- Move interference sources away from cabling. Change the location of microwave ovens, copiers, or other equipment, if possible.

- Install a better grade of cable. If using unshielded twisted-pair cable, install Category 5e or higher even if the application only requires Category 3.

- Maintain cable pair twist up to the point of termination. Category 5e, for example, requires that twists be maintained to within 13 mm (0.5 in) of the point of termination.

- Make sure there are no split pairs.

- Maintain separate cross-connect fields for voice and data.

- In extreme EMI cases, such as radar facilities, magnetic resonance imaging (MRI) rooms in hospitals, or some factory environments, consider the use of shielded cabling or optical fiber.

Locating cables.   If documentation is not complete, it may be necessary to follow the path of a cable to locate the ends. Many cable field test instruments have accessories available to help with these tasks.

Testing the system.   Once the repairs are finished, the final step is testing and documenting the system. First, complete a visual inspection to determine that the:

- Location has been correctly labeled.

- Correct patch cords/equipment cords have been used.

- Correct cable and termination devices have been installed in accordance with the applicable standard.

Having completed the verification phase (visual inspection) and determined that the trouble is within the cabling network, the fault isolation phase begins.

Fault isolation.   The troubleshooting flow chart (Figure 16.5) provides a comprehensive, organized approach to cable troubleshooting. However, before beginning this phase, wire properties should be reviewed.

To most residential cabling installers, a wire pair appears as follows:

Figure 16.3 Twisted-pair wire.

However, to an electron this same wire pair appears as a transmission line.

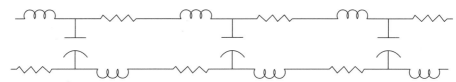

Figure 16.4 Schematic of a transmission line.

As seen in the above diagram, a transmission line is composed of resistors, inductors, and capacitors. When a second pair is added, the influence of these components is felt on the second pair. Crosstalk loss occurs when a signal traveling down a pair exceeds the proper limits and induces a current into an adjacent pair.

If the cable is stretched, the capacitance (the cable sheath) is changed; if stretched hard enough, the resistance (a change in wire gauge) is also changed. If a coil is placed in the link, an inductor is added. By placing one or more cables together for an extended length, there is the chance of an increase in crosstalk loss. If cable ties are cinched too tightly, the jacket of the cable is deformed, resulting in a change in capacitance and, therefore, a change in NEXT loss and return loss.

> **NOTE:** When dealing with ScTP or enhanced shielded twisted-pair (STP) cables, it is also important to verify the ground as well as the individual pairs.

Figure 16.5 provides the guidelines for successful problem isolation.

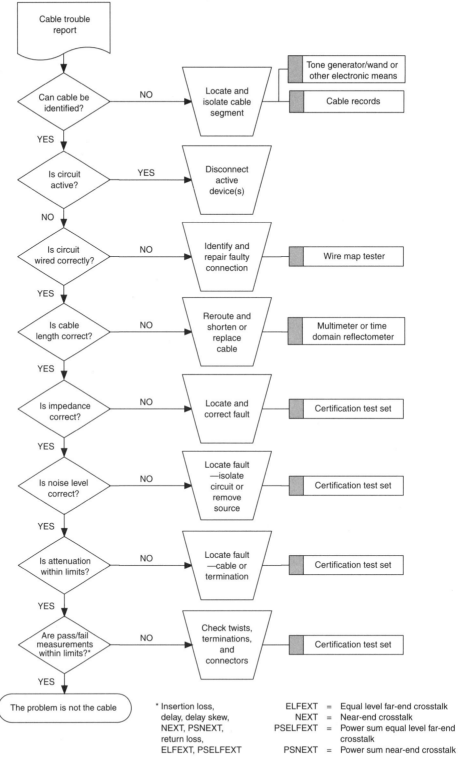

Figure 16.5 Troubleshooting flowchart.

After following the troubleshooting guidelines, the next objective is to match the test equipment with the situation.

TABLE 16.3   Field test instrument selection

Step	Media Type		
	Twisted-Pair	Unshielded Twisted-Pair	Screened Coaxial
Segment identified?	Documentation Tone generator/ wand	Documentation Tone generator/ wand	Documentation Tone generator/ wand
Segment active?	Telephone field test instrument	Telephone field test instrument	Telephone field test instrument
Wired correctly?	Wire map field test instrument	Wire map field test instrument	Multimeter
Length correct?	Multimeter Time domain reflectometer	Multimeter Time domain reflectometer	Multimeter Time domain reflectometer
Impedance correct?	Certification field test instrument	Certification field test instrument	Certification field test instrument
Noise level correct?	Certification field test instrument	Certification field test instrument	Certification field test instrument
Attenuation level correct?	Certification field test instrument	Certification field test instrument	Certification field test instrument
Crosstalk loss issues correct?	Certification field test instrument	Certification field test instrument	Certification field test instrument

NOTE: The range of NEXT and FEXT loss (including power sum) issues and tests are interrelated such that a certification field test instrument is required to perform the tests and display the results. Many extended crosstalk loss performance parameters are actually calculations made from one or more related tests. All crosstalk loss issues must be reviewed as part of a total performance parameter package, not a stand-alone test.

Up to this point, obvious faults have been described. What about the types of problems that are more difficult to isolate? Two of the most common are discussed below:

▪ Symptom: Certification field test instrument cannot locate the remote end.

– Possible Cause: Assuming both the field test instrument and injector is attached to the correct circuit, the most probable cause is an open in all four pairs of the cable.

- Isolation: Switch the field test instrument to the TDR function, and it will display the cable length. If the length is shorter than anticipated, check at that point for an open.

▪ Symptom: Unable to log onto the network, sluggish performance, disk errors, station freezes, reboots.

- Possible Cause: If the cable has been checked and eliminated as the problem, there is a good possibility that there is an intermittent noise problem. Noise is being induced into the system by an outside source such as an electric motor (e.g., an electric pencil sharpener), lighting, ac power lines, heaters, or other equipment. This effect only occurs when the device is activated and puts a "spike" on to the line.

  NOTE: Check for the location of a noise source (e.g., electric pencil sharpener).

- Isolation: Using the certification field test instrument, first check the average noise on the circuit. (Make sure two-way radios are turned off. Check for impulse noise [the spike].) If these levels are high and the field test instrument has an active time domain crosstalk loss capability, it should be able to determine where the noise spike is entering the line. If it does not, the line will have to be hand traced to locate the possible equipment that may be inducing the noise. If these efforts prove negative, other potential sources of the problem may be the grounding, bonding, or electrical system.

In this case, perform a visual inspection, beginning with the connection between the bonding conductor for telecommunications and the telecommunication main grounding busbar (TMGB) of the telecommunications grounding and bonding network (see Chapter 8: Electrical Protection Systems). If all connections are correct, use a multimeter to test the connections for a maximum value of 0.1 Ω. If these efforts prove negative, a strip-chart recorder may be placed on the circuit to the device. If power fluctuations are evident, the branch circuit serving the equipment may have to be moved to another phase of the three-phase power feeding the building or the unit may have to be placed on an uninterruptible power supply (UPS) system.

Once the problem has been successfully isolated, the final phase is to repair and test the circuit. In the case of a noise interference problem, it may not be possible to clear the problem by moving the cable or the interfering source. If this is true, consider placing a coupled bonding conductor (CBC). The CBC provides equalization like the telecommunications bonding backbone (TBB) and also provides protection through electromagnetic coupling with the telecommunications cables.

Another solution is to replace the twisted-pair cable with either ScTP or enhanced shielded twisted-pair cable to correct the situation. Once the repair has been completed, the circuit should be retested in accordance with the testing procedures provided earlier in this chapter.

Follow the steps below to troubleshoot twisted-pair cabling.

Step	Twisted-Pair Cable Troubleshooting
1	Disconnect active equipment so that it will not interfere with the testing. For an existing system, ensure that users have been contacted and a convenient time has been arranged to remove equipment from service. ▪ Determine which equipment is connected to the line being tested. ▪ Disconnect the equipment.
2	Connect the cable field test instrument and carry out the appropriate measurements. Some field test instruments require daily set reference. Set the reference according to the manufacturer's specifications. ▪ Set the field test instrument for the correct type of cable. ▪ Conduct the appropriate measurements.
3	If the field test instrument is not the type to return a pass/fail reading, compare the results with the specifications for the cable. Verify that the readings fall into the acceptable range for the cable. If the original cable certification test results are available, compare the current reading/test result with the result of the original test.
4	Record the results.
5	Identify failing measurement values. ▪ Determine which results do not fall into the normal range or are different from the original test results. If the results of a test are near the limit (other than length) for troubleshooting purposes, consider that test to indicate a failure. Properly installed and functioning cable should be well within limits. Marginal results indicate potential problems. ▪ Record any failing results.
6	Use measurement results to determine the source of the problem. ▪ Inspect hardware and connections. ▪ Visually inspect the cable for damage. ▪ Check the cable type. ▪ Verify test settings. ▪ Inspect the terminations. ▪ Examine the bend radius, pair twist, and the sheath. ▪ Observe color codes. ▪ Verify pin configuration.
7	Test specific channel components, the link, and each patch or equipment cord separately.
8	Perform additional tests as necessary.
9	Take corrective action for any observed conditions that do not meet cabling standards.
10	Retest the cable to verify it passes all relevant tests.

Step	Twisted-Pair Cable Troubleshooting
11	Record the:
	▪ Circuit identification number and location.
	▪ Customer name and telephone number.
	▪ Service type.
	▪ Test results.
	▪ Repairs made.
	▪ Passing test results.
12	Perform housekeeping as needed.
	▪ Replace any cables moved during the testing.
	▪ Ensure that all cables are properly routed.
	▪ Ensure that all cables maintain at least the minimum bend radius.
	▪ Ensure that all cables have the appropriate slack.
	▪ Leave a clean space when the installation is complete.

Troubleshooting power distribution

Troubleshooting begins with verifying that the 120 Vac source is working. Typically, 120 V ac power is converted to dc power using an ac/dc transformer inside the DD. In some cases, the ac/dc transformer may be a wall-mount version external to the DD.

For most systems, the dc power output is typically 12-18 Vdc unregulated. Amps range from −400 milliampere (mA) to −3 amp; above this, power issues are more complicated. As a result, manufacturers tend to limit the total ac power that is available in the DD.

The dc transformers are fused to meet safety requirements. Most fuses on dc transformers are nonresetting. Be sure to check voltage output during troubleshooting.

Ideally, dc is transformed one time and then distributed by a dc power distribution function/module to all modules requiring power in the DD. However, in some cases, dc may go straight to the modules, often using many 120 Vac receptacle/transformers (i.e., power strips).

Typically, only a limited number of functions (modules) in the DD require power. Today, dc plugs/jacks typically use the standard 5.1 mm or 5.5 mm plug connector. A mismatched 5.5 mm with a 5.1 mm connector will work with intermittent problems (i.e., causes troubleshooting problems).

If the DD is overdrawing the dc source, the voltage will typically drop and cause system problems before the dc source gives up or protects itself.

Video functions (amplifiers) are the most susceptible to voltage drops. Video performance will suffer and is sometimes difficult to troubleshoot. A dc voltage problem is often diagnosed as a video problem. Data products are less susceptible to voltage drop and may work fine until the dc source completely quits.

All modules that use power should have a power indicator for troubleshooting. Most DDs do not have power backup capability. When implemented, power backup capability is completed by backing up the 120 Vac with a typical UPS or by backing up the dc with a standard lead-acid battery.

It is important to note that many systems place dc on coaxial cable. All unused coaxial connectors that are not self-terminating require 75 Ω termination caps.

> **NOTE:** 75 Ω caps, without dc blocks, on a system with dc power on coaxial cables cause immediate and serious dc power problems. Check for this problem during troubleshooting.

Troubleshooting optical fiber cable

Troubleshooting is a process for locating a decreased performance condition (problem) that is not acceptable to the user of an optical fiber circuit. This condition may be minor in nature or a total outage. Restoration is the process of returning the circuit to an operational condition that is acceptable to the user.

The troubleshooting process for optical fiber cable is identical to that used for copper cable. The four steps of verifying, isolating, repairing, and testing are applicable regardless of the medium.

Verify the problem.   The questions asked of the user remain basically the same. It may also be helpful to take the user through the same telephonic verification as performed for copper cable.

- Answer the following questions:
  - Who is affected by the condition?
  - What is the nature, scope, and severity of the problem?
  - Where are the possible areas that the condition could occur?
  - When did the problem first occur? Does it correlate with any other activities or abnormal conditions?
  - Why does this condition exist?
  - How could this condition happen (e.g., electrical outage, weather conditions, people working in the area)?
- Upon arrival at the site, assess the situation. Determine the priority of the condition and the resources available to correct the problem. The following items should be considered:
  - What is the priority of the restoration? A circuit serving one computer is not as critical as a backbone cable in a hospital.
  - What test equipment is available?
  - What trained manpower is available?
  - What method or procedure will be used to find the problem?

Isolate the problem. This is the most interesting part of the troubleshooting process. The two common approaches for isolating a problem are:

- Start at the beginning of the circuit and work toward the other end. This is appropriate where there is only one link or one that is relatively short in length.

- Start in the middle of the circuit and determine in which direction the problem lies. Move to the middle of an identified defective section and again determine which section is defective. Continue this process until the defective link is found.

A thorough visual inspection of all areas is necessary. It is important that each connector be thoroughly cleaned before inspection. A dust particle on the end of a connector becomes an efficient attenuator.

Most problems in an existing circuit can be located in a visual inspection; for instance:

- Disconnected jumper.

- Jumper that has been incorrectly changed because of move, add, or change (MAC) activity.

- Jumper to a hub or router that has been unplugged.

- Hub or router that has been turned off. Look for the power-on indicator.

- Broken fiber inside a FDU or DD. The fiber inside a 900 μm buffer can be pinched or broken by a door being closed or a drawer being moved. The plastic buffer on the fiber may or may not show some damage.

> **NOTE:** The damaged area can usually be felt with your fingers because any compression severe enough to damage the fiber will also distort the plastic buffer.

- Damaged cable inside a DD, MDU-TR, ER, or entrance facility (EF).

Until this point all previously listed conditions would result in a dead fiber; that is, no light exits. There are other conditions, such as higher attenuation at 1300 nm than 850 nm or a greater loss in one direction than in the other. If the system had been tested at one wavelength and in one direction according to ANSI/TIA/EIA-568-B, *Commercial Building Telecommunications Cabling Standard*, the majority of these conditions could have existed when the fiber was originally placed but did not become evident until the fiber was activated.

ANSI/TIA/EIA-526-14-A states that "Bidirectional testing is a default requirement of this document as it is the most conservative." For example, at the time of original testing, the fiber was tested from the MDU-TR to the residence in the 850 nm window. The fiber is now to be activated and is to operate in the 1300 nm range. The loss reading at 1300 nm is higher than at 850 nm. Possible causes of the problem are excessive bending:

- At the connector.
- Along the route.
- At a splice point.

Lower wavelengths normally have a greater loss than higher wavelengths because of a higher scattering and absorption loss at the lower wavelength.

The second condition is at a greater loss at both wavelengths when tested from the other direction. In the original test, the light was transmitted from Connector 2 to Connector 1. In that case, all of the light traveled into Connector 1.

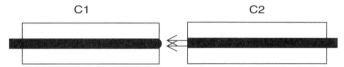

Figure 16.6   Connector loss, connector 2 (C2) to connector 1 (C1).

When the transmit and receive directions are reversed because of the dome-shaped polish on Connector 1, part of the light is allowed to escape, causing an increase in attenuation. This can be corrected by repolishing the connector.

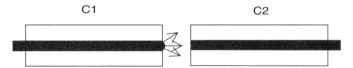

Figure 16.7   Connector loss, connector 1 (C1) to connector 2 (C2).

Another possible cause may be small, asymmetrical nicks in the fiber. When light is transmitted in one direction, the light is reflected into the fiber, while transmission from the opposite direction causes the light to be reflected out of the fiber.

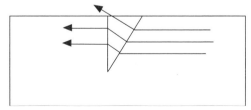

Figure 16.8   Reflective loss.

This condition causes some light to be reflected outside the fiber and lost. When the light is transmitted from the opposite end of the fiber, all light is reflected into the fiber.

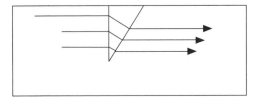

Figure 16.9   Reflective loss.

Repair the problem.   Once the problem has been isolated, the next action is to correct the problem and restore the circuit to its original working condition. This may involve the following:

- Cleaning a connector
- Reseating a connector
- Replacing a connector
- Installing a splice on a broken backbone fiber
- Replacing a broken splice
- Replacing a broken jumper
- Replacing a broken pigtail
- Installing a replacement section of either horizontal or backbone cable

Test the system.   Once repairs are finished, the final step is the testing and documentation of the repaired circuit. Follow the steps below for fault isolation.

Step	Isolating an Optical Fiber Fault
1	Perform a visual inspection of the fiber and connectors.
2	If physical damage is evident (e.g., chipped fiber at the connector, broken fiber at the panel or faceplate, etc.), make the necessary repair and proceed to Step 5. If physical damage is not evident, proceed to Step 3.
3	Clean the connectors by performing the optical fiber connector cleaning procedure detailed earlier in the chapter.
4	Check the fiber with the optical fiber flashlight. If light is visible, proceed with Step 5. If light is not visible, proceed to Step 12.
5	Set up and calibrate the light source and optical power meter.
6	Test the fiber from the origin to the extreme.

Step	Isolating an Optical Fiber Fault
7	Document results.
8	Test the fiber from the extreme to the origin.
9	Document the result.
10	Perform the mathematical averaging of the two readings (R1 + R2)/2.
11	If the results are within the standard values for the specific cable under testing, document the results and place the fiber back into service. If not, continue with Step 12.

NOTE: The maximum loss for a particular backbone cable must be calculated using the length of the cable and the number and type of splices.

At this point of the fault-isolation procedure, the fault is one of the following conditions:

- A broken fiber, which was not evident during the initial visual inspection (Steps 12 through 15).

- High attenuation from an unknown cause (Steps 16 through 20).

Step	Isolating an Optical Fiber Fault
12	If looking for a no-light condition, connect the visual fault locator or hot red light to one end of the fiber.
13	Perform a visual inspection of the fiber path. A red glow should be evident at any point where light is escaping from the fiber. Mark each location on the sheath. Note that there may be more than one fault in the cable. If no light is detected, proceed to Step 16.
14	If a broken cable is detected in the horizontal, replace the cable. If a broken cable is detected in the backbone, pull slack from the service loop and splice.
15	Once the repair has been completed, complete Steps 5 through 11.

At this point, there are two options: the fault locator or the OTDR. Each will find the cause of the no-light or the high attenuation condition. For the remainder of this checklist, the steps used will be those for the OTDR field test instrument. Prior to setting up the OTDR, obtain a copy of the original OTDR tests for the fiber in question. These will be invaluable in the interpretation of the new test results.

Step	Isolating an Optical Fiber Fault
16	Set up the OTDR. If the original traces are available, take special care to set the refractive index, pulse width, pulse duration, sampling rate, etc. to that original trace.
17	Connect the field test instrument and perform the test. Since the interest is in the total cable, it is imperative that both a launch and receive jumper of proper length be used for the test. As the interest is in the total cable, it is imperative that both a launch and receive jumper of proper length be used for the test.
18	Compare the new test with the original. Note any point of difference between the two tests. These differences are the faults, either broken or partially broken locations.
19	Make repairs as necessary.
20	Once the repairs have been completed, complete Steps 5 through 11 again.

Part 3

Multi-Dwelling Unit

Multi-Dwelling Unit
Structured Cabling Systems

General Considerations for Multi-Dwelling Units (MDUs)

Multi-Dwelling Unit (MDU) structures

MDUs include apartments, townhouses, condominiums, gated communities, and assisted-living facilities (ALFs). MDU facilities may be under a single roof or they may consist of multiple buildings in a residential campus. In addition, MDUs may include only residential units or they may have residential units along with commercial and retail spaces.

MDU low-voltage residential cabling infrastructure consists of multiple telecommunications cabling subsystems (one or more per residence or residences). The incoming telecommunications services (e.g., voice, Internet, and video) often enter the MDU at one or more common points of entry—individual residences may or may not have unique points of entry.

Some challenges associated with the design of an MDU residential structured cabling system (RSCS) are to:

- Establish multiple demarcation points (DPs), if needed.

- Implement backbone cabling and auxiliary disconnect outlet (ADO) cabling pathways within common areas to allow for proper maintenance of the residential cabling systems.

- Use adequate cabling and cabling pathways to allow for expansion and upgraded system performance.

- Provide the proper division of services to allow for the delivery and billing of applications to each residence.

▪ Provide secure access of equipment rooms (ERs), multi-dwelling unit telecommunications rooms (MDU-TRs), and distribution devices (DDs).

Low-Rise Multi-Dwelling Units (MDUs).   Low-rise MDUs are facilities in which each unit has access to the ground level and also has a roof line (e.g., town-houses). Low-rise units may have direct access to a DP on the exterior of the building—much as with single-family homes—and to roof top-positions for the mounting of satellite dishes.

Low-rise residential units are often built in a campus style, and backbone cabling paths are typically run underground between buildings. This limits the amount of backbone cabling run within common areas and, therefore, cabling system maintenance is more easily performed. This MDU design most closely resembles that of single-family homes, in that there is a limited number of shared resources.

Mid-Rise Multi-Dwelling Units (MDUs).   Mid-rise MDUs include duplexes and other building styles in which units are stacked upon one another. ADO cabling paths are short and generally installed to a common area in the MDU where each access provider (AP) has a DP. Mid-rise construction may require an increasing number of common telecommunications applications (e.g., inter-com systems to common entryways).

High-Rise Multi-Dwelling Units (MDUs).   High-rise MDUs most closely resemble commercial buildings. Few of the units have direct access to the roof line or ground floor. Backbone cabling is typically run from ERs through common areas between floors to MDU-TRs. ADO cable pathways (and sometimes outlet cable pathways) are typically run from MDU-TRs or ERs to residences through common areas.

Community systems and services

The residents within an MDU require the same basic services as the home-owner of a single-family home. However, due to the community infrastructure and the influence of homeowners' associations (or the building owner in the case of apartments and ALFs), the delivery of these applications and the scope of offered services may differ from the single-family home.

Telephone service is usually provided to the MDU through a common point of entry. This is composed of a cabling facility within a common area of the building or telecommunications cabling pedestals located externally to the building (particularly in the case of townhouses).

Dedicated Internet service (e.g., nondial-up cable modem or digital sub-scriber line [DSL]), is provided to the MDU through a high-bandwidth link (e.g., T1 line). The AP supplies this link to a dedicated router on the premises that provides shared bandwidth to each residence. Various levels of bandwidth are routed to each residence and billing is allocated accordingly.

Video service is provided by a cable television company through an underground cabling infrastructure or by a satellite television service provider (SP) through a satellite dish and headend. The headend accommodates splitting of the satellite signal and the subsequent amplification of the signal for transport to the individual units. In the case of a community antenna television (CATV) solution, the AP's central office is usually within range of the MDU and a separate headend is not required on site.

Other possible services include:

- Access control.
- Fire/life safety.
- Intercom.
- Closed circuit television.
- Building automation.

Cabling to support the required telecommunications services for the residents of an MDU is challenging. Video service may come from either ground level (CATV) or from the roof (satellite). Backbone cabling may come from below, above, or both directions.

Just as with single-family homes, the homeowner or resident within each unit of an MDU may desire internal telecommunications, entertainment, security, and control functions. One unique requirement of cabling a unit in an MDU is to determine what constraints applying to the installation of low-voltage cabling systems are put on the homeowner by the building owner or homeowners' association. This can become an issue when residential cabling or satellite dishes must occupy common facilities within and upon the building or when issues such as excessive noise caused by such services disturb other homeowners.

The individual unit's cabling system in an MDU—the cabling directly terminated at the telecommunications outlet/connectors—should be run from a DD located within the individual unit in a star topology per ANSI/TIA/EIA-570-A, *Residential Telecommunications Cabling Standard*. However, for a variety of reasons, including better economics and ease of maintenance, many multi-dwelling unit RSCSs serve telecommunications outlet/connectors within residences directly from a common distribution point in an MDU-TR or ER, and is not recommended because:

- The DP that defines ownership of the RSCS is blurred.
- System maintenance and upkeep is difficult.
- There may be security issues.
- The intraresidential unit capabilities (e.g., internal video, and networking) are limited.
- The MDU resident will not have access to the ADO in the MDU-TR or ER.

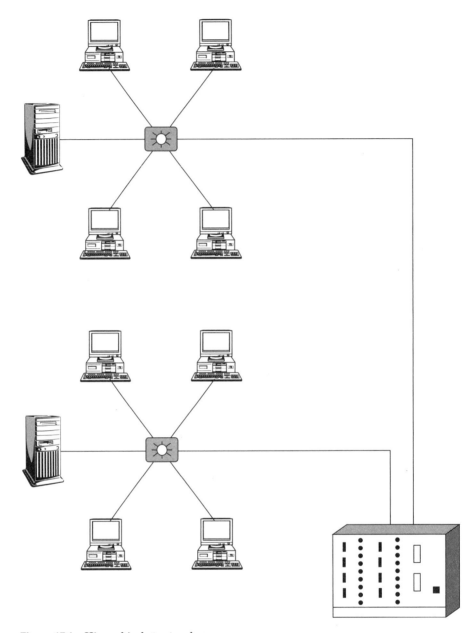

Figure 17.1   Hierarchical star topology.

To take advantage of these services, the hierarchical star should be used. The hierarchical star topology is an extension of the star topology. In this configuration, network devices in specific locations (e.g., the home office) are connected to a hub or switch as in a star topology (see Figure 17.1). These hubs or switches are then connected to each other via a central hub, also following a star configuration. This is the recommended topology for backbone RSCSs in large buildings and in campus environments.

Fire resistance requirements for telecommunications cables

The cable running in the ceilings, floors, and walls of a building can contribute to the spread of a fire and can worsen the damage it causes. The pathways and spaces occupied by telecommunications cabling and other utilities (e.g., electrical wiring) provide ideal conduits for the spread of fire. The cable itself can contribute fuel to the fire. In addition, the noxious fumes and smoke resulting from burning cable sheathing and other plastic components can be more deadly to humans than the fire itself.

Firestopping, through the establishment and maintenance of firewalls and other structural barriers to impede the spread of fires once they start, is a vital component of all commercial and most residential construction. It is important to firestop wall, ceiling, and floor penetrations when cabling both new construction and retrofit projects.

In addition to installing and repairing firestops in a range of different construction situations, it is the responsibility of the residential cabling installer to pull cable that conforms to local fire codes. A fire-rating classification system has been established for communications cable (see the *National Electrical Code® [NEC®, 2002]*). In the early 1990s, this fire-rating classification system was harmonized with that in the *Canadian Electrical Code® (CEC®)* by a binational technical committee appointed to review differences between the two national electrical codes.

These fire-rating systems establish a hierarchy of classifications that place limits on the fuel and noxious smoke and gases that cabling can contribute to a building fire. The limits for different classifications differ in their stringency, based on where the cable is placed. For example, cable to be placed in a plenum, or air-handling space, must meet the requirements of the most stringent classification. Such cable is commonly known as plenum-rated cable. Riser-rated cable must meet slightly less stringent requirements, because cable in risers has been judged to be less conducive to building fires than has cable in plenums.

The fire ratings of the various cabling products on the market are verified by independent testing laboratories using standardized tests of varying stringency.

Cabling products that are submitted to an independent testing laboratory, tested using one of these procedures, and meeting the requirements of the procedure for combustion characteristics and products are announced to the cabling industry as being approved for use, usually via a directory or catalog published by the testing laboratory. Products may be listed as "qualified," "tested," "listed," "classified," and "approved." Different terms are used by different laboratories, but all are essentially synonymous.

Articles 800.50 and 800.51 of the *NEC* describe the fire-resistance ratings of copper-based communications cables. Articles 770.50 and 770.51 provide the corresponding ratings for optical fiber cables. The ratings for copper-based cables installed in the United States are outlined in the following table.

TABLE 17.1   Cable markings for copper-based cables

Cable Marking	Type
MPP	Multipurpose plenum
CMP	Communications plenum
MPR	Multipurpose riser
CMR	Communications riser
MP, MPG	Multipurpose general-purpose
CM, CMG	Communications
CMX	Communications, limited use
CMUC	Undercarpet

MP-type cables are multiconductor or coaxial cables that must satisfy *NEC* requirements in addition to the basic C-type requirements. For example, MPR cable must meet CMR requirements and additional requirements applied either to multiconductor or coaxial cables.

The cable markings in Table 17.1 are listed in a hierarchy, from top to bottom based on the stringency of the fire-resistance test the cable has passed. For this reason, a cable marking may be substituted for any cable marking listed below it in the table. For example, CMP plenum-rated cable may be substituted for CMR riser-rated cable in an installation. MP-type cables are listed before C-type cables because they meet all C-type requirements, as well as their own additional requirements. MP-type cable may be substituted for the corresponding C-type cables listed below it (e.g., MP may be substituted for CM). The fire ratings for optical fiber communications cables in the United States are outlined in the following table.

TABLE 17.2   Cable markings for optical fiber cables

Cable Marking	Type
OFNP	Nonconductive plenum
OFCP	Conductive plenum
OFNR	Nonconductive riser
OFCR	Conductive riser
OFN, OFNG	Nonconductive general-purpose
OFC, OFCG	Conductive general-purpose

The optical fiber cable markings listed in Table 17.2 are given in a descending hierarchy, based on the severity of the fire-resistance testing the cable type has passed. As a result, a cable marking may be substituted for any cable marking below it in the table. Generally, nonconductive optical fiber cables can be substituted for their conductive counterparts.

Cost is the major obstacle to using higher-rated cables in general-purpose applications. The higher-rated cables, and especially plenum cable, tend to be more expensive, primarily because of the requirement for specialized jacketing materials.

It is the responsibility of the residential cabling installer to:

- Identify the function of the space through which a cable run passes (e.g., plenum, riser, etc.).

- Ensure that all cable pulled in that space is labeled as meeting the minimum fire rating applicable for that space, as determined by the *NEC* or local fire and building codes.

- Ensure that all cable pulled in that space is listed, marked, or otherwise labeled by an approved third-party independent testing laboratory as attaining the fire-rating classification stamped on the cable.

> **NOTE:** Once the cable pull is completed, it is also the responsibility of the residential cabling installer to ensure that all fire barriers that have been penetrated are properly sealed or restored.

Firestopping in the residential environment

Firestopping or plugging the holes bored in headers and footers in the residential environment has been overlooked. Local building codes may require the holes in headers and footers (between floors) used for cabling to be firestopped. This movement is prevalent in both single and multi-family dwellings. Building codes vary from area to area and state to state. Knowledge of local building codes is a necessity if firestopping is required.

According to the *NEC*, communications circuits are required to firestop any penetration through floors or ceilings. Article 800.52, *Installation of Communications Wires, Cables, and Equipment, (B) Spread of Fire or Products of Combustion*, states "Installations in hollow spaces, vertical shafts, and ventilation or air-handling ducts shall be made so that the possible spread of fire or products of combustion is not substantially increased. Openings around penetrations through fire resistance-rated walls, partitions, floors, or ceilings shall be firestopped using approved methods to maintain the fire resistance rating. The accessible portion of abandoned communications cables shall not be permitted to remain."

Follow the *NEC* to properly firestop any penetration in a header or footer. Check with the local authority having jurisdiction (AHJ) to determine if the local building code mandates firestopping.

Components of a Multi-Dwelling Unit (MDU) Residential Structured Cabling System (RSCS)

The basic components of an MDU RSCS described below are standard for any type of MDU building, including townhouses, apartment complexes, and high-rise condominiums. The cabling requirements for such a structure are outlined in ANSI/TIA/EIA-570-A, in a section following the cabling information on single-family residences. The information in this section is brief and should be supplemented with an understanding of the two main commercial building cabling standards:

- ANSI/TIA/EIA-568-B.1, *Commercial Building Telecommunications Cabling Standard, Part 1; General Requirements,* 2001.
- ANSI/TIA/EIA-569-A, *Commercial Building Standard for Telecommunications Pathways and Spaces,* 1998.

The two standards differ as follows: ANSI/TIA/EIA-568-B.1 addresses cabling, while ANSI/TIA/EIA-569-A covers the architectural pathways and spaces into which cabling and components are installed. Both components of an MDU RSCS and the pathways and spaces housing the components are covered in this chapter.

MDU residential structured cabling installations include most or all of the following components:

- Entrance facilities
- Main terminal space
- Equipment rooms (ERs)
- Multi-dwelling unit telecommunications rooms (MDU-TRs)
- Backbone cables
- Auxiliary disconnect outlet (ADO) cables
- Distribution device (DD)
- Outlet cables
- Telecommunications outlet/connectors

See Chapter 2: Residential Structured Cabling Systems, for information on ADO cables, DDs, outlet cables, and telecommunications outlet/connectors.

Entrance Facilities (EFs)

The EFs includes the cabling components needed to provide a means to connect the outside services facilities to the premises cabling. This area can include:

- Service entrance pathways.
- Cables.

- Connecting hardware.
- Grounding and bonding.
- Primary (electrical) protection devices.
- Transition hardware.
- Firestopping.
- Water blocking.

The AP is generally responsible for the installation of the items listed above to a specified DP, which is the interface between the AP's facility and the customer. For purposes of this manual, the residential cabling installer is responsible for extending services from the DP to the RSCS of the resident and to the resident's telecommunications outlet/connector.

An EF is meant to provide the following:

- DP between the APs and customer premises cabling (if required).
- Primary (electrical) protection devices governed by the applicable electrical codes.
- Space to house the transition between cabling used in the outside plant (OSP) to cabling approved for intrabuilding use. This usually involves transition to a Listed cable per the *NEC*.

An EF must enter and terminate within the building at the most suitable location needed to serve the occupants of the building. The EF includes the:

- Route these facilities follow on private property.
- Entrance point (EP) to the building.
- Termination point within the building.

The structure of the EF depends on the:

- Type of entry being used (e.g., underground, tunnel, buried, or aerial).
- Route for the facility.
- Building architecture.
- Aesthetic considerations.

Although optical fiber cables are specified in many situations, the most common medium for providing connections to the AP is copper cable.

There are four principal types of EFs:

- Underground
- Tunnel

- Buried
- Aerial

Underground entrances. Underground entrances use conduit or other types of mechanical pathways to provide out-of-sight service to a building (see Figure 17.2). The pathway for an underground service:

- Is usually provided by the building owner from the building to the property line (or to another customer-owned building on the same property).
- Runs from the building entrance location to a pole, pedestal, or maintenance hole (MH) provided and maintained by the AP.

The advantages of underground entrances are that they:

- Preserve the aesthetic appearance of the building.
- Are adaptable for future facility placement or removal.
- Are economical over the life cycle.

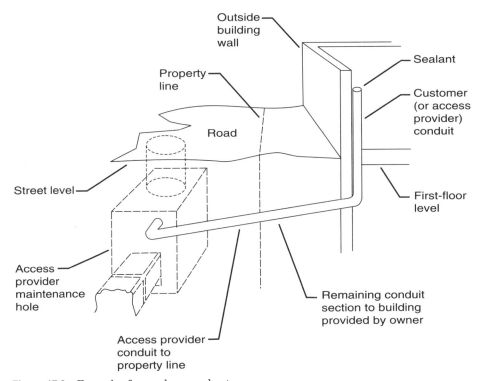

Figure 17.2 Example of an underground entrance.

- Provide the security of additional physical cable protection.
- Minimize the need for possible subsequent repairs to the property when growth is required for existing facilities.

The disadvantages of underground conduit entrances are that they:

- Have a high initial residential cabling installation cost.
- Require careful route planning.
- Provide a possible path for water or gas to enter buildings if improperly sealed.
- Usually take more time to install.

Buried entrances

Buried entrances differ from underground entrances in that they are trenched or plowed rather than running in conduit. They are a more economical means of providing out-of-sight service to a building without conduit (see Figure 17.3). The trench:

- Is usually provided by the building owner from the building to the property line (or to another customer-owned building on the same property).
- Runs from the building entrance location to a pole, pedestal, or maintenance-hole (MH). From here, the AP takes it to a pole or MH in the provider's system.
- Requires a sleeve or entrance conduit through the perimeter wall. This sleeve or entrance conduit shall be properly firestopped and water-blocked.

The advantages of buried entrances are that they:

- Preserve the aesthetic appearance of the building.
- Usually have a lower initial cabling installation cost than an underground installation.
- Can easily bypass obstructions, compared with underground installations.

The disadvantages of buried entrances are that they:

- Are inflexible for future service reinforcements or changes.
- Do not provide the same physical protection for the cable sheath as underground systems.
- May be difficult to locate unless metallic warning tape or other means are colocated, especially in the case of optical fiber without a metallic member.
- Discourage accurate route planning and recordkeeping.

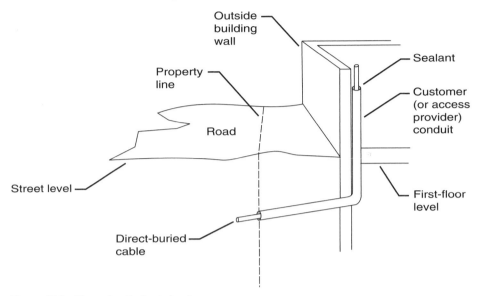

Figure 17.3   Example of a buried entrance.

Aerial entrances

Aerial entrances are another means of providing service to a building (see Figure 17.4). Aerial refers to cables placed overhead.

The advantages of aerial entrances are that they:

- Usually provide the lowest residential cabling installation cost.
- Are readily accessible for maintenance.

The disadvantages of aerial entrances are that they:

- Affect the aesthetic appearance of the building.
- Are subject to traffic and pedestrian clearances.
- Can damage a building's exterior.
- Are susceptible to environmental conditions (e.g., falling tree limbs, wind, lightning, and ice loading).
- Are usually joint-use cabling installations with the electric company, CATV company, and telephone or data SPs.

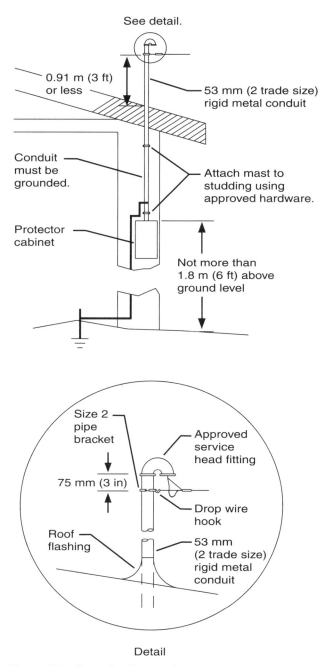

Figure 17.4   Example of an aerial entrance.

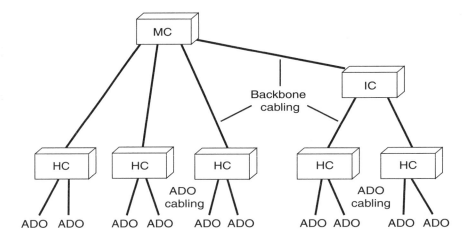

ADO = Auxiliary disconnect outlet
HC = Horizontal cross-connect
IC = Intermediate cross-connect
MC = Main cross-connect

Figure 17.5   Example of structured backbone cabling installation.

Backbone cabling

The term backbone is used to describe cables that handle the major network traffic (see Figure 17.5). There are two types of backbone cables:

- Interbuilding—Defined as a cable that handles signaling between buildings.
- Intrabuilding—Defined as a cable that handles traffic between MDU-TRs and ERs in a single building.

Backbone cabling is unlikely to be an issue in single-family residential construction, but intrabuilding backbone cable will play a role in MDUs. Interbuilding backbone cabling may be required for multi-building developments.
    The main components of backbone cabling are:

- Cable pathways—Shafts, conduits, raceways, and floor penetrations (e.g., sleeves or slots) that provide routing space for the cables.
- Firestopping.
- Grounding and bonding.
- Cables—Optical fiber, twisted-pair copper, coaxial copper, or some combination of these cables. Refer to Chapter 7: Cabling Media, for recognized backbone cables.
- Connecting hardware—Connecting blocks, patch panels, interconnections, cross-connections, or some combination of these components.

Certain considerations should be addressed when planning a backbone cabling system:

- The useful life of the backbone cabling system is expected to consist of several planned growth periods (typically three to 10 years). This is shorter than the overall life of the premises cabling system.

- Prior to the start of a planning period, the maximum amount of backbone cable for the planning period should be projected; growth and changes during this period should be accommodated without installing additional cabling.

- Planning the routing and support structure for copper cabling should avoid areas where potential sources of electromagnetic interference (EMI) may exist.

These pathways must be sized such that all necessary cabling is accommodated. In addition:

- Spare pathway capacity shall be planned for future media additions or modifications.

- A minimum of one 103 mm (4 trade size) conduit or sleeve shall be provided for each pathway for cable extending from the main terminal space to an MDU-TR.

- A minimum of one 41 mm ($1^1/_2$ trade size) conduit or sleeve shall be provided where cable bundles of 25 mm (1 in) or less extend through each apartment or housing unit.

Interbuilding backbone pathways provide the means to connect main terminal spaces with other main terminal spaces.

Equipment Rooms (ERs) and Multi-Dwelling Unit Telecommunications Rooms (MDU-TRs)

An ER is a special-purpose room that provides space and maintains a suitable operating environment for large telecommunications equipment. ERs are generally considered to serve an entire building (or even a campus), whereas a MDU-TR should serve no more than three floors of an MDU.

Equipment rooms:

- Provide for termination and cross-connection of backbone and ADO cables.

- Provide work space for service personnel.

- Are designed according to specific requirements associated with the cost, size, growth, and complexity of the equipment.

- Can also serve as a portion of an EF or as an MDU-TR.

- House large pieces of common control equipment (e.g., VDV, fire/life safety, security, building automation).

Although an ER usually serves an entire building, occasionally buildings use more than a single ER to provide one or more of the following:

- Separate facilities for different types of equipment and services
- Redundant facilities and disaster avoidance

Certain considerations should be kept in mind for an ER:

- It must be versatile. An ER must be designed to accommodate both current and future applications. It must have provisions for growth and the ability to

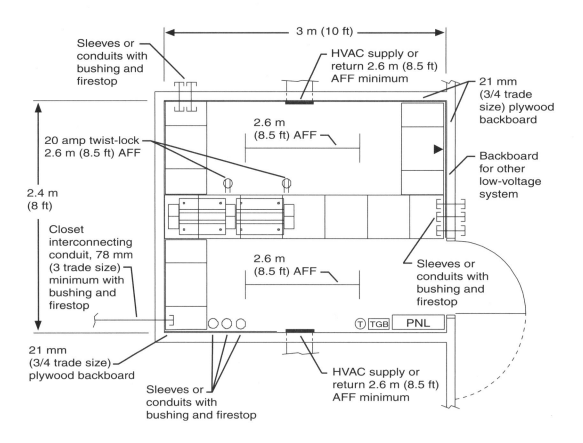

Figure 17.6   Example of an equipment room.

go through numerous equipment replacements and upgrades during its life, with minimal service disruption and cost.

- It must meet lighting, air-conditioning, floor-loading, electrical, and minimum-space requirements.

MDU-TRs differ from ERs and EFs in that they are generally considered to be floor-serving as opposed to building-serving facilities.
These MDU-TRs:

- Serve as a point of termination for ADO and backbone cables on compatible connecting hardware.

- House the horizontal cross-connect—A collective reference for the connecting hardware, jumpers, and patch cords used for completing cross-connection and/or interconnection of ADO and backbone cable terminations.

- May contain intermediate cross-connect points for different parts of the backbone cabling system.

- Provide a controlled environment for the telecommunications equipment, connecting hardware, and splice closures.

Consider the following in planning MDU-TRs:

- The size of the building, the usable floor space served, the occupant needs, and the telecommunications service to be used

- The need to optimize the ability of the MDU-TR to accommodate change

- Lighting, air-conditioning, floor-loading, electrical, and space requirements

Observe all applicable codes and plan for all proposed uses of MDU-TRs during construction.

TABLE 17.3   Minimum space for multi-dwelling unit telecommunications room.

Space	Grade 1	Grade 2
Minimum space for first five resident units	375 mm (15 in) wide 610 mm (24 in) high	775 mm (30.5 in) wide 610 mm (24 in) high
Minimum additional space per resident unit	32 000 mm^2 (50 in^2)	64 500 mm^2 (100 in^2)

If active equipment is placed within the MDU-TR, a dedicated, unstitched 15 ampere, 120 volt alternating current (Vac) nominal outlet shall be provided within 1.5 m (5 ft) of the MDU-TR. The height of the electrical outlet should be suitable for the MDU-TR being installed and shall be in compliance with applicable codes.

Main terminal spaces

The main terminal space may be colocated with the EF. It may also be used to house active equipment. The main terminal space may house the DP, ADO cable, and backbone cable.

The main terminal space is the point of entrance into the building for most services. Each AP shall be provided space for its respective network interface devices (NIDs) and DPs. This facility contains pathways that link to either MDU-TRs or each residence. Primary protection should be provided within this facility if required.

A dedicated, nonswitched 15 ampere (120 Vac) duplex electrical outlet shall be provided within the main terminal space. The circuit shall be terminated in a duplex electrical outlet. All equipment within the main terminal space shall be grounded.

Cable Support Systems

Cable support accessories are designed to route, manage, and protect telecommunications cabling. These products are widely used in routing cable between floors and units within an MDU, or within the MDU-TR inside an MDU. They may also be used to conceal cable upon the exteriors of buildings or within a building or home.

Cable trays and ladder racks

Cable trays and ladder racks are available in many sizes and configurations. Cable trays and ladder racks are manufactured from steel and aluminum. Typically, cable trays provide a pathway to support installed cables. They are also used to support cables from one wall to another within a MDU-TR or ER. Cable trays and ladder racks are available in various styles such as:

- Tubular ladder rack.
- Rod stock (basket).
- Center rail.

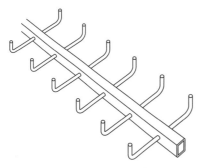

Figure 17.7   Center rail cable tray.

Conduits

Conduits are raceways that house backbone cables and ADO or outlet cables. Conduits that enter an MDU-TR or ER should be terminated at specific locations on the wall to enable orderly routing of the cables to terminating connecting blocks in the room. Alternatively, they may be routed through the room to another location. It is vital to adhere to a minimum cable bend radius of not less than 10 times the outer diameter of the cable sheath (see Figure 17.8).

Surface-mounted raceway

A surface-mounted raceway is installed on the surface of walls, ceilings, and floors. It is available in both metallic and nonmetallic versions. The metallic version is commonly used to conceal electrical cable. The nonmetallic version is most often used to conceal low-voltage cabling. It can, however, be used to conceal electrical cable as well. Metallic surface raceway should be bonded and grounded when used for electrical or low-voltage cabling.

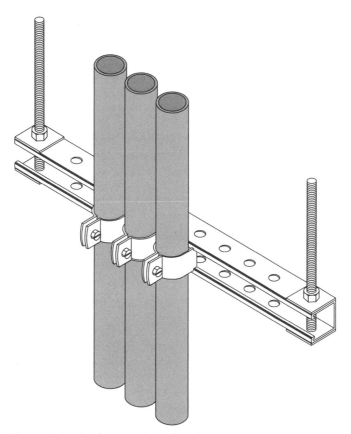

Figure 17.8   Conduits on channel stock.

Fire protection

Fire protection and fire safety are important issues for any residential construction project, whether single-family or MDU. It should be noted that residential telecommunications cabling is subject to inspection by building and electrical inspectors, as well as fire marshals and other locally appointed officials. Consult the local fire code AHJ, usually the fire marshal or building inspector, to obtain the firestopping and fire safety procedures applicable to a construction project.

Multioccupant Cabling Layout

Typical layouts for cabling to the DP or ADO in MDUs are shown (see Figures 17.9–17.15). These diagrams are typically the designs used in distribution systems up to the ADO that is colocated with the DD.

NOTE: The DP may be located at the protected terminal, the MDU-TR, the ADO, or in another building or structure.

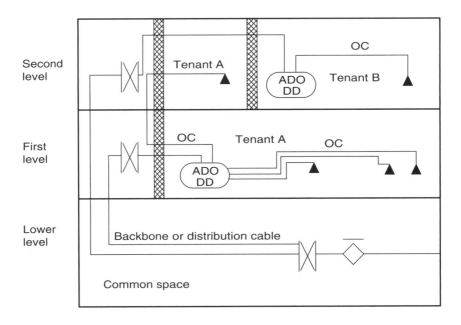

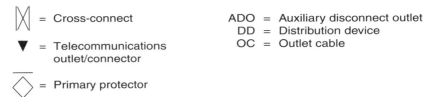

Figure 17.9 Multioccupant cabling layout.

Apartment buildings with central and multiple backbones

Figure 17.10 illustrates an example of telecommunications backbone and ADO cabling layout for an apartment building with a central backbone.

NOTE: Residential units are not similar in design and are not stacked one above the other.

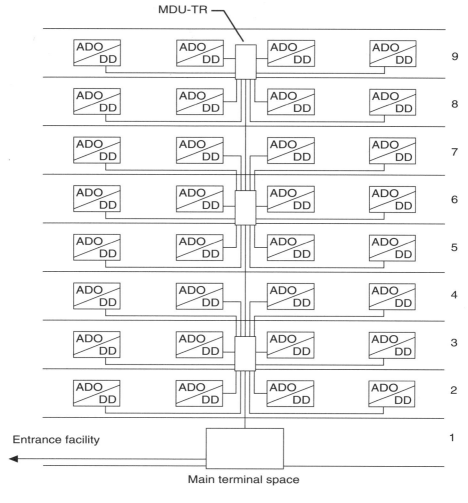

ADO = Auxiliary disconnect outlet

DD = Distribution device

MDU-TR = Multi-dwelling unit telecommunications room

Figure 17.10 Telecommunications backbone and ADO cabling layout for an apartment building with a central backbone.

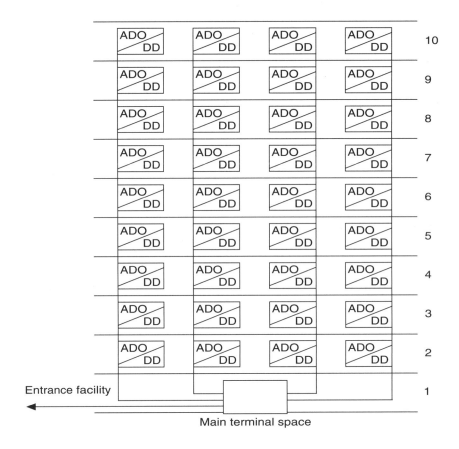

ADO = Auxiliary disconnect outlet

DD = Distribution device

Figure 17.11   Telecommunications backbone and ADO cabling layout for an apartment building with multiple backbones.

Figure 17.11 illustrates an example of telecommunications backbone and ADO cabling layout for an apartment building with multiple backbones.

NOTE: All residential units are of similar design and stacked one above the other.

Townhouses and side-by-side duplexes

Figure 17.12 illustrates an example of conduit distribution for a seven-unit townhouse. Distribution is in:

- Metallic conduit in the ceiling.
- Rigid polyvinyl chloride (PVC) in the slab.

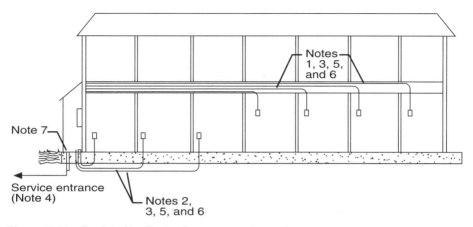

Figure 17.12 Conduit distribution for a seven-unit townhouse.

NOTES: 1. One 27 mm (1 trade size) electrical metallic tubing (EMT) conduit should be run to each unit. Conduit may be in slab or ceiling space.

2. One 27 mm (1 trade size) rigid PVC Type 2 or metallic conduit should be run to each unit.

3. All conduit ducts should be limited to two 90-degree sweeping bends.

4. Size and location of conduit ducts should be determined by the AP.

5. Provide and install all ducts and conduits with a pull string.

6. CATV should be in a separate system.

7. One 27 mm (1 trade size) PVC conduit should run to the accepted ground grid.

Figure 17.13 illustrates an example of cabling distribution for a side-by-side duplex residence. Distribution is in:

- Metallic conduit to Unit 1.
- Open cabling through the wood-frame construction in the wall cavity.

NOTES: 1. Location of the service entrance (drop) is determined by the AP. This location may be considered as the NID placement area by the AP. If the drop is to be concealed in the wall cavity and there are two floors or more, supply and install a

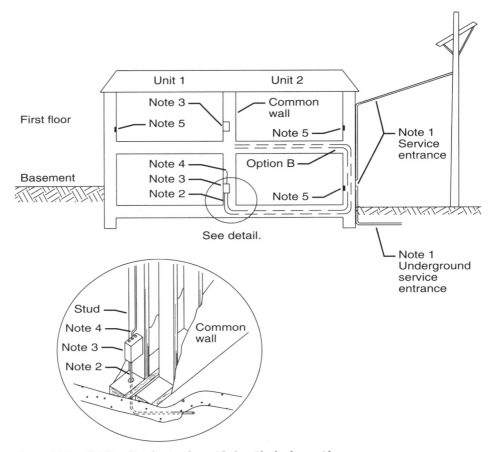

Figure 17.13   Cabling distribution for a side-by-side duplex residence.

53 mm (2 trade size) metallic conduit in the wall cavity. If underground service is required, provide a 53 mm (2 trade size) conduit.

2. Provide and install a 27 mm (1 trade size) rigid PVC conduit in slab or, as an alternative, a 27 mm (1 trade size) metallic conduit in ceiling space. Provide a waterproof pull cord in the conduit. CATV is to be in separate system.

3. Note 3 indicates an outlet box.

4. Note 4 indicates station cabling or distribution cabling installed in wall cavity.

5. Note 5 indicates a low-voltage mounting bracket.

Frame apartment and apartment complex projects

Figure 17.14 illustrates an example of cable distribution for frame apartment projects. Distribution through the wood-frame construction in the wall cavity is in:

▪ Metallic conduit.

▪ Open cabling.

> **NOTES:** 1. Size and location to be determined by the AP.
>
> 2. Telephone backboard size to be determined by the AP.
>
> 3. One 27 mm (1 trade size) conduit to the ADO on the first floor from the terminal room.

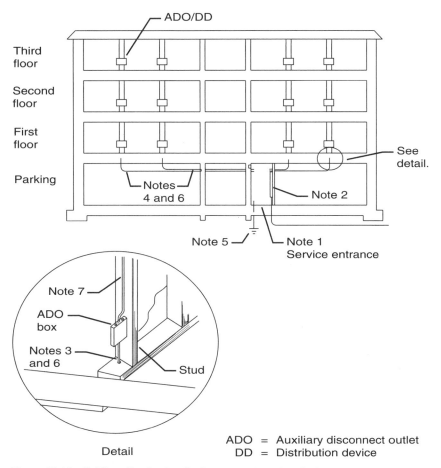

Figure 17.14 Cabling distribution for frame apartment projects.

4. One 27 mm (1 trade size) conduit.

5. One 27 mm (1 trade size) conduit to an approved ground.

6. CATV to be in a separate system.

7. ADO cable is installed in the wall cavity within an individual unit.

Figure 17.15 is an example of an apartment complex with backbone cable extending across a campus.

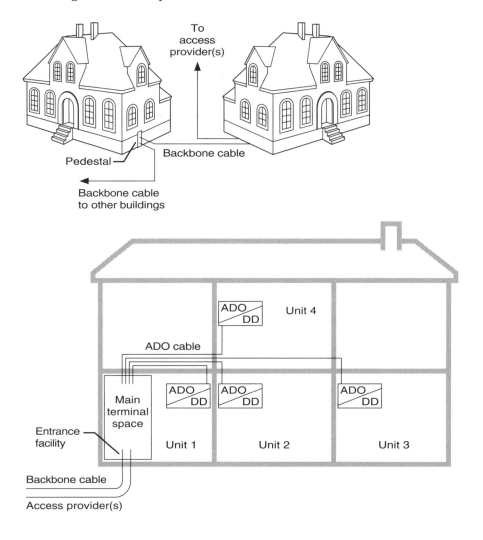

ADO = Auxiliary disconnect outlet
DD = Distribution device

Figure 17.15   Apartment complex with backbone cable.

Chapter 18

Multi-Dwelling Unit Structured Cabling Installation

Overview

Multi-dwelling units (MDUs) can be viewed as essentially commercial enterprises. For the most part, a landlord or building owner rents, leases, or sells some or all of the individual units in a particular complex, although the landlord or building owner may occupy some part of the premises.

A building owner cables MDUs for commercial reasons, while a homeowner does so to suit personal and family tastes and perceived needs for communications and entertainment. These differing rationales for cabling the residence call for different cabling schemes, a distinction recognized in ANSI/TIA/EIA-570-A, *Residential Telecommunications Cabling Standard*. Section 5 of the standard focuses on cabling systems for single residential units, while Section 6 covers multi-dwelling and campus infrastructure.

This chapter reviews the MDU requirements of the cabling standard. The following unique residential installation issues and procedures that apply to MDUs will be described: setting up a multi-dwelling unit telecommunications room (MDU-TR), wiring cross-connects, pulling backbone cable, and more.

Cabling Standard Requirements for Multi-Dwelling Units (MDUs)

The residential structured cabling system (RSCS) for a multi-dwelling building is defined as the infrastructure connecting the demarcation point (DP or demarc) with the MDU-TRs, and the MDU-TRs with telecommunications outlet/connectors in individual dwelling units.

ANSI/TIA/EIA-570-A makes the following points:

- The DP in MDUs may be located at a minimum point of entry or individual tenant spaces may have their own.
- Access to shared-use space is controlled by the building owner or agent.
- Where the total length of cabling from the DP to the farthest outlet exceeds 150 m (492 ft), the access provider (AP) must be notified at the design stage so that transmission requirements can be accommodated.
- Grounding and bonding is to be performed in accordance with local electrical codes. Additional information on this subject is provided in ANSI/TIA/EIA-607, *Commercial Building Grounding and Bonding Requirements for Telecommunications.*

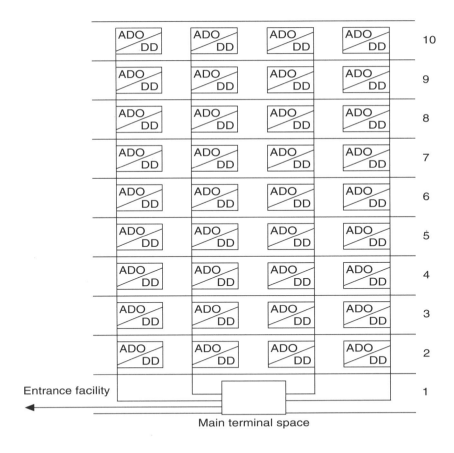

ADO = Auxiliary disconnect outlet

DD = Distribution device

Figure 18.1   Typical backbone cabling system for a multi-floor building (stacked).

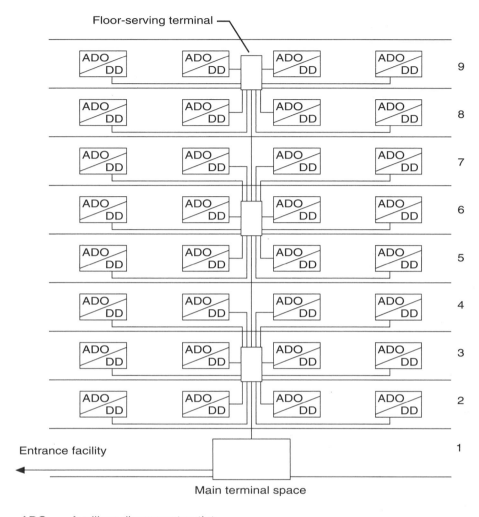

ADO = Auxiliary disconnect outlet

DD = Distribution device

Figure 18.2 Typical backbone cabling system for a multi-floor building (MDU-TR).

Adjoining Walls in Multi-Dwelling Units (MDUs)

Building code controls adjoining or common walls (the wall located between any two separate living units) in multiple dwelling units. Building code varies from area to area and state to state. For example, in some areas, the adjoining or common wall must be dual construction, contain an air gap with offset studs, and extend between the floor and shingled roof. In other areas, the adjoining or common wall must extend beyond the shingled roof and the adjoining wall must be masonry.

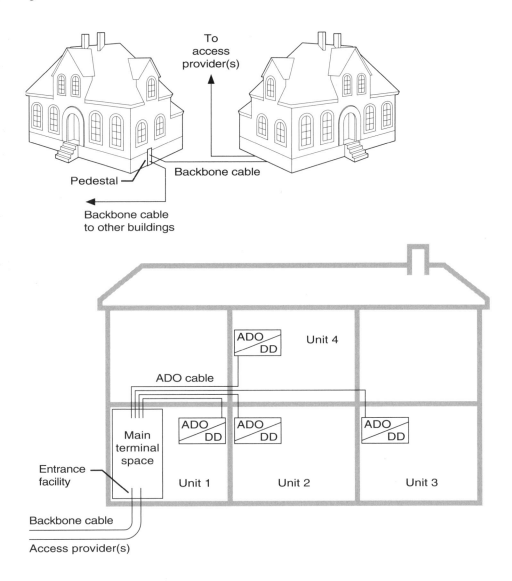

ADO = Auxiliary disconnect outlet
DD = Distribution device

Figure 18.3   Typical backbone cabling system for a multi-unit building or campus environment.

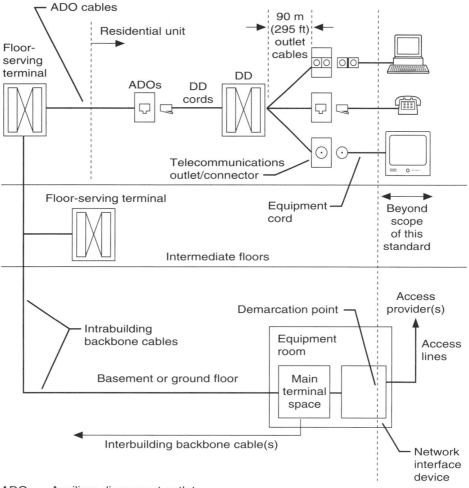

ADO = Auxiliary disconnect outlet
DD = Distribution device

Figure 18.4   Typical cabling system components of a multi-dwelling or campus environment.

The same building code controls the location of outlet placed in the adjoining or common wall. Typically, the outlets in these walls must be offset and separated by at least a given distance (in one area, this distance is 400 mm [16 in]).

When locating telecommunications outlet/connectors in the adjoining or common wall of an MDU, one must adhere to the local building code. To gain this knowledge, one must contact the authority having jurisdiction (AHJ).

Backbone specifications

Inside multi-dwelling buildings, the telecommunications distribution designer or residential cabling installer should consider establishing spare pathway capacity in the form of extra conduit to accommodate future media changes or additions that otherwise would be difficult or impossible to cable.

Intrabuilding telecommunications backbone pathways should use conduits, sleeves, slots, or cable trays to route backbone cable. In addition:

- The residential cabling standard specifies a minimum of one 103 (4 trade size) conduit or sleeve for each backbone pathway between the main terminal space and MDU-TRs.

- Cable bundles with a diameter of 25 mm (1 in) or less that extend through apartment closets should be accommodated in a backbone pathway containing a minimum of one 41 ($1^1/_2$ trade size) conduit or sleeve.

- More information on intrabuilding backbone pathways may be found in ANSI/TIA/EIA-569-A, *Commercial Building Standard for Telecommunications Pathways and Spaces*.

In a campus or multi-building environment, interbuilding telecommunications backbone pathways are required. These pathways:

- Consist of underground, buried, aerial, or tunnel constructions.

- Are described in more detail in ANSI/TIA/EIA-758, *Customer-Owned Outside Plant Telecommunications Cabling Standard*.

Backbone cabling media recognized by the residential standard are:

- 100 Ω twisted-pair.

- 50/125 μm multimode fiber.

- 62.5/125 μm multimode fiber.

- Singlemode fiber.

- Trunk distribution feeder coaxial cable.

- Series 6 and Series 11 coaxial cable.

A star topology should be used for twisted-pair and optical fiber backbone cabling. Coaxial backbone cabling should be cabled using a star or bus topology.

When cables are connected with interbuilding telecommunications backbone cabling, the applicable fusing and voltage-protection codes are to be followed.

Setting Up the Multi-Dwelling Unit
Telecommunications Room (MDU-TR)

There are two types of telecommunications spaces where the termination of cabling permits cross-connection and interconnection:

- Equipment rooms (ERs)
- MDU-TRs

These two spaces share some of the same basic purposes. They both support the installation of cables, connecting hardware, cross-connects, and electronic equipment. For instance, while an MDU-TR is generally floor serving, an ER is generally building serving. The primary difference is in the nature of what they serve.

ERs are designed to house large equipment items such as telephone cabinets, data-processing mainframe computers, uninterruptible power supplies (UPSs), and video headend equipment. The floor loading of ERs must be rated higher than for MDU-TRs because of the anticipated high concentration of equipment in a confined space. The heating, ventilating, and air conditioning (HVAC) requirements for these spaces are also greater.

MDU-TRs are designed only for limited equipment and floor loading. They may house splice cases, termination hardware, and relay racks. Small items of equipment (e.g., hubs, switches, multiplexers, and key telephone systems) can also be found in MDU-TRs.

Cross-connects between ADO and backbone systems are found in MDU-TRs. Cabling between MDU-TRs is considered to be backbone cable.

Additional space may be available in a building that is currently being used as a makeshift MDU-TR. The proper equipment must be installed and provisions made for the cable and connecting hardware to upgrade the space to a true MDU-TR. If the existing MDU-TR is shared with electrical, plumbing, or HVAC facilities, then this choice for an MDU-TR must be reconsidered.

IMPORTANT: Never house telecommunications facilities in a space that handles other building utilities.

It is recommended that MDU-TRs have 19 mm (0.75 in) thick plywood backboards installed on at least two walls. The backboard should be painted with two coats of nonconductive, fire-retardant paint of a light color. The plywood provides a space for wall-mounting connecting hardware. A 300 mm (12 in) wide cable tray or ladder rack should be mounted on the same wall as the backboard. Good planning dictates that all walls of an MDU-TR be equipped with plywood and cable tray/ladder rack to facilitate future growth.

The designer's documents will indicate the size, location, quantity, and nomenclature of the equipment to be installed in the MDU-TR, along with a routing diagram of the cables to be installed or that pass through the room. They will also show the location of the pathways entering or leaving the space and who is responsible for installing them. If equipment racks are to be installed, a plan view of the space will indicate where their respective footprints are located and how they relate to other equipment being installed. In most instances, the designer's documents will indicate where voice, data, and video (VDV) cables are to be terminated. The designer may be an architect, consulting engineer, or a Registered Communications Distribution Designer (RCDD®).

Always ensure that appropriate clearances are maintained around all pieces of equipment. In the absence of a specified distance, plan for a minimum of 1.0 m (3.3 ft) of work and aisle space. If an MDU-TR layout is not provided, prepare one.

Plywood backboards

Plywood backboards are installed on walls in MDU-TRs. Plywood is available in two types: interior and exterior, and in four grades—A, B, C, and D.

Sheets of plywood are normally sized 1.2 m (4 ft) wide by 2.4 m (8 ft) high. The thickness is variable, but, for the purposes of this manual, only two will be considered: 19 mm (0.75 in) and 25 mm (1 in). Plywood that is too thin allows the screws used by residential cabling installers to penetrate completely through the plywood and sometimes does not offer enough strength to ensure that mounted hardware is securely anchored. Plywood sheets thicker than 25 mm (1 in) are not usually required. This is contingent on the sheet of plywood being properly attached to the building structure.

The finishing grade of plywood (A, B, C, or D) describes the quality of the surface (i.e., degree of knotholes or blemishes).

- Grade A is the highest grade and is without any surface blemishes.
- Grade B has the knotholes cut out and replaced with a patch of clean wood.
- Grade C contains some blemishes and an occasional small knothole.
- Grade D contains knotholes without any repair or corrective action by the manufacturer.

Grading of a sheet of plywood may result in a different grade for each of the two sides. For instance, a sheet of plywood could be graded A/B—one side is A and the reverse side is B.

Plywood should be void free. This means that the space in each layer inside the plywood where the knotholes are removed is completely filled with replacement wood patches. Voids inside the sheet of plywood may create a weak spot where the attachment hardware (e.g., screws and toggle bolts) cannot hold fast.

For telecommunications use, grade A/C plywood should be used. The A side is exposed to the interior of the MDU-TR, and the C side is placed against the building structure or cabinet wall. Generally, start mounting the plywood 150 mm (6 in) above the finished floor. Although not required by TIA standards or the *National Electrical Code® (NEC®)*, BICSI strongly recommends that plywood be painted with two coats of fire-retardant paint in the color specified by the designer.

> **NOTE:** Some local codes may require the use of treated plywood. If this applies, do not paint the plywood. The effects of the treatment will cause the paint to crack, deteriorate, and peel off the plywood backboard. In addition, the saline solution can cause the metal hardware mounted to the plywood to corrode.

Installing plywood

Plywood sheets used for backboards should be installed with the longest dimension reaching from the floor level up toward the ceiling. In the case of a 2.7 m (9 ft) or 3 m (10 ft) ceiling, do not yield to the temptation to raise the bottom of the plywood to split the difference. This will raise the working height level such that a ladder may be required to work on equipment mounted at the top of the plywood.

Plywood should be installed properly, so there is no separation between adjacent sheets. When installing plywood in a corner, the plywood backboard can be installed plumb and adjacent to the edge of one side of the wall at the corner with the sheet on the intersecting wall butted up against the first sheet to form a smooth, 90-degree corner (see Figure 18.5).

The plywood backboard must be secured on top of the existing Drywall or to the studs in the perimeter walls of the room. When installing the plywood on bare studs (no Drywall), screws a minimum of 19 mm (0.75 in) longer than the depth of the plywood backboard must be used.

When installing plywood on Drywall that has already been installed on the studs, always verify the load rating of the wall prior to installing the plywood. If the load rating will permit this type of installation, use toggle bolts (butterfly bolts) to ensure the stability of the installation. These toggle bolts should be a minimum of 6.3 mm (0.25 in) in diameter and must be sufficient in length to allow the bolt to seat behind the Drywall after installation (see Figure 18.6).

For this Drywall application, toggle bolts should be installed at approximately 610 mm (24 in) spacing around the entire perimeter of the plywood board. If desired, recess the bolt heads to allow for use of the entire area on the plywood. Never recess the bolt head on any plywood less than 25 mm (1 in) thick. This affects the ability of the plywood to hold the desired load of equipment and termination hardware. The toggle bolts should be installed 50 mm (2 in) from the edges of the sheet of plywood on approximately 610 mm (24 in) centers. Locate the toggle bolts using care to avoid the studs when drilling. The studs will prevent the toggle bolt's wings from opening behind the Drywall (see Figure 18.7).

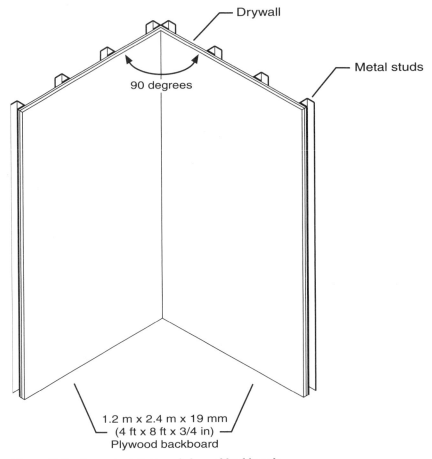

Figure 18.5   Corner installation of plywood backboards.

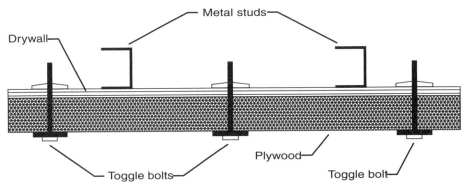

Figure 18.6   Installation using toggle bolts in Drywall construction.

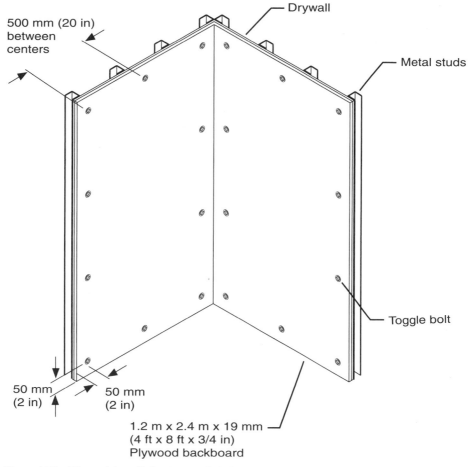

Figure 18.7 Plywood installed using toggle bolts.

The following checklist should be used for an MDU-TR.

TABLE 18.1   Checklist for a telecommunications room installation

✔	Tasks
	1. Obtain blueprints/specifications.
	2. Determine equipment layout.
	3. Verify type of wall construction and load capacity.
	4. Install plywood backboard per designer's drawings.
	5. Install vertical and auxiliary disconnect outlet (ADO) pathways.

continued on next page

TABLE 18.1   Checklist for a telecommunications room installation

✔	Tasks
	6. Install relay racks.
	8. Install terminal blocks and/or patch panels.
	9. Install D-rings and other wall-mounted hardware.
	10. Install telecommunications grounding busbar (TGB) or telecommunications main grounding busbar (TMGB) at the location shown on the designer's drawings.
	11. Install grounding and bonding conductors.
	12. Install optical fiber connecting hardware.
	13. Label all pathways and hardware in accordance with the designer's drawings.
	14. Label all terminal blocks, patch panels, and other connecting hardware in accordance with the designer's drawings.
	15. Label all grounding and bonding conductors in accordance with their originating and terminating locations.
	16. Update design documents to reflect any changes required by field conditions.
	17. Verify that the electrical panels, electrical outlets, and lighting fixtures are adequate for the MDU-TR requirements.
	18. After the work is complete, remove all trash, wire clippings, boxes, packing crates, excess materials, and any other equipment or cable that does not permanently reside in the MDU-TR. Make the space ready for installation of the electronic equipment.

Layout

If possible, install plywood backboards around the entire perimeter of the MDU-TR. This enhances the use of the wall space in the MDU-TR and allows cables to be installed around the walls to where terminal equipment will be located, now or in the future. It also facilitates attaching cables that pass through vertically to MDU-TRs located on floors above or below.

Installing Backbone Pathways

Overview

When installing backbone pathways, it is important to ensure that the route for the pathways is verified prior to actually installing the support mechanisms for the pathways. Fire- and smoke-rated barriers have to be penetrated. Obstructions (e.g., HVAC ducts, large pipes, and structural beams within the building) have to be accommodated. Be sure that the chosen route will provide a clear path.

Conduit materials

Conduit is primarily used for backbone pathways but can also be used as an ADO pathway. There are three types of conduit used inside commercial buildings today:

- Electrical metallic tubing (EMT)

- Intermediate metallic conduit (IMC)

- Galvanized rigid conduit (GRC)

EMT is available in 16 mm ($^1/_2$ trade size) through 103 mm (4) trade size); IMC and GRC are available in 16 mm ($^1/_2$ trade size) through 155 mm (6 trade size). Both the outside diameter (OD) and inside diameter (ID) of steel conduit are larger than the trade size designation (e.g., 16 mm [$^1/_2$] trade size). EMT has an OD of 18 mm [0.7 in] and the ID is about 17 mm [0.67 in]). The metric designators were developed to aid in using design specifications for government jobs. They are designators only and not actual measurements in millimeters. The metric trade size designators for electrical conduit differ from pipe sizes to differentiate between conduit and pipe/tubes for other purposes.

Polyvinyl chloride (PVC) conduit should not be used inside MDU buildings. When subjected to extreme temperatures during fires, this type of conduit emits toxic smoke and gases that can cause extensive injury or death to those persons attempting to escape the fire. The products of combustion can also be severely corrosive to telecommunications equipment and can disrupt service to critical telecommunications equipment. PVC conduit can be used under slab-on-grade construction (where codes permit).

Electrical Metallic Tubing (EMT)

EMT is thin wall metal tubing. The ends are not threaded. It is used widely today in electrical distribution systems and as a pathway for telecommunications cabling. EMT is supplied in standard 3 m (10 ft) lengths. Each length of tubing is referred to as a section.

Couplings. EMT uses two types of couplings to connect sections of conduit together. Compression couplings are installed by first slipping the fitting's nut and then its compression ring onto the conduit. The conduit is then slipped into the main coupling body and the compression ring is compressed as the nut is threaded and then tightened onto the coupler body. This is accomplished using two pipe wrenches or slip-joint pliers. The other type of coupling is a set-screw coupling. It is installed by inserting the conduit into the fitting and tightening the set screws with a screwdriver. Both couplings are used extensively throughout the industry.

Be sure conduit ends are securely placed in the coupling or connector. This is important to ensure that the set screws have room to bite into the conduit

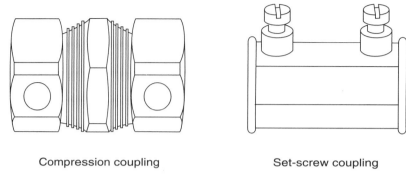

Compression coupling   Set-screw coupling

Figure 18.8   Electrical metallic tubing couplings.

for a secure joint. Correct measurement, including length of the elbow legs, is necessary to have sufficient conduit length for secure joints.

Intermediate Metallic Conduit (IMC)

IMC has a heavier wall than EMT. Although the wall is less than that of GRC, it is even stronger than EMT. Like EMT, it is available in 3 m (10 ft) lengths (sections). Each section of IMC is factory-threaded on both ends. When a section of this conduit is cut to length, the residential cabling installer must then install new threads on the cut end. The threads are required to have a 19 mm (0.75 in) per foot taper and must comply with ANSI/ASME B.20.1. This provides a secure joint when made up with the standard threaded couplings, which have straight-tapped threads.

These conduits can be purchased with an attached coupling, which can be secured without turning the pipe, making installation easier. Elbows of this integral type are also available and are very useful in tight applications. Compression-type connectors and couplings are also available (see Figure 18.9).

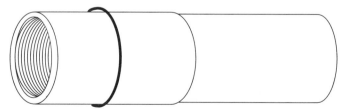

Figure 18.9   Intermediate metallic conduit.

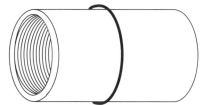

Figure 18.10 Intermediate metallic conduit coupling.

Couplings. Straight-tapped screw-on couplings are used to connect sections of IMC together. When standard couplings are used, the conduit, not the coupling, is turned. This generally means the conduit on the right and left of the coupling are turned in opposite directions. Pipe wrenches are required to tighten the conduit into the couplings. If they are not sufficiently tightened, the connection will not be airtight and will loosen over time. This could result in the separation of the conduit section and the coupling. Integral-coupling-type IMC is available in 53 mm (2 trade size) through 103 mm (4 trade size); for this type only, the coupling must be turned to make up the joint. This is especially useful in tight spaces. Couplings used on IMC are the same as those used on GRC. There is nothing to prohibit joining a stick of IMC to a section of GRC (see Figure 18.10).

Galvanized Rigid Conduit (GRC)

GRC is thicker and heavier than IMC. Like IMC, it is factory threaded and joints are screwed together. The same couplings and thread forms are used on IMC and GRC. GRC is available in 3 m (10 ft) lengths (sections). The sections of conduit are already threaded on both ends. As with IMC, when a section of GRC is cut to length, the residential cabling installer shall then install new threads on the cut end. As with IMC, threads are required to have a 19 mm (0.75 in) per foot taper and shall comply with ANSI/ASME B.20.1 (see Figure 18.11).

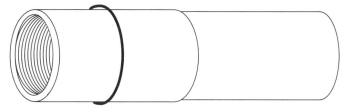

Figure 18.11 Galvanized rigid conduit.

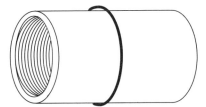

Figure 18.12   Galvanized rigid conduit coupling.

Couplings.   Straight-tapped screw-on couplings are used to connect sections of this conduit together. The threads of GRC are deeper than IMC. This is made possible by the thickness of the conduit wall. When standard couplings are used, the conduit, not the coupling, is turned. This means the conduit on the right and left of the coupling are turned in opposite directions. Pipe wrenches are required to tighten the conduit into the couplings. If they are not sufficiently tightened, the connection will not be airtight and can loosen over time. This could result in the separation of the conduit section and the coupling. Integral coupling-type GRC is available in 53 mm (2 trade size) through 103 mm (4 trade size). For this type only, the coupling shall be turned to make up the joint. This is especially useful in tight spaces.

Proper support, good tight joints, and bonding for an assured ground are the important elements of a good installation of EMT, IMC, and GRC. Grounding and bonding shall comply with Article 250 and Chapter 8 of the *NEC*. Good workmanship in these areas will provide a backbone pathway that will serve well for many years (see Figure 18.12).

Support requirements

The support requirements of *NEC* Article 348 for EMT, Article 345 for IMC, and Article 346 for GRC shall be followed. Where out-of-the-ordinarily heavy loads or abuse are anticipated, residential cabling installers may choose to add extra supports to assure joints remain secure. EMT, IMC, and GRC shall not be supported by the ceiling grid or by ceiling support wires. Separate support shall be provided. This is particularly important when the ceiling is fire-rated, since any extra load could compromise the rating.

All exposed raceways are to be run as near parallel or perpendicular to walls and ceilings as the residential cabling installer can achieve.

Do not use raceways as support for equipment. Provide separate support, unless otherwise permitted by the *NEC*.

Install the complete raceway before installing the cables. Be sure joints are tight and the raceway is securely terminated and held firmly in place.

Cutting and threading Intermediate Metallic Conduit (IMC) and Galvanized Rigid Conduit (GRC)

A roll-type cutter should not be used on EMT. Using a hack saw or band saw will permit reaming of the EMT without flaring the ends. Be sure to make a square cut. It is best to ream EMT with a tool that is designed for that purpose. This will make fittings install easier and better. If other tools are used (e.g., pliers), special care must be taken to not flare the ends.

Be sure to measure the exact length of conduit needed. If it is too short, good thread engagement cannot be made with IMC and GRC, or set screws and rings of compression fittings will not engage with EMT.

Use a standard 19 mm (0.75 in) per foot taper die that is sharp. Cut full, clean threads. A worn die or poor threading practices can result in ragged and torn threads.

Adjust the threading dies and use a factory-threaded piece to set the die; lock the dies so they are firmly held in the head by tightening the screws or locking collar. A proper thread will usually be one thread short of flush with the thread gauge. This is within permitted tolerances.

IMC and GRC can be cut with a saw or roller cutter. It is important to make a straight cut and to start the die on the pipe squarely. Wheel-and-roll cutters must be revolved completely around the pipe, tightening the handle about one-quarter turn each time it is rotated.

Ream interior edges after cutting and before threading.

One of the most important aspects of good threading is to use cutting oil freely. Apply the oil after the die has taken hold. Keep the conduit well lubricated throughout the entire threading process.

It is a good practice to thread one thread short to prevent butting of conduit in a coupling and allow the coupling to cover all of the threads on the conduit when wrench tight.

After the die is backed off, clean the chips and lubricant from the thread.

Joint make-up

Always read and follow packaging instructions for any fittings. Use fittings specifically for the raceway type and size being installed. This information is generally found on the container.

Expansion fittings are seldom needed for steel conduit installed in buildings. Expansion joints for the raceway shall be evaluated if large temperature extremes are expected, or building expansion joints are in the pathway. The coefficient of expansion to be used is $2.83 \times 10 - 4$ mm/°C ($6.5 \times 10 - 6$ in/°F).

The need for square cut ends has been discussed where threadless fittings are used. These ends are to be clean and assembled flush against the fitting's end stop. Threadless fittings for IMC or GRC should not be used unless the conduit manufacturer specifically recommends that application. Threadless fittings do not give a gas-tight and oil-tight seal.

All threaded joints have to be wrench tight. Engage at least five full threads, but be careful not to overtighten.

EMT is joined by set screw or compression fittings. There are a variety of designs, but there is no one method of tightening that applies to all. It is important not to overtorque or overtighten screws while assuring they are firmly secured. Most bad joints are due to lack of attention to workmanship and failure to set the screws or compression glands.

For good joints and terminations, it is important to measure the length of conduit needed and select the right elbows.

Conduit elbows and bends

Factory-manufactured bends are recommended for conduit installation. Bends can be manufactured in the field, but specialized equipment must be used to accomplish the required angle of bend. Field-manufactured bends typically result in an elliptically shaped cross-section, rather than being completely round. This is caused by the action of the tools used to bend the conduit. A hand bender can be used in the field to bend 16 mm ($^1/_2$ trade size) up to 35 mm ($1^1/_4$ trade size). Conduits are specified by their trade size.

If the conduit has an internal diameter of 50 mm (2 in) or less, the bend radius must be at least six times the internal conduit diameter. Bends are available in standard radii or sweeps. Recommended conduit bends are commercially available beginning at 21 mm ($^3/_4$ trade size).

Conduit bends are readily available in various bend configurations. Elbows and bends of 11.25 degrees, 15 degrees, 22.5 degrees, 30 degrees, 45 degrees, and 90 degrees are available through distributors. These bends are available in all conduit trade sizes. Special radius and long sweeps are available but may have to be special ordered.

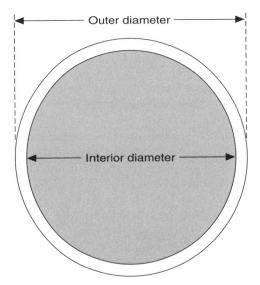

Figure 18.13   Cross-section of conduit, outer diameter vs. inner diameter.

If odd bend degrees are needed, various configurations of factory-manufactured bends can be used to accomplish the specific requirements. For example, one 45-degree bend and one 15-degree bend can be coupled together to form a 60-degree bend; two 30-degree bends can also be used to accomplish this. However, it is preferable to use one elbow of the proper degree for easier pulling of the conductors. Long sweeps are also available, but may need to be special ordered. Conduit bends are commercially available with varying leg lengths on each side. A leg is the extension of the conduit bend past the point where the bend stops and straightens out.

Most elbows are furnished with a standard leg length for each size. Nonstandard leg lengths are also available, but may have to be special ordered. Leg lengths may vary from one manufacturer to another. The residential cabling installer should confirm these lengths in order to maintain a uniform installation with accurate termination points on the conduits.

Whenever possible, conduit sweeps should be used rather than bending the conduit in the field using mechanical methods.

Hangers

Many different types of hangers are available to support the installation of this type of conduit. Because of this, a limited number of options are described in this section.

Pipe hangers are used to support conduits. The hanger is a pear-shaped device that is attached to an all-threaded-rod (ATR). An ATR is a length of rod stock that has been threaded for its entire length at manufacture. ATR is available in various lengths and thickness.

A pipe hanger is suspended from the building structure by an anchor and a section of ATR. The anchor is installed in the concrete structure of the floor or beam. When concrete is not available and steel trusses are installed in the building, beam clamps can be used to support the ATR and hanger. The selection of anchors and beam clamps should be determined by the load weight of the conduit and cable to be supported. The ATR is attached to the pipe hanger with nuts and lock washers. The assembly is then capable of supporting the conduit that is installed through the pipe hanger (see Figure 18.14).

Another type of conduit support is a trapeze. This device is made by using two ATRs and a section of channel stock. The ATR is suspended from the building structure and attached to each end of the channel stock using appropriate nuts and washers. The conduits are then attached to the channel stock with pipe clamps and locked in place.

A third type of conduit support is a one-piece conduit hanger. This device is manufactured in such a way that the compression bolt is part of the hanger itself and, when loosened, will not come off the hanger. This type of hanger can be mounted directly to the building structure, to red iron, or to a specialized hanger mount.

Other types of conduit hangers are available and should be researched by the residential cabling installer prior to performing the work.

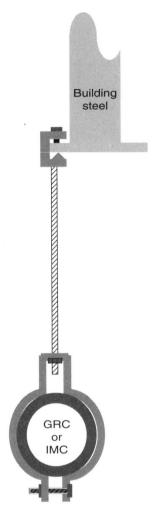

Pipe hanger assembly

GRC = Galvanized rigid conduit
IMC = Intermediate metallic conduit

Figure 18.14   Pipe hanger.

Conduit terminations

Backbone conduits should be terminated where they enter or leave an MDU-TR or ER. The recommended location for terminating the conduit is in a horizontal plane where the conduit penetrates the wall of the MDU-TR or ER. Allow no more than 50 mm (2 in) of conduit and bushing to extend into the room. Conduits that enter an MDU-TR should terminate near the corners to allow for proper cable routing. Terminate these conduits as close as possible to the wall where the backboard is mounted (to minimize the cable route inside the room).

Conduits should be reamed in accordance with procedures outlined in the section on cutting and threading. Conduits should have a bushing installed to reduce cable sheath damage during the pulling operation.

Conduits shall be equipped with grounding bushings. Grounding bushings are installed on the end of the conduit. With EMT conduit, a set-screw grounding bushing is placed on the end of the conduit and tightened using the appropriate tool. On IMC and GRC conduit, a grounding bushing is screwed onto the threaded end of the conduit. Tighten the bushing until it is secure. A threadless grounding bushing is also available that can provide a more convenient means for locating the ground lug. These are installed with set screws.

Another method for securing conduits is employed when conduits are turned down the wall and terminate at the top of the plywood or turned up and terminate at the bottom of the plywood.

Terminate conduits that protrude through the structural floor 25–75 mm (1–3 in) above the surface. This prevents cleaning solvents or other fluids from flowing into the conduit.

When conduits are turned down in a room, terminate them above the plywood backboard. This allows full use of the plywood backboard for termination and routing of cabling. If this method is used, channel stock can be used to attach the conduits in a fixed manner to the wall. Each conduit can be attached to the channel stock with a pipe clamp. If the pipe clamps are equipped with "teeth" that bite into the conduit and the channel stock is not painted, a grounding bushing is not required on every conduit. A single conduit in each run of channel stock can be equipped with a grounding bushing. The entire section of channel stock can effectively be grounded using a single bushing and ground wire. An alternative method is to install a listed grounding lug onto the channel stock and then route the ground wire to the appropriate ground bar.

NOTE: Backbone conduits should never be turned down in an MDU-TR, as it adds an extra bend to the conduit and increases the coefficient of friction during the pulling operation.

The following table provides guidelines for adapting designs to conduits with bends.

NOTE: Consider an offset as equivalent to a 90-degree bend.

TABLE 18.2   Adapting designs

If a Conduit Run Requires...	Then...
More than two 90-degree bends	Provide a pull box between sections with two bends or less.
A reverse bend (between 100 and 180 degrees)	Insert a pull point or pull box at each bend having an angle from 100 to 180 degrees.
More than two 90-degree bends between pull points or pull boxes	For each additional bend, derate the design capacity by 15 percent or use the next larger size of conduit.

No more than two 90-degree bends are allowed in any one conduit segment. A third bend may be acceptable in a conduit segment without derating the conduit's capacity if:

- The run is not longer than 10 m (33 ft), or

- The conduit size is increased to the next trade size, or

- One of the bends is located within 300 mm (12 in) of the cable feed end. (This exception only applies to placing operations where cable is pushed around the first bend.)

> **NOTE:** In practice, a third bend should not be used unless there is no way to avoid it.

Securing conduit formations

Backbone conduits shall be secured on each end and throughout the entire route to prevent swinging and swaying during cable placement. When large, high-pair-count cables are installed in backbone conduits, the pulling of the cable places significant tension on the conduit. Winches are employed in this placing operation.

Sometimes the winch is anchored to the building or attached directly to the conduit. This tension causes swinging and swaying of the conduit and its hangers. Excessive movement of the conduit can cause the hangers to loosen and possibly come free. If this happens, the conduit and its cable could fall to the floor resulting in damage to the cable, the building, and possibly the residential cabling installers.

Conduits must be secured at each end in such a manner that they do not move. Cross braces can be used throughout the route to stabilize the conduits and prevent movement. This can be accomplished by the use of conduit clamps, channel stock, or ATR placed at opposing angles (180 degrees opposite from each other). This helps prevent lateral movement of the conduit during placing operations. The same anchoring mechanisms can be used to secure the clamps and cross braces as are used to hang the conduits from the building structure.

Pathway Preparation

It is important to determine the entire route of a backbone pathway prior to installation of the supporting hangers, ATRs, or other support mechanisms. The entire route should be planned ahead of time to ensure that the conduit can be installed without unforeseen obstacles. This is especially true when having to penetrate fire- or smoke-rated walls and floors. If the penetration cannot be established, then all the work done to install the support hardware may have to be repeated at another location.

Always make penetrations through fire- or smoke-rated walls and floors prior to installing the hangers, clamps, and trapezes. Once the conduit is installed, firestop the penetrations using approved methods.

Installing Auxiliary Disconnect Outlet (ADO) Pathways

Stub-up/stub-out conduits

The terms stub-up and stub-out imply that a section of conduit is used to provide a pathway in a vertical and then horizontal direction from a point of termination. While similar in many ways, they are significantly different from a residential cabling installation perspective.

Stub-ups are usually single sections of small diameter (not less than 21 mm [$^3/_4$ trade size]) metallic conduit. They originate at a single- or double-gang box installed in Drywall or paneling. The stub-up continues vertically through the wall cavity where it penetrates the wall cap and stubs up into the suspended-ceiling area. It terminates at that point and is usually equipped with a conduit bushing and a pull string. Sometimes the stub-up is equipped with a 90-degree bend that is turned back into the room, especially when installed in fire- or smoke-rated walls.

Stub-outs are usually short runs of small-diameter metallic conduit. They originate at a single- or double-gang box installed in Drywall or paneling. The stub-out continues vertically through the wall cavity, where it penetrates the wall cap, and continues into the suspended ceiling area. In a typical residential cabling installation, the conduit continues out of the room and into an adjacent hallway. The conduit may terminate as it exits the wall of the hallway or may continue to another type of supporting structure (e.g., a cable tray or ladder rack). It terminates at that point and is usually equipped with a conduit bushing and sometimes a pull string.

NOTE: Conduit sizes larger than 32 mm (1.25 in) ID are not generally employed in this type of residential cabling installation because the knockouts on standard receptacle boxes do not accept box adapters for larger conduits.

Surface-mounted raceway

Surface-mounted raceway is installed on the surface of walls, ceilings, floors, modular furniture panels, and modular furniture. It is available in both metallic and nonmetallic versions. Metallic surface raceway should be bonded and grounded when used for electrical or low-voltage cabling.

Metallic surface raceway.   This type of raceway is available in many sizes and configurations. Do not confuse it with wireway or hinged-cover duct systems.

This raceway is available in two sections: a base and a cover. The base is installed on the wall surface using fasteners. The cabling is installed and then the cover is installed over the top of the cabling. The cover is held in place by snapping it over the base. The entire assembly is secured to the surface by the use of two-hole clamps that are specifically designed to fit over the installed assembly. The two-hole clamps are anchored to the wall by fasteners.

Nonmetallic surface raceway.   Like metallic raceway, nonmetallic raceway is available in many sizes and configurations. Two types of nonmetallic raceway are available: noncategorized installation raceway and Category 5-type installation raceway. The primary differences between the two are the elbows and fittings used to couple sections of the raceway together. Nonmetallic raceways, being nonconductive, avoid most of the problems associated with bonding and grounding of pathway components.

Nonmetallic surface raceway is available with a single channel or divided channels. When divided channels are required, they are available in two- and three-channel versions. The dividers used to create the channels are removable. They can be removed by breaking them out of the raceway section. By flipping the divider back and forth in a 180-degree arc, the attachment to the base is weakened, allowing the divider to be pulled out of the section of raceway.

This type of raceway is available in two component sections or a single-component version. Some manufacturers refer to the single-component version as a latching duct. This version is manufactured as a single assembly and, once anchored in place, the cover simply snaps onto the opposite sidewall, latching the cover to the base. Other manufacturers offer a hinged version. The hinged version is a two-piece assembly. One side of the cover is equipped with a round projection that is snapped into a cavity on one side of the base. When closed, the opposite side latches onto the other side of the base.

These raceways are available in a wide range of colors. Most of the colors offered by manufacturers match the National Electrical Manufacturers Association® NEMA® colors (ivory, black, grey, red, white, brown, and almond). Most manufacturers offer special colors at additional cost. In most instances, these special orders require additional lead time for delivery. Some manufacturers also allow their raceways to be painted after installation. This increases the difficulty of reentering the raceway after the paint has cured and may not be aesthetically desirable.

This type of raceway is also available with or without adhesive backing. If adhesive backs are used on a project, be aware that some of the adhesive backing will not fully adhere to all types of wall finishes. The backing will hold for a limited time, after which the raceway will fall off the structure. This may necessitate use of an additional adhesive, anchors and screws, or some other type of fastener to ensure the stability of the installation.

Various connectors are also available for surface raceway. Splice connectors are used to join two sections of the raceway in a straight line. Internal, external, and flat elbows are used to change direction with the raceway. Where joining three sections of the raceway together creates a T, a T cover is used to conceal the joint. End caps are used at the termination of a raceway where a surface-mount box is not used. Conduit adapters, reducers, and ceiling fittings are also available for surface-mount raceway.

Three hand tools are essential for the proper installation of surface raceway. They are a level, a straight edge, and a ruler. A PVC cutter is used to cut sections of the nonmetallic raceway to the desired length. A compound power miter saw is helpful in cutting difficult angles. With the proper blade, it provides an exact fit between the two sections of raceway. By using this power tool, any angle of cut is obtainable to exacting measurements, creating an almost imperceptible joint after installation of the raceway.

Surface-mount boxes

Surface-mount boxes are available in single- and double-gang versions. They are also available in shallow, standard, and deep versions. The single-gang box measures approximately 50 mm (2 in) wide by 100 mm (4 in) tall by 54 mm (2.12 in) deep. The double-gang version measures approximately 100 mm (4 in) wide by 100 mm (4 in) tall by 54 mm (2.12 in) deep.

Surface-mount boxes are available in a wide range of colors. Most of the colors offered by manufacturers match the NEMA colors (ivory, black, grey, red, white, brown, and almond). Most manufacturers offer special colors at additional cost. Some manufacturers also allow their boxes to be painted after installation.

Surface-mount boxes are available with or without adhesive backing. If adhesive backs are used on a project, be aware that some of the adhesive backing will not fully adhere to all types of wall finishes. The backing will hold for a limited time, after which the raceway will fall off the structure. This may necessitate use of an additional adhesive, anchors and screws, or some other type of fastener to ensure the stability of the installation.

Pulling Cable

Pulling cable for telecommunications residential cabling installations inside of buildings requires using many techniques. Some cables are large and heavy, disguising the fact that the pairs inside the cable are actually fragile. Be prepared to learn the correct way to handle the cable so that the residential cabling installation will meet specifications.

Setting up, or getting ready, is the proper way to ensure a smooth job of pulling cable. Precable pulling procedures are essential for a successful cable pull:

- The work area must be secured (cleared of pedestrian traffic) for safe conditions to exist for everyone.
- Equipment for supporting the cable reels must be in place.
- Equipment for pulling the cable must be available.
- Equipment for temporarily holding the cable in place must be available.
- Correct lengths of cable must be on hand.
- Before the pull begins, locations for accessing and pulling the cable should be identified.
- Adequate workers must be available.
- Equipment allowing residential cabling installers to communicate must be on hand.

ADO cabling begins at the MDU-TR on each floor and ends at the distribution device (DD). Types of support structures include:

- Conduit.
- Cable trays.
- J-hooks.
- Modified types of bridle rings.
- Beam clamp.
- Innerduct.
- Building red iron or roof-support structures.

Conduit provides a good pathway for the cable. Conduit may be made of:

- EMT.
- Rigid metal.
- Rigid PVC.
- Fiberglass.

Conduit made of flexible metallic tubing should not be used unless it is the only practical alternative. Check local codes to ensure flexible tubing is permitted.

Conduit provides a safe environment that prevents cable from being accidentally cut or damaged. Normally, conduit is installed from the ER to each of the MDU-TRs in what is usually called a home run.

Moving the cable through the conduit from one end to the other requires some type of object (pull string or rope) to precede the cable through the conduit. This is known as fishing a conduit.

Methods for installing the pull string or rope through the conduit includes using:

- The manual fish tape method.
- Air-propelled or vacuum methods.

A fish tape is a steel or fiberglass wire-like device that is rigid enough to be pushed all the way through the conduit, or in longer runs, to a pull box. It is used to retrieve the pull string from the far end.

Air-propelled methods include:

- A vacuum on one end and a foam "rat" or ball attached to a pull string on the other end.
- A compressed air bottle or mechanical blower used to propel a pull string attached to a foam ball or rat.
- Other devices either purchased or made up on the job.

ADO cabling in an open ceiling uses a different pulling method from those described above. Various ways are used to pass the pull string through the trusses or other structural elements. Cables should have a straight and smooth path; this is essential for a good cable pull. There are various methods used for placing pull string, including attaching:

- A ball to a pull string and throwing it through the open ceiling space.
- A fiberglass pole to the pull string.
- Remote-controlled appliances to the pull string.

Cables should always be installed parallel or perpendicular to exterior and interior walls.

Backbone cabling provides interconnections between the MDU-TRs, ERs, and entrance facilities. For vertical backbone cabling pathways, the cabling is more easily pulled from the top down than from the bottom up because gravity helps with the pull. It is the preferred method if the reels of cable can be moved to the top floor. Sometimes this cannot happen because the reels are too large to fit through doors or on a freight elevator. When necessity demands it, backbone cable is pulled from the bottom up. In both cases, special cable-handling devices are required, including:

- Winches.
- Cable reel brake.
- Temporary take-up devices.
- Bullwheels.
- Pulleys.
- Mesh grips.
- Bull lines (pull ropes).

The type of equipment needed depends on the direction of the pull. If pulling from the bottom up, a winch may be needed. If pulling from top down, a reel brake may be needed. A reel brake is a mechanical device used to stop or slow a freewheeling reel, thus keeping it from unreeling too fast due to gravity.

Communication between coworkers is essential in every cable pull. Everyone should be prepared to alert the person pulling the cable, with or without power equipment, to ensure the cable is traversing the route smoothly without twisting, kinking, or getting bound up in some way. Two-way communication is also normally used to monitor the progress of the job.

Adhering to fire and building codes and standards gives residential cabling installers the ability to minimize the likelihood of the spread of fire and smoke in a building. All penetrations through fire- or smoke-rated walls and floors need to be firestopped with approved methods to reduce the chance of spreading fire and smoke. Firestop materials are available as:

- Putty.
- Pillows.
- Caulking compounds.
- Cementitious compounds.
- Blankets.
- Wrap strips.
- Collar devices.
- Composite sheets.
- Sprays.
- Mechanical systems.

All firestopping solutions are a combination of firestopping materials, holding devices, packing materials, and other devices that make up a listed (approved) system. Do not use unapproved methods to firestop a penetration. Always use an engineered and approved system to properly firestop a penetration. Contact an appropriate firestop manufacturer for any situations that are not addressed specifically by the manufacturer's listed methods.

Cable pulling setup

There are many preparation tasks that affect safety. Once the job of pulling cable begins, everything and everyone should be in the right place. This can only occur with proper planning. A good residential cabling installer is efficient and organizes the work.

A good cable setup means all materials are in place so the cables can be handled properly. A residential cabling contractor may need specialized equipment capable of holding large reels. Smaller equipment (e.g., cable tree) may be brought in to handle the many rolls containing low-pair-count cable. If all preparations are made correctly, time will be saved once pulling begins.

The job site should be secured from occupants of the building. Other contractors as well as building occupants need to know that an installation is going on. The safety of everyone involved is important. Securing the area is a way to let other people in the building be aware that an installation is underway. The equipment in use may create a hazard where people walk.

Prepare for starting the cable into the conduit by:

- Securing the area.
- Setting up the cable.
- Setting up the pull string.
- Selecting and identifying the cable-labeling system to be used.
- Identifying pull points.

Following are steps in a cable pulling setup.

Step	Cable Pulling Setup
1	Secure the area (see Figure 18.15).
	▪ Set up cones, signs, barricades, and/or caution tape in work areas to alert everyone of danger in the area.
	▪ Place caution tape across the entrance to the area to restrict access to anyone other than residential cabling installers or other authorized personnel.
	▪ Notify appropriate personnel that work is beginning.
2	For large cable reels, set up a jackstand or reel dolly. A reel dolly is basically a jackstand with wheels that allows easy relocation of the assembly (see Figure 18.16).
	▪ Find out if the cable reel being installed is so large and heavy that a jackstand or reel dolly is required.
	▪ Many contractors use jackstands for holding the reels of cable off the floor. Sometimes this equipment is made from homemade supports and a pipe is used to appropriately suspend the reel.

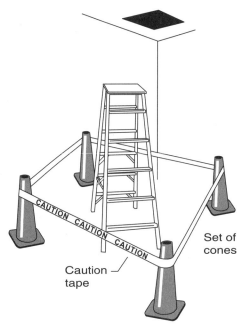

Figure 18.15 Secured area with safety cones and caution tape.

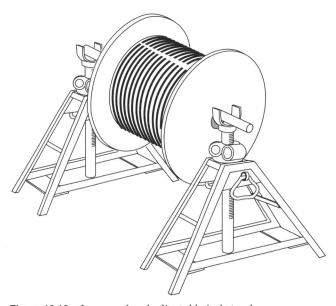

Figure 18.16 Large reel and adjustable jackstands.

Step	Cable Pulling Setup

- Select a location that is large enough for the number of spools needed.
- Set up the jackstand inside the MDU-TR if there is room or set it up in the area just outside the room.
- Place a pipe or crossbar (also known as a mandrel) through the center hole in the reel. Get help from coworkers to lift the spool onto the jackstand. Jackstands used for placing larger cables usually have a mechanical ratcheting or hydraulic lift mechanism to assist in lifting the reels into place for pulling.
- Make sure the mandrel can support the weight of the cable reel(s).
- In some pulling operations, reel brakes may be needed to control the pay out of the cable.

3   For smaller reels, set up a cable tree (see Figure 18.17).

- Cable trees are used when there are multiple small spools of cable being pulled at the same time.
- Select the spools of cable that will be needed for the job. Cable can be ordered with various put-up lengths.
- Select a location near the cable feed point that is large enough for the cable tree.
- Bring in the cable tree and set it up inside the MDU-TR if there is space, or in an area outside the MDU-TR (e.g., in the hallway).
- Mount the spools on the cable tree.
- Find the end of each cable reel and bring them together, through the guide hole on the pay-out end of the cable tree.
- Select one of the cables in the group as a guide cable.

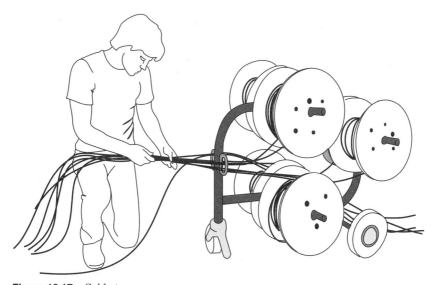

Figure 18.17   Cable tree.

Step	Cable Pulling Setup

- Take the next two cables and tape them to the guide cable approximately 75 mm (3 in) behind the guide cable. Continue attaching cable to the bundle in groups of two until all cables are attached to the bundle.

- Prepare the guide cable for attachment to the pull string.

4   For even smaller amounts of cable, set up a box of cable.

- Knock a hole in the top of the box or remove the plastic insert to expose the cable inside.

- Cable is shipped in several different containers.

- The boxes are convenient and require no rack to hold them. For pulling multiple strands of cable, the boxes can be stacked and numbered sequentially for proper labeling later.

- Find the end of the cable and thread it out through the hole in the top of the box.

- Keep the cable free from any obstruction that would crimp or bend it, causing damage.

5   Set up pull string or pull rope. If no pull string or pull rope exists, and it is required in the conduit system, one must be installed.

6   Identify pull points for each ADO cable run.

- Determine the distance of the complete cable run. Ensure the distance does not exceed 90 m (295 ft). Take care to identify the changes in direction, vertically or horizontally. These changes in direction may cause the run to exceed the 90 m (295 ft) limit.

- Take note of the number and locations of bends and identify where to feed the cable around sharp bends or turns. Pulley hangers may be used at some points to save manpower (see Figure 18.18).

7   Allow no more than two 90-degree bends per 30 m (100 ft) pull.

- Divide the total cable run distance into segments where each segment is less than the maximum of 30 m (100 ft), including end measurements.

- Identify a pull point, where necessary, for accessing and handling the cable.

- Where possible, minimize the number of pull points, since each pull point requires an additional person, a pulley, or a separate pull.

Figure 18.18 Bullwheel and pulley hangers.

Step	Cable Pulling Setup

NOTE: When personnel are not available to be at all the pull points, the cable will have to be pulled to the farthest manned pull point and then placed in a figure-eight on the floor. To continue the pull, workers are relocated to the next vacant pull points until the cable pull is complete.

Pulling Auxiliary Disconnect Outlet (ADO) cable in conduit with fish tape

ADO cable is installed between the MDU-TR and the DD.

Conduit installations are usually designed to be parallel or perpendicular to the external walls of the building. No one conduit segment should be more than 30 m (100 ft) in length and no more than two 90-degree bends are allowed in one segment. Exceeding these limits increases the coefficient of friction and can possibly stretch the cable and damage it.

A pull string is a thin cord used to pull cable through the conduit. The pull string must extend through the entire length of the conduit before cable can be pulled. This can be done by a manual method using fish tape or by pressurized air methods using a foam rat or a foam ball attached to a pull string. Fish tape passes through the conduit to reach the pull string attached to the cable at the far end. Some fish tapes have swivel-type clips on the end for attaching to objects that need to be pulled back through the conduit. Fish tape is used to pull the string through the conduit, not for pulling the actual cable. Some pull string is designed to break if excessive tension is placed on the cable.

ADO cabling in conduit may require the use of a lubricant. Many products are available for lubricating the cable as it enters the conduit to assure a smooth pull.

Always label cables and their reels prior to pulling cable into place. It is easier to identify and label the cables before they are pulled through the conduit.

The building plans or blueprints are the record of what is in the building. After the installation is complete, the building owner has the as-built plans to refer to for future work. It is a lasting record of cable information that documents the placing of cables.

Following are steps for pulling ADO cable in conduit with fish tape.

Step	Pulling ADO Cable in Conduit with Fish Tape
1	Estimate length of run by walking off the distance or using a measuring wheel.

- Find the area on the floor that is just below the location in the ceiling where the conduit run is located.
- Walk the distance of the cable run, noting where the conduit bends and pull boxes are.

Step	Pulling ADO Cable in Conduit with Fish Tape
	■ Estimate the distance walked and allow for all changes in elevation. (e.g., include the distance from conduits to floors or ceilings).

> **NOTE:** A common general practice is to count ceiling tiles of known lengths, such as 0.6 m (2 ft) by 0.6 m (2 ft) lay-in or 0.6 m (2 ft) by 0.6 m (2 ft) decorator tile.

■ Ensure each cable dispenser has enough cable to reach the full length of the run.

■ Place an identification label on each cable end.

■ Label each cable dispenser and cable a few meters (feet) back from the cable's end. Do not put the label too close to the end or the tape will cover the label.

2   Determine the length of fish tape required. After estimating the length of the run, add a meter (a few feet) to each end to be sure enough fish tape is available.

3   Feed fish tape through conduit.

■ Put on all necessary personal safety equipment.

■ Feed the fish tape into the end of the conduit.

■ Do not force the fish tape if the conduit contains existing cables.

■ If there is a metallic tab on the end of the fish tape, cover it with electrical tape to prevent snagging other cable already in the conduit.

■ Never feed a fish tape into a conduit without knowing where it may go. It could accidentally be fed into a live electrical junction box.

■ Push the fish tape through the conduit until it comes out the far end of the conduit.

4   Attach pull string to fish tape.

■ Go to the far end of the conduit where the end of the fish tape is now located.

■ The end of the fish tape will have a hook, a ring, or some type of attachment device.

■ Attach the end of the fish tape to the pull string to be pulled through the conduit securely to prevent it from coming off during the pulling operation.

■ If the clip at the end of the tape has been removed, secure the attachment by wrapping a piece of electrical tape around the pull string and the fish tape.

5   Pull the fish tape out of the conduit.

■ Go back to the other end of the conduit.

■ Slowly rewind the fish tape reel to retrieve the pull string from the far end of the conduit, storing the fish tape in the holder as it is taken up.

■ Detach the fish tape from the pull string.

■ If there will be a delay prior to cable-pulling efforts, tie off the pull string at both ends. This will prevent accidentally pulling the string back into the conduit and the need to refish the conduit.

■ Put the fish tape away.

6   Overcoming pulling problems.

■ Feed out a few meters (feet) of cable to prepare for entering the conduit.

■ Arrange the cable(s) to form a smooth transition from a single cable to a larger bundle. Tape cables as necessary to hold them in place while the pull string is attached to the cable bundle.

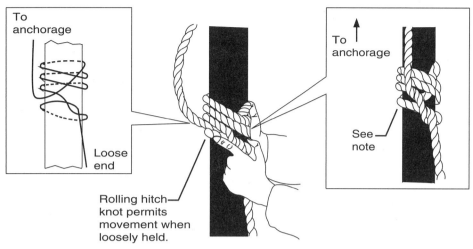

Figure 18.19   Rolling hitch knot.

Step	Pulling ADO Cable in Conduit with Fish Tape

- Attach the pull string using a rolling hitch (three half hitches and a clove hitch). Each half-hitch will bite into the cables and prevent them from slipping. The clove hitch is of little importance as long as a tight knot is placed around the cables after each half-hitch. Place electrical tape over the knots to prevent them from slipping or becoming caught during the pull (see Figure 18.19).

- Attach a trailer string to the cables to provide a string for future pulls.

- As the cable is being pulled through the conduit, keep slack available at the entrance so the cable can flow freely into the conduit.

- Guard against developing excessive bends or kinks as cable is led to the conduit.

- Some conduits may require the use of pulling lubricant to prevent excessive pulling tension on the cable(s).

7   Pull cable.

- Maintain proper bend radius of four times the diameter of the 4-pair cable.

- Monitor pull force closely during installation to be sure that the manufacturer's specifications are never exceeded.

  NOTE: Maximum pull force for one 4-pair, 24 AWG [0.51 mm 0.020 in)] cable is 110 newton (N) (25 pound-force [lbf]). Bends and lubricants also affect performance.

- Avoid excessive tension and deforming of the cable when going around corners or bends.

- Consider using a 11 kg (25 lb) breakaway swivel or a 11 kg (25 lb) fishing line attached between the pull rope and the unshielded twisted-pair (UTP) cable to ensure that no more than 11 kg (25 lb) of tension is exerted on the pulled cable.

- Leave enough slack in the MDU-TR to reach the farthest corner and add the distance from floor to ceiling. This allows service slack to be stored within the MDU-TR for any possible mistakes or future reconfigurations.

Step	Pulling ADO Cable in Conduit with Fish Tape

> **NOTE:** Most optical fiber distribution cabinet manufacturers require 3 m (10 ft) of slack within their cabinets in addition to the service slack described above.

- Allow sufficient slack for copper and optical fiber cables at the DD.
- Tie off both ends of the trailer string.

8   Identify cables.

- Once the amount of required service slack has been determined, mark the cable prior to cutting it off the cable dispenser.
- The cable label shall be a unique identifier that is clearly visible on each end of the cable after the pull is complete. If this is a temporary label, ensure that the person who follows can understand it.

9   Document cable information.

- Obtain a copy of the plans or blueprints.
- On the plan show clearly what type of cables were installed.
- Document the origination and termination point of each cable.
- On the plans show clearly which conduit was used.
- Describe the application of the installed cables.
- Install a label on the cable at each end that conforms to the labeling scheme to be used on the project. The label should identify the cable(s) so that the residential cabling installers will be able to identify the cables for termination. When terminating the cable, permanent labels should be installed.

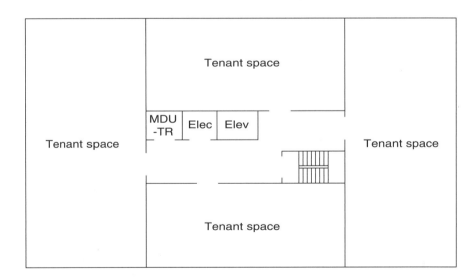

Elec = Electrical room
Elev = Elevator
MDU-TR = Multi-dwelling unit telecommunications room

Figure 18.20 Example of marked job floor plans with common symbols.

Step	Pulling ADO Cable in Conduit with Fish Tape
10	Job site clean up.

Keeping a job site clean:

- Prevents development of safety hazards.
- Reflects on the professionalism of the residential cabling installer and the cabling installation company.

General housekeeping specifics include but are not limited to:

- Picking up used pull strings and pull ropes immediately after use.
- Disposing of removed sheath and wire scrap from cables terminated in MDU-TRs.
- Storing cable reels and boxes when pulling function is complete.
- Placing termination scrap in an appropriate container while terminating cables.
- Vacuuming residue from work areas.
- Disposing of all personal use items (e.g., luncheon materials, coffee cups) and used cleaning supplies.
- Storing tools and equipment properly at the end of the work day.

When the fish tape method is not practical (i.e., because of excessive conduit lengths), the following alternative methods may be employed.

Alternative methods.   Alternatives to using fish tape include using an air bottle and foam ball, or a vacuum cleaner. These methods produce the intended result of retrieving the pull string at the far end.

Step	Alternative Method 1—Blow String through Conduit
1	Select a propellant object that is lightweight, such as a plastic bag crumpled up or a piece of foam rubber designed for the purpose.
2	Tie a pull string to the object.
3	Place the propellant object in the conduit.
4	Place the grommet on the conduit and attach the air hose.
5	Using pressurized air, blow the ball attached to the string through the conduit until it reaches the other end.
	WARNING: Never look into the end of a conduit when a pull line is being blown through it.
6	Go to the opposite end of the conduit and tie off the pull string, separating it from the ball (see Figure 18.21).

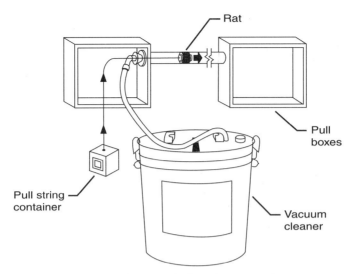

Figure 18.21   Vacuum blowing a ball or a bag.

Step	Alternative Method 2—Vacuum Pull String through Conduit
1	Attach a vacuum hose to the conduit to pull the object forward from one pull point to the next.
2	Retrieve the object ball with the pull string attached.
3	Secure the string at the receiving end of the conduit (see Figure 18.22).

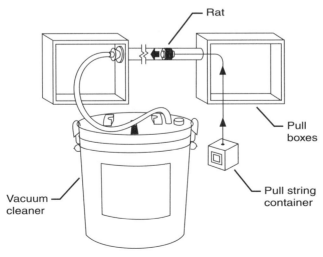

Figure 18.22   Vacuuming a ball.

NOTE: If it is necessary to pull heavy cable, attach the pull string to a heavier string or rope strong enough to pull the weight of the cable in the conduit.

Pulling Auxiliary Disconnect Outlet (ADO) cable in open ceiling

The procedure for residential cabling installations in open ceilings is different from that in conduits. Cables may be supported by beam clamps that hold bridle rings of various sizes to accept the cable. J-hooks may also provide cable support. Beam clamps and J-hooks are mounted every 1.2–1.5 m (4–5 ft) and at each change of direction; the specific distance is such that each J-hook supports less than 11 kg (25 lb). The minimum bending radius for high-performance cables is four times the OD of the cable. Always use supports that provide support for minimum bending radii.

Since there is no conduit, the weight of the cable is supported by mounts above the ceiling. The cable path must be determined before the support devices or J-hooks are mounted. Alternatively, the cable is manually pulled through the red iron supports.

Similar to conduit pulls, a pull string with adequate tensile strength or pull tape may be used to pull cable into place in open ceiling runs by first threading it through the supports. A gopher pole, which telescopes to various lengths, is helpful in threading the string through the supports. Use of the gopher pole limits the number of ceiling tiles that must be moved to provide access to the cable. It also reduces the number of times a ladder has to be moved.

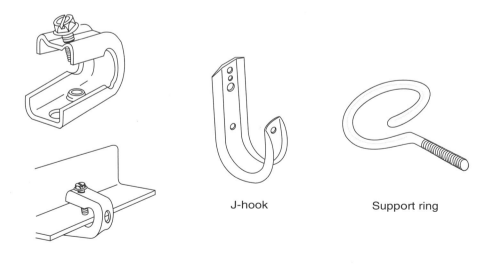

J-hook

Support ring

Beam clamp

Figure 18.23   Beam clamp, J-hook, and support ring.

Other devices available for installing pull strings in open-ceiling installations include a small crossbow or a remote-controlled toy.

NOTE: Do not attach tools to pull strings and throw them through the ceiling. This presents a safety hazard and may cause damage to the ceiling system, tiles, or other utilities contained in the ceiling space.

Each 90-degree bend identifies a pull point, which requires a residential cabling installer at that spot to assist in pulling the cable. Using a pulley can help minimize labor.

Maintain specified distances from possible sources of electromagnetic interference (EMI).

TABLE 18.3 **Minimum separation distances from possible sources of electromagnetic interference**

	Minimum	Separation	Distance
Condition	< 2 kVA	2–5 kVA	> 5kVA
Unshielded power lines or electrical equipment in proximity to open or nonmetal pathways	125 mm (5 in)	305 mm (12 in)	610 mm (24 in)
Unshielded power lines or electrical equipment in proximity to a grounded metal conduit pathway	64 mm (2.5 in)	150 mm (6 in)	305 mm (12 in)
Power lines enclosed in a grounded metal conduit (or equivalent shielding) in proximity to a grounded metal conduit pathway	—	75 mm (3 in)	150 mm (6 in)
Electrical motors and transformers	—	—	1220 mm (48 in)

NOTE: kVA = Kilovoltampere

For both safety and performance purposes, keep power cables physically separated from telecommunications cables.

NOTE: Route open or nonmetallic telecommunications pathways a minimum of 125 mm (5 in) away from fluorescent fixtures.

Following are steps for pulling ADO cable in open ceiling.

Step	Pulling ADO Cable in Open Ceiling
1	Determine tensile strength of string.
	▪ Pull string has a tensile strength rating which appears on the box.
	▪ Tensile strength ratings range typically from 330 N to 880 N (75 lbf to 200 lbf).
2	Verify manufacturer's specifications for cable tensile strength limitations.
	▪ Read manufacturers' guidelines for instructions as to cable strength limitations.
	▪ Excessive tension on the cable stretches the conductors inside the cable, degrading the cable's performance.
	▪ Pay close attention to the requirements for maximum bend radius, which for 4-pair UTP is equal to four times the diameter of the cable.
	▪ For extensive quantities of cable in a ceiling environment, use cable trays or ladder racks.
	▪ Avoid selecting paths that require exceeding the maximum cable bend radius.
3	Route string through the ceiling area.
	▪ Select a gopher pole that will reach the required distance. Telescoping poles reach from 1.83 to 7.62 m (6 to 25 ft).
	▪ Attach the pull string to the hook on the end of the gopher pole.
	▪ Maintain minimum separation distances from possible sources of EMI.
	▪ Keep sufficient distance from sources of heat (e.g., hot water pipes).
4	Attach string to cable.
	▪ Tape the pull string to the lead end of the cable, along with an additional pull string for future use.
	▪ Manually lift the cable into a position to follow the string through the open ceiling area.
5	Place cable in support devices.
	▪ Lift the cable(s) and place them in the cable supports. Once all cables have been installed into a support, close the support, if required.
6	Precautions.
	▪ Pull cable, being careful not to snag or pull the string against sharp objects (e.g., ceiling obstructions [HVAC duct, etc.]) or ceiling grid stringers.
	▪ Avoid creating friction that causes burns or tears in the jacket of the cable.
7	Job clean up.
	Keeping a job site clean:
	▪ Prevents development of safety hazards.
	▪ Reflects on the quality of the residential cabling installer.
	General housekeeping specifics include but are not limited to:
	▪ Picking up used pull strings and pull ropes immediately after use.
	▪ Disposing of removed sheath and termination scrap from cables terminated in MDU-TRs.

Step	Pulling ADO Cable in Open Ceiling
	▪ Storing cable reels and boxes when pulling function is complete.
	▪ Placing termination scrap in an appropriate container while terminating cables.
	▪ Vacuuming residue from work areas.
	▪ Disposing of all personal use items (e.g., lunch materials, coffee cups) and used cleaning supplies.
	▪ Replacing all ceiling tiles after use.
	▪ Storing tools and equipment properly at the end of the work day.

Pulling backbone cable in a vertical pathway from the top down

Backbone cable in the vertical riser shaft may be high-pair-count cable, which is extremely heavy, or high-strand-count optical fiber cable. Therefore, several considerations must go into deciding the best method for making this kind of installation. The cable may be placed vertically directly in an open riser shaft, through cores, sleeves or slots, or within a large conduit.

The residential cabling installer must determine the size and type of reel onto which the cable is to be loaded. These reels may be made of steel that must be returned to the factory, or of wood, which is disposable. The size of the reel is dependent on the size and quantity of the cable. High-pair-count cable can be loaded onto reels that are up to 2.1 m (7 ft) high and 1.5 m (5 ft) wide.

Cables should be ordered with a factory-equipped pulling eye. If this is not possible, a substitute can be installed prior to placing the cable.

When cable is received at the job site, cable length should be verified. Although cable is shipped from the manufacturer with a cable-run label that indicates the length of the cable, do not take this for granted. Inspect both ends of the cable to verify the footage markings on the outer jacket and compute the actual length of the jacket. This is the only accurate method for verifying cable length. Verify and note the weight of the cabled reel. When shipped from a distributor, cable is usually taken from a larger reel and spooled onto a smaller reel.

Determine if a loading dock is available to off-load the reel upon delivery to the site. An elevator of sufficient size and load capacity must be available to transport the reel to the top floor. Many commercial buildings have freight elevators that are designed specifically for this type of operation.

Once a reel is brought to the top floor, it must be set up for the placing operation. A reel dolly or jackstand is necessary to lay out the cable. It may be necessary to use a reel brake to help control the descent of the cable as it is pulled off the reel.

During pulling operations, a reel must be situated in a location that provides enough slack for managing the cable length within the MDU-TR once the placing operations are complete.

Pulleys may be needed to handle the cable from the reel location to the point where it will be dropped down to lower floors. These pulleys can be attached to the overhead structure in the building and provide a pathway for the cable as it is being placed.

In the room where the cable will enter the vertical pathway, a bullwheel will be required to ensure that the jacket is not damaged as it enters the pathway. Situate the bullwheel so that it allows the cable to drop vertically into the pathway. Bullwheels must be attached to the building structure, thereby preventing any lateral movement. Any lateral movement will cause the cable to become misaligned with the vertical pathway and will damage the outer jacket.

Residential cabling installers should be located on each floor through which the cable will pass, as well as on the floor where it will terminate. They must also be equipped with the proper tools to perform critical tasks during the placing operation. Some of the tools include:

- Personal protection equipment (e.g., safety glasses, hard hats, and gloves).

- Communication equipment (e.g., portable, handheld radios).

- Temporary restraining devices for the cable. Cable type and size will determine the restraining devices. One common method of temporarily restraining cable is the use of a large hemp or manila rope equipped with an eyelet. The eyelet is attached to some secure structure in the room. A half-hitch is made in the rope, and the rope is pulled on by the residential cabling installer. The amount of tension placed on the rope will control how quickly the cable slips through the half-hitch. This method allows the cable to be halted and restarted during residential cabling installation without losing control of the cable as it descends (with or without a reel brake).

A guide rope should be installed throughout the cable route. This rope is used to guide the cable through the various spaces rather than pull the cable.

If a pulling eye was not factory installed, install a device to substitute. An alternate method of pulling is the use of a pulling grip, which is a flexible wire-mesh device that clamps down on the cable as tension is applied. These devices are available in various sizes based on cable diameter. Ensure that it is properly sized for the cable. Place it over the cable jacket and then tape it to the jacket. This will prevent it from coming off should the pull be halted and the tension relaxed on the grip.

If a wire-mesh grip is not available and the pathway is large enough, the pull rope can be attached over the jacket in a series of half-hitches. Vinyl tape is placed over the guide rope, securing it to the cable.

If the pathway is too small to use half-hitches, and a pulling grip is not available, a pulling eye can be created on the cable using the conductors. To do this, remove a section of the outer jacket, shield, and inner jacket. Separate the conductors into two equal groups. Weave the two groups back onto themselves, forming an eye on the end of the cable. Once the eye is formed, vinyl tape should be placed over its entire length, including a section of the outer jacket. By using this device, tension will be evenly dispersed over each wire in the cable as well as to the jacket.

The coefficient of friction may be one of the obstacles to installation of cables, especially in conduit pathways. The use of cable lubricants can significantly reduce friction and speed residential cabling installation. Select the lubricant to be used based on the type of pull, type of conduit, cable jacket, and size of the cable. Apply the lubricant as the cable is placed.

Once everyone and everything are in place, the cable can be pulled. Communicate with each member of the team, and advise them that the pull is starting. Begin to lower the cable and monitor its progress throughout the pull. Continue the placing operation until the cable is completely placed.

When the cable reaches the bottom floor, the residential cabling installer must temporarily restrain the cable. Beginning on the bottom floor, the residential cabling installer should pay out the additional slack needed in each MDU-TR. Once the slack is brought into the room, it should be secured in its final location.

Securing the cable on each floor ensures that the weight of the cable will not become excessive at any one point. There are many different devices that can be used to secure the cable. The most common one used today is the split grip, which is a wire-mesh grip that is open on one side. The wire-mesh grip is placed around the jacket of the cable. A pin is then woven through the open sides of the wire-mesh grip, securing them around the cable jacket.

A section of steel rod, a minimum of 13 mm (0.5 in) thick, is inserted through the loop in the end of the wire-mesh grip. The cable is then allowed to slide through the opening in the vertical pathway until the steel rod lies on the top of the pathway.

Another method of securing the wire-mesh grip is to attach it to the building or a plywood backboard with an anchoring device.

The cable should then be identified, labeled, and the documents updated to reflect the operation.

After completing the pulling operation, the cable reel and any remaining cable can be returned to storage or prepared for salvage. If the cable was shipped on a wooden reel, the reel can be dismantled and disposed of. If the cable was shipped on a metal reel, it should be returned to the manufacturer for credit.

Following are steps for pulling backbone cable in a vertical pathway from the top down.

Step	Pulling Backbone Cable in Vertical Pathway—Top Down
1	Verify proper cable length and path.

   - Examine the vertical backbone pathway for the installation and check for a clear pathway. Avoid, if possible, a cable path with obstacles or transitions. Check vertical pathways visually and, if necessary, with a flashlight.

   - Determine the actual length of the cable run, based on information from the blueprint or by installing a pull tape equipped with sequential footage markings. This pull tape, if sufficiently sized, can also be used as the guide rope.

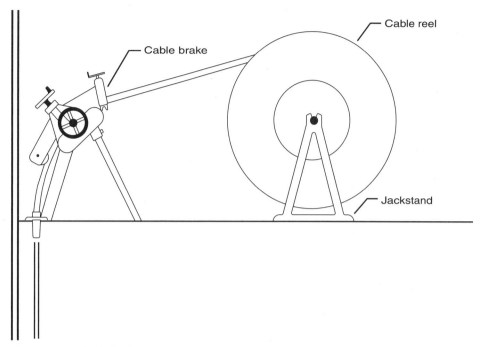

Figure 18.24   Cable brake attached to a cable reel.

Step	Pulling Backbone Cable in Vertical Pathway—Top Down
2	Move cable reels to the upper floor of the vertical feed point.
3	Set up the cable pulling area and put reels on a jackstand or reel dolly. Secure the work area.
4	Attach a cable brake to the cable, if needed (see Figure 18.24).
5	Install and attach a guide rope to cable lead. This pull rope helps to guide the cable as it descends.

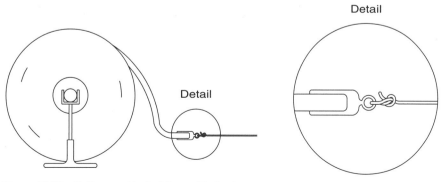

Figure 18.25   Pull rope attached to a cable lead.

Step	Pulling Backbone Cable in Vertical Pathway—Top Down
6	Attach pulleys, if required.

- Decide whether the weight (based on the size and length of cable) requires the use of pulleys and a brake to help control the gravitational pull.

- If the cable is on a wooden reel, remove the spike that holds the cable to the reel prior to beginning cable placement.

- Feed the cable off the reel and onto the pulleys mounted above the route to the pathway to support the cable to the point of entry into the vertical pathway.

- Mount the pulley on any appropriate superstructure available at the location or on overhead steel. If none is available in the place where needed, set up a swing-set type of pulley mechanism.

- Use a bullwheel, if necessary, to provide a sweeping arc of the cable down into the vertical shaft.

- Use a bullwheel to direct the cable into the vertical pathway, keeping the cable from scraping the edge of the floor and chafing the jacket, which damages the integrity of the cable (see Figure 18.26).

NOTE: This operation may require that the bullwheel be attached to the building in three places. The three attachments will support the bullwheel and prevent it from pulling away from the pathway entrance and from moving laterally.

7	Communicate with coworkers for routing.

- Establish locations for coworkers to clearly view the progress of the cable descending.

- Have constant communication between floors to notify the person controlling the cable brake of the progress being made.

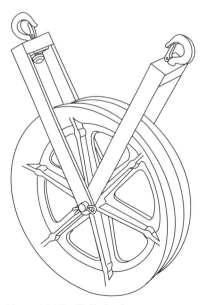

Figure 18.26 Bullwheel.

Step	Pulling Backbone Cable in Vertical Pathway—Top Down
	▪ Listen carefully for a report from a coworker who may spot a problem in the cable path that no one else can see.
8	Drop until finished.
	▪ Watch and observe the functions of any pulley in use.
	▪ Maintain control of all pull strings or guide ropes in use.
	▪ Stay in position as a communicator with coworkers until the pull or drop is finished.
	▪ Be prepared to temporarily take up the cable should a problem develop in the placing operation and prior to permanently securing the cable on each floor.
	▪ Secure all cables at each floor level. Cables that are extremely heavy may fall if they are not properly installed. Free-hanging cables create a serious hazard. See Figure 18.27 for a channel with straps.
9	Route and secure cables in vertical pathway.
10	When the vertical pathway enters an MDU-TR:
	▪ Make a sweep of the cable toward the termination point. Do not exceed the proper bend radius of the cable. Multipair cables require a bend radius of 10 times their diameter.
	▪ Secure the cable in D-rings, with cable ties, straps, split, wire-mesh grips, or clamps and place in the provided cable trays (see Figure 18.28).
	▪ Mount the cable to the wall using D-rings, cable clamps, or special-purpose cable ties designed to support the weight and type of cables to be installed (see Figure 18.29).

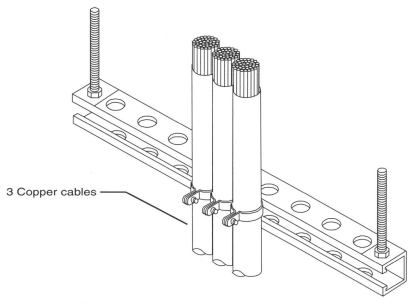

3 Copper cables

Figure 18.27   Channel with straps.

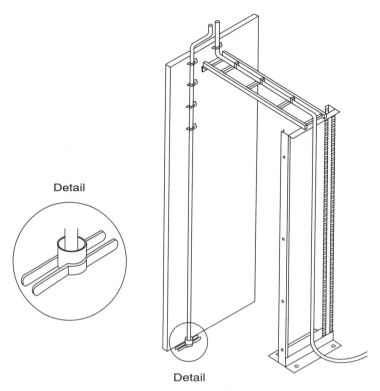

Detail

Detail

Figure 18.28   Cable on tray from vertical pathway.

Step	Pulling Backbone Cable in Vertical Pathway—Top Down
11	Identify cable.
	▪ Place a unique identification number (from the installation drawings) at each end of the cable showing the point of origin and destination. Also, place a label on the cable where it enters or leaves an MDU-TR on each floor.
12	Document cable information on the posted floor plans.
	▪ Show clearly what kind of cable was installed.
	▪ Show clearly which conduit was used.
	▪ Describe the purpose of the installed cables.
13	Perform job wrap-up tasks.

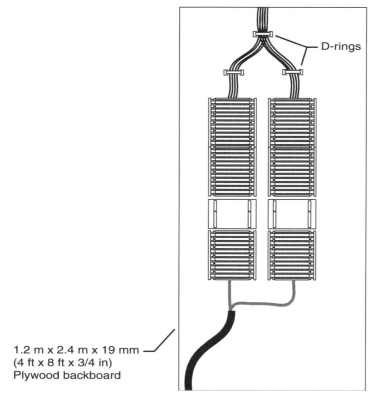

1.2 m x 2.4 m x 19 mm
(4 ft x 8 ft x 3/4 in)
Plywood backboard

Figure 18.29 Backboard layout with D-rings.

Pulling backbone cable in a vertical pathway from the bottom up

Conditions that exist when pulling backbone cable from the top down are the reverse of those that exist when pulling from the bottom up. The first choice is to pull the cable from the top down. Sometimes, however, it is not possible to get the large spools of cable up to the top floor of a building. In those situations, specific equipment must be brought in to handle the task of hauling the cable up through the riser to the necessary heights. Because of gravity, the heavy weight of the cable makes a bottom-up installation more difficult.

The residential cabling installer must determine the size and type of reel onto which the cable is to be loaded. These reels may be made of steel that must be returned to the factory, or of wood, which is disposable. The size of the reel is dependent on the size and quantity of the cable. High-pair-count cable can be loaded onto reels that are up to 2.1 m (7 ft) high and 1.5 m (5 ft) wide.

Cables should be ordered with a factory-equipped pulling eye. If this is not possible, a substitute can be installed prior to placing the cable.

When cable is received at the job site, cable length should be verified. Although cable is shipped from the manufacturer with a cable-run label that indicates the length of the cable, do not take this for granted. Inspect both ends of the cable to verify the footage markings on the outer jacket and compute the actual length of the jacket. This is the only accurate method for verifying cable length. When shipped from a distributor, cable is usually taken from a larger reel and spooled onto a smaller reel.

Determine if a loading dock is available to off load the reel upon delivery to the site.

Before pulling the cable, determine the required cable length and how the weight will be distributed. The number of floors and MDU-TRs that must be accessed are factors in the decision process.

Once the reel is brought to the bottom floor, it must be set up for the placing operation. A reel dolly or jackstand is necessary to allow for paying out the cable. The reel must be situated in a location that will provide enough slack for the bottom floor to be cabled once placing operations are completed.

Cable sheaves may be necessary to handle the cable from the reel location to the point where it will be pulled to upper floors. These cable sheaves can be laid on the floor along the route to the vertical pathway and provide a pathway for the cable as it is being placed. In the room where the cable will enter the vertical pathway, a bullwheel will be required to ensure that the jacket is not damaged as it enters the pathway. The bullwheel should be located so that the cable ascends vertically into the pathway. Bullwheels must be attached to the building structure so that they do not move laterally. Any lateral movement will cause the cable to become misaligned with the vertical pathway and damage the outer jacket. This operation may require that the bullwheel be attached to the building in three places. The three attachments will support the bullwheel and prevent it from pulling away from the pathway entrance and also will keep it from moving laterally.

Residential cabling installers should be located on each floor through which the cable will pass as well as on the top floor where it will terminate. These individuals must be equipped with the proper tools to perform critical tasks during the placing operation. Some of the tools include:

- Personal protection equipment (e.g., safety glasses, hard hats, and gloves).

- Communication equipment (e.g., portable, handheld radios).

- Temporary restraining devices for the cable. Cable type and size will determine the necessary restraining devices. One common method of temporarily restraining cable is the use of a large hemp or manila rope equipped with an eyelet. The eyelet is attached to some secure structure in the MDU-TR. A half-hitch is made in the rope, and the residential cabling installer pulls on the rope. The amount of tension placed on the rope will control how quickly the cable slips through the half-hitch. This method allows the cable to be halted and restarted during placing operations without losing control of the cable if it falls (with or without a reel brake).

The pull rope chosen for this operation is critical. Manila or hemp rope should be used because it does not stretch when placed under tension. It does not deteriorate quickly when subjected to friction and heat. It does, however, have a tendency to twist as tension is applied, requiring the use of a swivel between the cable and the pull rope. The swivel prevents the cable from twisting during the placing operation and thus damaging the jacket, the shield, and the conductors.

If a pulling eye was not factory installed, install a device to substitute. An alternate method of pulling is the use of a pulling grip that is a flexible wire-mesh device that clamps down on the cable as tension is applied. These devices are available in various sizes based on cable diameter. Ensure that it is properly sized for the cable. Place it over the cable jacket and then tape it to the jacket. This will prevent it from coming off should the pull be halted and the tension relaxed on the grip.

If a wire-mesh grip is not available and the pathway is large enough, attach the pull rope over the jacket in a series of half-hitches. Place vinyl tape over the guide rope, securing it to the cable.

If the pathway is too small to use half-hitches, and a pulling grip is not available, a pulling eye can be created by using the conductors. To do this, remove a section of the outer jacket, shield, and inner jacket. Separate the conductors into two equal groups. Weave the two groups back onto themselves, forming an eye on the end of the cable. Once the eye is formed, vinyl tape should be placed over its entire length, including a section of the outer jacket. By using this device, tension will be evenly dispersed over each wire in the cable as well as to the jacket.

Attach the swivel to the pull rope, then attach the cable to the swivel. Ensure the swivel is rated for the weight of the pull.

To lift a cable of significant weight, a winch is required. A winch is an electrical motor equipped with a capstan. It is secured to the building structure on the top floor. The pull rope is wound around the capstan. When the residential cabling installer places tension on the rope, it tightens around the capstan. Friction created by the tightening causes the rope to advance toward the residential cabling installer. The residential cabling installer can control the rate at which the rope advances, thus controlling the advance of the cable along its route.

WARNING: It is absolutely critical that the winch be securely attached to the building structure. If the winch is not properly anchored and comes loose, it becomes a projectile and may cause damage.

The coefficient of friction can be an obstacle when installing cable, especially in conduit pathways. The use of cable lubricants can significantly reduce friction and speed cable installation. A lubricant should be selected based on the type of pull, type of conduit, cable jacket, and cable size. The lubricant should be used as the cable is placed. When placing the cable from the bottom up, use a gel-type lubricant.

Once everyone and everything are in place, the cable can be installed. Communicate with each member of the team and advise them that the pull is ready to begin. Begin pulling the cable up through the vertical pathway. Monitor its progress throughout the pull. Continue the placing operation until the cable is completely placed. Ensure that enough slack is pulled to the top floor so that it can be lowered back down to the individual floors, where it may be required before it is permanently secured.

Once the cable reaches the top floor, the residential cabling installer must temporarily restrain the cable. Beginning on the bottom floor, pay out the additional slack that is needed in each MDU-TR. Once the slack is brought into the room, the cable can be secured in its final location.

Securing the cable on each floor ensures that the weight of the cable will not become excessive at any one point. Many different devices can be used to secure the cable. The most common one used today is the split grip, which is a wire-mesh grip that is open on one side. The wire-mesh grip is placed around the jacket of the cable. A pin is then woven through the open sides of the wire-mesh grip, securing them around the cable jacket.

A section of steel rod, a minimum of 13 mm (0.5 in) thick, is inserted through the loop in the end of the wire-mesh grip. The cable is then allowed to slide through the opening in the vertical pathway until the steel rod lies on top of the pathway.

Another method of securing the wire-mesh grip is to attach it to the building or a plywood backboard with an anchoring device.

The cable should then be identified, labeled, and the documents updated to reflect the operation.

After completing the cable-pulling operation, the cable reel and any remaining cable can be returned to storage or prepared for salvage. If the cable was shipped on a wooden reel, the reel can be dismantled and disposed of. If the cable was shipped on a metal reel, it should be returned to the manufacturer for credit.

Following are steps for pulling backbone cable in a vertical pathway from the bottom up.

Step	Pulling Backbone Cable in Vertical Pathway—Bottom Up
1	Verify proper run and cable length.
	▪ Examine the conduit/vertical pathway for the planned installation and check for a clear pathway. Avoid, if possible, a cable path with obstacles or transitions.
	▪ Use the blueprints to estimate the actual length of the cable run or verify by installing a pull tape equipped with sequential footage markings. This pull tape, if sufficiently sized, can also be used as the guide rope.
2	Move cable reels into place and secure the pulling area. Place the reel on a reel dolly or jackstands. If necessary, place cable sheaves along the floor to the entrance of the vertical pathway.

Step	Pulling Backbone Cable in Vertical Pathway—Bottom Up
3	Determine appropriate rope size and strength.

- A critical factor in the success of the job depends on the strength of the rope used in the pull.
- Check manufacturer's specifications about the length and weight of the cable that will be installed to ensure the rope meets the strength requirements.

4	Install the pull rope.

- Check to see that the pull rope, which must hoist the cable to the upper floors, is lowered down to the proper building level so that it can be attached to the cable.
  - Bring the pull rope to the upper floor.
  - Secure one end at the top.
  - Lower the rope to the bottom floor or vertical feed point.

5	Attach the cable to the rope.

- Check the pulling end of the cable to see if it is equipped with a factory-installed pulling eye. Attach the pull rope to a swivel. Attach the swivel to the cable.

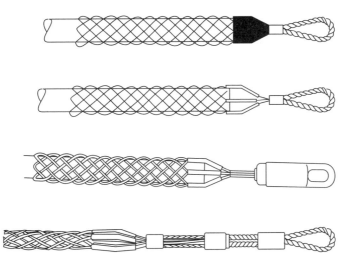

Figure 18.30   Wire-mesh grips.

- Use a wire-mesh grip, if a pulling eye is not provided on the cable.

6	Use a bullwheel, if required.

- Use a bullwheel to maintain the proper bend radius as the cable comes off the reel and is fed up into the opening.
- The arc formed by pulling the cable under the bullwheel protects the integrity of the conductors.
- Unlike pulleys, bullwheels do not alter the pulling force required.

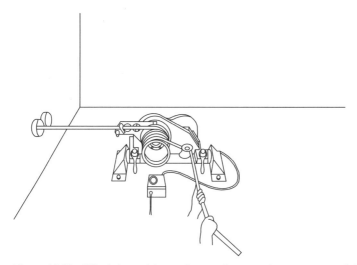

Figure 18.31 Winch in position and properly secured to a concrete slab.

Step	Pulling Backbone Cable in Vertical Pathway—Bottom Up
7	Attach winch and pulleys, if required; verify they are anchored and secured (see Figure 18.31).

- Confirm that the winch is at the required location.

- Verify that the winch is solidly anchored and firmly secured. Adjust winch pulley wheels to proper tension to begin tugging.

- Verify that the short, tugging motions that the winch makes will not move the equipment due to vibration.

- Verify that each pulley is positioned in a secure location and that it is firmly anchored.

- Pulling equipment must be anchored to the structure of the building (e.g., concrete anchors, steel cables, bolts, etc.). See Figure 18.32 for a properly secured pulley.

| 8 | Lubricate cable/pipe. |

- Obtain an approved lubricant that is not abrasive or damaging to the cable.

- Apply a liberal amount of the lubricant to the open end of the conduit.

- Apply the lubricant to the end of the cable where it is attached to the pull rope. Some dispensers are commercially available for automatically putting the lubricant on the cable as it enters the conduit.

| 9 | Communicate with coworkers. |

- As the cable pull begins, communicate with coworkers to pass on information about how things are going.

- Ask all coworkers to call out the job status—if everything is OK or if there is a problem.

- Encourage everyone to take responsibility for the job and to report even the slightest appearance of a problem. It is better and safer to prevent the problem than to correct or repair one.

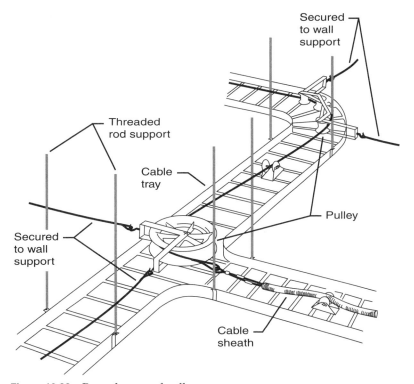

Figure 18.32   Properly secured pulley.

Step	Pulling Backbone Cable in Vertical Pathway—Bottom Up
10	Start winch slowly; notify personnel at the other end.

- Make sure a coworker is at each end of the pull and at all midway positions. Otherwise, workers will have to move between floors. The pulling operation may have to be temporarily halted for them to relocate.

- Notify coworkers of the conditions at the end where cable is being fed into the riser shaft.

- Start the winch slowly, so that the tugs do not come too fast and everyone can observe what is going on.

| 11 | Maintain the operation until the cable is in place. Keep coworkers in position and watching for problems until the pull is complete. |
| 12 | Route and secure the cable, as necessary. |

- Secure the cable in riser shaft.

- Bring the cable into MDU-TR.

- Bring the cable toward the termination point.

- Continue securing the cable with plastic cable ties or other appropriate devices to the backboard, tray, or rack.

- Prepare the service slack, leaving slack in the cable for termination. Mount the cable to the wall, using D-rings, clamps, or ladder rack and cable ties.

Step	Pulling Backbone Cable in Vertical Pathway—Bottom Up
	NOTE: The service slack is used to help relieve tension on the cable and provide slack for future moves and changes.
	▪ The service slack should be long enough to reach the other side of the MDU-TR in case of relocated or additional equipment. Keep in mind that these cables usually require a minimum bending radius of 10 times the OD of the cable for multipair copper cables and 15 times the OD for optical fiber cables.
13	Identify cable.
	▪ Place a unique identification number (from the installation drawings) at each end of the cable showing the point of origin and destination. Also, place a label on the cable where it enters or leaves the MDU-TR on each floor.
14	Document cable information on the posted floor plans.
	▪ Show clearly what kind of cables were installed.
	▪ Show clearly which conduit was used.
	▪ Describe the purpose of the cables installed.
15	Perform job wrap-up tasks.

Pulling optical fiber cable

Introduction.   Optical fiber is one of three types of cable suitable for backbone systems. Optical fiber cable is more rugged than generally perceived; however, just as with copper cable, care must be taken when pulling optical fiber so as not to exceed the cable bend radius or the manufacturer's recommended pulling tension. Most optical fiber cable manufacturers provide maximum pulling tension based on cable construction.

Within buildings, the predominant placement of optical fiber cable is in the backbone between MDU-TRs and ERs.

Regardless of whether the application is backbone or ADO cabling, optical fiber cables may be installed in innerducts as an indicator that fiber cable is present. This method of installation also tends to reduce the pulling tension required, especially when multiple innerducts are installed in conduit.

Plenum- and nonplenum-rated innerducts are available in a variety of colors. Though usually purchased with a pull rope preinstalled inside for attaching the fiber cable to be pulled, innerducts are also available without pull ropes. The pathway of a fiber cable must be free of sharp bends and turns. Normally, innerducts are placed inside conduit, through sleeves, or placed in cable trays. Care should be taken to ensure that the properly fire-rated innerduct is being installed.

NOTE: Innerduct is not a recognized pathway but rather a pathway within a pathway. All cables installed in innerduct must have the proper sheath rating for the area in which they are installed.

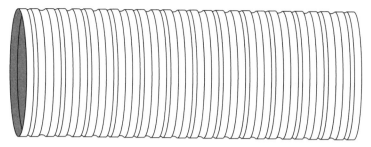

Figure 18.33 Innerduct.

As in copper cable installations, communication among coworkers is essential to ensure the pulling function progresses properly and without excessive stress or tension on the cable.

Following are steps for pulling optical fiber cable.

Step	Pulling Optical Fiber Cable
1	Ensure that only tested and accepted lengths of optical fiber cable are installed.

■ Verify that there is fiber continuity while the cable is on the reel. In order to do so, the cable must be ordered with access to both ends of the fiber on the reel. Maintain the test results for inclusion in the client's documentation package.

■ Test the optical fiber cable for continuity. This can be done by shining a light source (such as a flashlight) into the fiber cable or using a power meter and light source or optical time domain reflectometer (OTDR). A bare-fiber adapter or mechanical reusable splice is necessary to connect either the power meter and light source or OTDR to the unterminated fiber on the reel.

WARNING: Never look directly at the end of an installed optical fiber cable because of the possibility that a laser light source (which is not visible to the human eye) may be present. This can cause permanent damage to the retina of the eye.

2   Install and secure innerducts.

■ Install innerduct for optical fiber cable where necessary and appropriate.

■ Verify the application (plenum or nonplenum) and place the innerduct(s), depending on the specific location (vertical shaft, cable tray, or open ceiling) with the appropriate support. If secured with plastic cable ties in a plenum area, use plenum-rated plastic cable ties.

■ Innerduct can be ordered with or without a pull rope already placed inside. If ordered without a rope, follow the procedure for installing a pull string in a conduit.

■ If the requirement is to place innerducts within a conduit, determine the size and number of innerducts permitted. Innerducts are classified by OD, whereas trade size conduits are classified by ID. This fact allows a total of four 25 mm (1 in) innerducts to be placed inside a 103 mm (4 trade size) conduit.

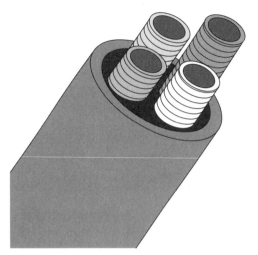

Figure 18.34 Four-inch conduit with four one-inch innerducts.

Step	Pulling Optical Fiber Cable

- 25 mm (1 in) innerduct is the usual size placed within buildings; however, 32 mm (1.25 in), 38 mm (1.5 in), and 50 mm (2 in) innerducts are available for larger fiber cables.

- Attach the optical fiber cable(s) to the pull rope for pulling of the optical fiber cable through the innerduct. There are two methods of connection commonly used. The most common is to remove the sheath for about 300 mm (12 in) length from the fiber cable and place the exposed aramid yarn strength member through a loop in the rope. Secure it with tape, as shown in Figure 18.35.

- The second method is to place the optical fiber cable(s) in a multiweave, wire-mesh grip of the appropriate size that has a swivel pulling eye to which the pull rope is attached as in Figure 18.36.

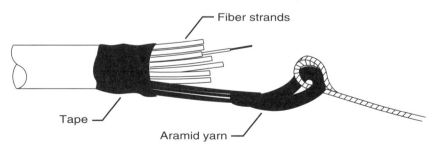

Figure 18.35 Connecting aramid yarn.

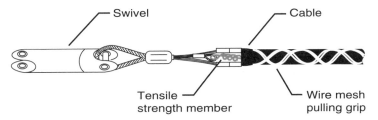

Figure 18.36 Multiweave wire-mesh grip with swivel pulling eye.

Step	Pulling Optical Fiber Cable
3	Install without innerduct.
	■ When optical fiber cable is installed without innerduct, care must be taken to ensure that the run is as straight as possible.
	■ At transitions, be sure that a coworker is placed at such points to relieve excessive tension during the pull. This ensures that the optical fiber cable does not rub against obstructions that can nick or cut the jacket.
4	Pull fiber.
	■ Feed the optical fiber cable into the innerduct. Normally the optical fiber cable can be pulled by hand.
	■ Do not exceed the recommended pull tension or the recommended bend radius of the optical fiber cable.
5	Always leave a service loop.
	■ The service loop is used to relieve all tension on the optical fiber cable and provide slack for future moves or changes.
	■ Always bring the optical fiber cable with the innerduct into a point of termination using the grommets provided in the optical fiber cabinet. Secure the optical cable to the cabinet, ladder rack, cable tray, or backboard. The optical fiber strands should be stored and dressed inside the cabinet.
	■ The optical fiber cable service loop should be long enough to reach the other side of the MDU-TR in case relocated or additional equipment is encountered in the future and the fiber must be relocated within the MDU-TR. The service slack stored inside the cabinet should be 3 m (10 ft).
6	Secure the cable to the cabinet per the manufacturer's specifications.
	■ All fiber cables should be anchored in the termination cabinet by using the aramid yarn of the cable to act as a strain relief, as stated in the manufacturer's instructions.
	■ Care should be taken to not overtighten plastic cable ties around optical fiber cable when dressing. Keep them loose enough that they can be turned from side to side using finger pressure.
	■ Maintain proper bend radius and pull. For instance, the bend radius requirements of some optical fiber cables are 10 times a cable's OD when the cables are at rest and 20 times a cable's OD while the cables are under stress. These values may be 15 times and 30 times the cable's OD for some cables.

Step	Pulling Optical Fiber Cable
7	Identify cables.
	▪ Place a unique identification number (from the installation drawings) at each end of the cable showing the point of origin and destination.
8	Document the job on the posted floor plans.
	▪ Indicate clearly what type of cables were installed (i.e., 24-strand, plenum-rated multimode fiber).
	▪ Show clearly which conduit was used and which innerduct, if applicable.
	▪ Describe on the floor plan the purpose of the installed cables (e.g., backbone from main cross-connect to third floor MDU-TR).
9	Perform job wrap-up tasks.

Glossary

Numbers

0 dBm Standard-level ratio of measurement that represents a zero-decibel reading from one milliwatt of power at a frequency of 1000 hertz across an impedance of 600 ohms.

Definitions

A

access line* A telecommunications circuit provided by a service provider at the demarcation point.

access provider (AP)* The operator of any facility that is used to convey telecommunications signals to and from a customer premises.

adapter* A device that enables any or all of the following:

- Different sizes or types of plugs to mate with one another or to fit into a telecommunications outlet.

- Interconnection between cables.

- Large cables with numerous wires to fan out into smaller groups of wires.

- The rearrangement of leads.

adapter; optical fiber duplex* A mechanical device designed to align and join two duplex optical fiber connectors (plugs) to form an optical duplex connection.

administration* The method for labeling, identification, documentation, and usage needed to implement moves, additions and changes of the telecommunications infrastructure.

---

*Terms marked with an asterisk are reprinted with permission of Telecommunications Industry Association (TIA). Complete copies of all TIA standards can be purchased through Global Engineering Documents at 800-854-7179 or 303-397-7956.

aerial entrance   Entrance facility where the cables providing service to a building are placed overhead.

alternating current (ac)   Current carried by voltage that alternates between positive and negative values; the standard commercial power frequency in the United States is 60 hertz (or cycles) of ac, so the voltage completes 60 sine waves per second.

American wire gauge (AWG)   A system used to specify wire size. The greater the wire diameter, the smaller the AWG value.

ampere (A)   Unit of electrical current. One ampere is equal to the current produced by one volt acting through a resistance of one ohm.

amplifier   An electronic device that takes an incoming signal and increases the signal strength so that the signal can transmit a greater distance.

analog   A format that uses variables (e.g., voltage amplitude or frequency variations) to transmit information.

antenna   A usually metallic device (e.g., a rod or wire) for radiating or receiving radio waves.

antenna entrance*   A pathway facility from the antenna to the associated equipment.

architectural drawing   Architectural and structural (engineering) drawings; part of the building drawing set.

asbestos   Fibrous mineral substance used in buildings as insulation between the mid-1940s and 1978 and later found to be carcinogenic.

as-built drawing   **1.** Documentation that indicates cable routing. **2.** A final form of drawing in electronic or hard-copy format inclusive of all "record copy" information and notes.

aspirator   An apparatus for producing suction; moving or collecting materials by suction.

asynchronous transmission   Data is transferred as a series of bits separated by start and stop bits, not related to time. See synchronous transmission.

attenuation   The decrease in magnitude of transmission signal strength between points, expressed in decibel as the ratio of output to input signal level. See also insertion loss.

attenuator   A device used to reduce the amplitude of voltage, current, or power (signal strength) without appreciable distortion.

audio   Something that is heard or having to do with something that is heard; in audio/video, having primarily to do with the process of reproducing sounds through a system of electronic components.

aural ear protectors   Resemble ear muffs; available in passive or active models.

auxiliary disconnect   See auxiliary disconnect outlet and auxiliary disconnect outlet cable.

auxiliary disconnect outlet (ADO)*   A device usually located within the tenant or living unit used to terminate the ADO or backbone cable.

auxiliary disconnect outlet (ADO) cable*   In residential applications, the cable from the auxiliary telecommunications disconnect outlet/connector or the distribution device in a customer's premises to the backbone facility or point of demarcation.

B

backboard   A panel (e. g., wood or metal) used for mounting connecting hardware and equipment.

backbone   A facility (e.g., pathway, cable, or conductors) between telecommunications rooms or multi-dwelling unit telecommunications rooms, the entrance facilities, and the equipment rooms within or between buildings.

backbone cable*   **1.** Term used to describe cable that handles major network traffic. **2.** See backbone.

backbone cabling   See backbone.

balun   A balanced to unbalanced circuit coupling device, used to convert from unbalanced to balanced transmission, and provides impedance matching for connecting twisted-pair to coaxial cable.

bandwidth   The information-handling capability of a medium, expressed in units of frequency (hertz).

base   Describes the number of signals traveling over media at the same time. Base is a shortened form of baseband, which has only one carrier channel and no multiplexing.

baseband service   A method of communication in which all of the bandwidth is dedicated to a single communications channel, and only one single message transfer can occur at a time.

baud   **1.** The signal-change rate on the line. **2.** The rate at which the signal changes from one state to the other.

bel   A measure of analog signal strength; named in honor of telephone pioneer Alexander Graham Bell.

bend radius   Radius of curvature that a media can bend without signal degradation.

bit   **1.** The basic unit of digital information, used to indicate the existence of one of two binary states or conditions, such as current flow or no current flow, on or off. **2.** A digital pulse representing a "1" or a "0;" contracted from the phrase "binary digit."

bit error rate (BER)   The fraction of bits transmitted incorrectly.

blueprint   A reproduction of an architectural plan and/or technical drawing that provides details of a construction project or an existing structure. These drawings are printed on special paper that allows graphics and text to appear as blue on a white background.

BNC connector   Bayonet Neil-Concelman connector that has a center pin that must be installed over the coaxial cable's center conductor.

bonding*   The permanent joining of metallic parts to form an electrically conductive path that will assure electrical continuity and the capacity to conduct safely any current likely to be imposed.

broadband   Commonly used to refer to high-speed high-bandwidth digital circuits where the communications channel is capable of transporting the data streams simultaneously.

bucket lift   Personnel lift composed of a fiberglass bucket mounted on the end of an extendable arm in which the user stands.

building code   Construction requirements that are adjudicated into law by the local authority having jurisdiction.

building plan   The representation of a building project.

building specification   The detailed written description of a building project.

bullwheel   A large wheel used to maintain an arc when feeding large cables into a backbone pathway.

bundled cable*   An assembly of two or more cables continuously bound together to form a single unit.

buried entrance   An entrance facility in which underground cables are trenched or plowed rather than running in conduit, providing a more economical but still out-of-sight means to connect outside service facilities to the premises cabling.

bursty   A sudden filling of cable bandwidth, although not necessarily to full capacity.

bus topology   A linear configuration where all network devices are placed on a single length of cable. It requires one backbone cable to which all network devices are connected.

C

cabinet (telecommunications)*   An enclosure used for terminating telecommunications cables, wiring and connection devices with a hinged cover, usually flush-mounted in the wall.

cable*   **1.** An assembly of one or more insulated conductors or optical fibers within an enveloping sheath. **2.** The act of installing cable.

cable chase   A pathway used for telecommunications cable.

cable modem   A device that converts broadband signaling to a computer.

cable run*   A length of installed media, which may include other components along its path.

cable sheath*   A covering over the optical fiber or conductor assembly that may include one or more metallic members, strength members, or jackets.

cable tray   A support mechanism used to route and support telecommunications or power cable. Typically equipped with sides that allow cables to be placed within the sides over the tray's entire length.

cabling   A combination of all cables, wires, cords, and connecting hardware.

campus*   The buildings and grounds having legal contiguous interconnection.

capacitance   **1.** The tendency of an electronic component to store electrical energy. **2.** Resistance to change in voltage.

carbon block   A protection device that conducts surge current to a grounding conductor.

cardiopulmonary resuscitation (CPR)   The emergency procedure used on a person who is not breathing and whose heart has stopped beating (cardiac arrest).

Category   The rating system imposed by the Telecommunications Industry Association on copper cabling to specify its transmission performance.

catwalk   Elevated walkway provided in some large buildings to help workers reach utilities.

CEBus   Protocol for home automation cabling.

center channel   Third front audio channel (in addition to main stereo left and right channels) found in surround sound audio systems with the primary task of reproducing movie dialogue (what the actors are saying) thus locking the voices to the screen for all listeners.

channel*   The end-to-end transmission path between two points at which application-specific equipment is connected.

characteristic impedance   The input impedance of a uniform analog transmission line of infinite length.

chase   See pathway.

circular mil   Measuring unit used to specify copper conductors.

cladding   The transparent outer concentric glass layer that surrounds the optical fiber core and has a lower index of refraction than the core.

Class C fire extinguisher   Fire extinguisher that does not have any chemicals with conductive properties so it can be used for electrical fires.

closed-circuit television (CCTV)   A private television system, typically used for security purposes, in which the signal is transmitted to a limited number of receivers.

coaxial cable   Copper cable deriving its name from the fact that there is a centrally located insulated conductor surrounded by an overall metallic shield enclosed in a jacket.

code   A rule intended to ensure safety during the installation and use of materials, components, fixtures, systems, premises, and related subjects. Codes are typically invoked and enforced through government regulation.

community antenna television (CATV) system   A system of television reception in which signals from distant stations are picked up by a master antenna and sent by cable to the individual receivers.

conductance   The ease with which electrical current flows through a substance. Uniformly distributed along the substance length, conductance varies as a function of a conductor's geometry as well as dielectric properties of the materials surrounding the conductor.

conductor   A material or object that permits an electric current or light wave to flow easily.

conduit*   **1.** A raceway of circular cross-section. **2.** A structure containing one or more ducts.

confined space   The work space defined by the Occupational Safety and Health Association as one a worker can enter and work in but that has limited or restrictive means of entry or exit and that is not designed for continuous occupancy (e.g., maintenance holes, splice pits, crawl spaces, and attics).

connecting hardware*   A device providing mechanical cable termination.

connector (plug)   A junction that allows an optical fiber or a cable to be repeatedly connected or disconnected to a device (e.g., a source or detector).

connector (plug), duplex; optical fiber* A remateable device that terminates two fibers and mates with a duplex receptacle.

contract Written document covering the entire understanding between the customer and the contractor.

control panel A device that arms, disarms, and supervises an alarm system at the user's premises.

cord (telecommunications)* A cable using stranded conductors for flexibility, as in distribution cords or line cords.

core The inner portion of a glass optical fiber used to carry light pulses.

crawl space Space where a residential cabling installer cannot stand upright.

cross-connect* A facility enabling the termination of cable elements and their interconnection or cross-connection.

cross-connection* A connection scheme between cabling runs, subsystems, and equipment using patch cords or jumpers that attach to connecting hardware on each end.

crosstalk Unwanted transfer of signal from one or more circuits to other circuits.

current (I) The flow of electrons in a conductor measured in amperes.

D

daisy-chain topology Devices are connected in series, one after the other, and the transmitted signals go to the first device, then the second, and so forth.

data communication The transmission and reception of electronically coded information.

data rate The number of bits of information that can be transmitted per second.

decibel (dB) A logarithmic unit for measuring the power or strength of a signal. See bel.

delay skew The difference in propagation delay between pairs in a cable.

demarcation point (DP)* **1.** The point of interface between application providers and the customer facilities. **2.** The point where the operational control or ownership changes.

detail drawing Large-scale drawing with details of construction required; often not drawn to scale.

detection badge See exposure monitor.

dielectric A nonconductor of direct electric current.

digital Term referring to a signal that is a discontinuous, changing from one state to another in a limited number of discrete (single) steps; in its simplest form, there are two states or signal levels—on and off, where the "on" translates to a digit "1" and "off" is seen as a zero level corresponding to the digit "0"—in digital transmission; the two states can be represented by positive and negative voltages.

digital subscriber line (DSL) A type of digital technology designed to use existing (legacy) cabling to support high data transfer rate applications.

direct broadcast satellite (DBS) A service that uses satellites to broadcast multiple channels of television programming directly to home-mounted small-dish antennas.

direct current (dc) A steady current value that does not change in direction of voltage or current flow, as does alternating current; a battery is an example of a dc source.

dispersion **1.** The widening or spreading out of the modes in a light pulse as it progresses along an optical fiber. **2.** The characteristics of the sound coverage field of a speaker.

distribution device (DD)* A facility located within the dwelling unit for interconnection or cross-connection.

distribution device cord* A telecommunications cord that extends between the distribution device and the auxiliary disconnect outlet.

drawing set A set of drawings that shows the dimensions and the relationship between components of a building project.

duplex connector See connector (plug), duplex; optical fiber.

E

earplugs Disposable foam or reusable rubber hearing protection that can be slipped into the ear canal.

electrical metallic tubing (EMT) Thin-wall metal tubing that does not have threaded ends, which is widely used in electrical distribution systems and as a pathway for telecommunications cabling.

electromagnetic interference (EMI)* Radiated or conducted electromagnetic energy that has an undesirable effect on electronic equipment or signal transmissions.

electromagnetic pulse (EMP) High-intensity electromagnetic radiation generated by a nuclear blast high above the earth's surface and held to disrupt electronic and electrical systems.

elevation drawing A drawing showing the exterior faces of a structure to be built.

end user* The owner or user of the premises cabling system.

entrance bridge* A terminal strip that is an optional component in a network interface device and is provided for the connection of the auxiliary disconnect outlet cable.

entrance facility (EF [telecommunications])* An entrance to a building for both public and private network service cables (including wireless) including the entrance point at the building wall and continuing to the entrance room or space.

entrance point (EP [telecommunications])* The point of emergence for telecommunications cabling through an exterior wall, a floor, or from a conduit.

equipment cord A terminated cable assembly that connects customer premises equipment to a telecommunications outlet(s)/connector(s).

equipment room (ER [telecommunications])* An environmentally controlled centralized space for telecommunications equipment that usually houses a main or intermediate cross-connect.

Ethernet A local area network protocol using a logical bus structure and carrier sense multiple access with collision detection.

exposure monitor   Device worn by the residential cabling installer to ensure that exposure to a hazardous substance does not exceed safe level.

F

F connector   Used to terminate 75 ohm coaxial cable.

faceplate   A protective plate for a device.

far-end crosstalk (FEXT)   The signal level at the near end of the disturbing pair compared to that transferred to the disturbed pair and measured at the opposite (far) end of the line.

fiber optic   See optical fiber.

fiber optic cable   See optical fiber cable.

filter mask   Protective breathing apparatus worn whenever harmful dust, gas, smoke, chemical vapor, or some other pollutant is present at the work site.

firestop   The installation of qualified firestop materials in holes made through fire-rated floors, walls, or ceilings for the penetration of pipes, cables, or other items.

firestopping   The process of installing special materials into penetrations of fire-rated barriers to reestablish the integrity of the barrier.

Firewire™   IEEE 1394 protocol for high-speed communications.

first aid   The emergency aid or treatment given before medical services can be obtained.

fixed device*   Any low-voltage device permanently affixed to a surface for purposes of security, fire detection or other control, data, or entertainment applications.

floor plan   Plan showing the layout of a building floor.

floor serving terminal   See multi-dwelling unit telecommunications room.

frequency   The number of cycles or sine waves occurring in a given time; if the unit of time is one second, the frequency is stated in hertz (Hz); one Hz is equal to one cycle per second.

frequency-division multiplexing (FDM)   In optical networks, a system that allows the transmission of more than one signal over a common path, by assigning each signal a different frequency band.

full-face shield   Protective eyewear worn where there is danger of splashing chemicals (e.g., when working with batteries).

fusing   The process of joining fibers together by fusion or melting.

G

gas tube   Protector (arrester) technology that basically operates in the same way as carbon blocks, but with a wider gap and higher reliability. See carbon block.

gated community   A community, composed of individual houses, duplexes, or apartment buildings, surrounded by a secured fence or other barrier allowing limited access through a secure gate.

gigahertz (GHz)   Unit of frequency denoting one billion cycles per second (hertz).

glass fiber   See optical fiber.

Grade 1 residential telecommunications cabling   Cabling that provides a generic cabling system that meets the minimum requirements for telecommunications services (e.g., telephone, satellite, community antenna television, and data services).

Grade 2 residential telecommunications cabling   Cabling that provides a generic cabling system that meets the requirements for current needs as well as for developing basic, advanced, and multimedia telecommunications services.

grommet   A protective edging placed around a hole.

ground*   A conducting connection, whether intentional or accidental, between an electrical circuit (e.g., telecommunications) or equipment and the earth, or to some conducting body that serves in place of the earth.

ground fault circuit interrupter (GFCI)   A device intended for the protection of personnel that functions to deenergize a circuit or portion thereof, within an established period of time, when a current to ground exceeds some predetermined value that is less than that required to operate the overcurrent protective device of the supply circuit.

ground grid   The conducting connection, intentional or accidental, between an electrical circuit or equipment and the earth or some conducting body that serves in place of it.

grunt sack   A bag that is raised or lowered on a rope to provide a means of safely passing tools and other small materials between individuals working in a construction environment.

H

hard hat   Protective headgear.

hard-wire system   A connection in which a device is either connected directly to the wires within a cable by screw lugs or a splice.

hertz (Hz)   **1.** A unit of measure used to express the range of frequencies associated with a given signal or communications channel. This range is also referred to as bandwidth. **2.** A unit of frequency equal to one cycle per second. A commonly used rate is megahertz.

hierarchical star topology   An extension of the star topology using a central hub.

home automation   The interoperation of devices and subsystems in the home (based on events, schedules, user actions, etc.) to increase the safety, convenience, or comfort of the home.

home run   A pathway or cable between two locations without a splice or intermediate termination points in between.

HomePNA™   The Home Phoneline Networking Alliance (HomePNA™) is an association of networking industry companies working to ensure adoption of a single, unified phoneline network standard and to bring to market interoperable home networking products compliant with that standard. The organization currently offers 1 megabit per second (Mb/s) and 10 Mb/s standards that support a range of residential applications using existing wiring in the home.

home theater   A system that is generally confined to a single room and is typically used in combining video with life-like audio soundtracks that enhance one's viewing pleasure.

hybrid cable   An assembly of multiple cables of the same or different types or categories covered by one sheath.

I

impedance   The total opposition offered to the flow of an alternating current, measured in ohms.

index sheet   A part of the drawing set, maintained by the architect and providing general information, including the site address, owner's name, and architect's name.

inductance   **1.** The property of an electrical force field built around a conductor when current flows through it. **2.** The resistance to change in current.

inductive coupling   The transfer of energy from one circuit to another (e.g., power lines on a utility pole can inductively couple a power surge onto telephone cables).

infrastructure (telecommunications)*   A collection of those telecommunications components, excluding equipment, that together provide the basic support for the distribution of all information within a building or campus.

insertion loss   A measurement of loss on a transmission line applying to all terminations other than the special case when load and source impedances match (or equal) the characteristic impedance of a wire pair; in this special case, the proper term of measurement is attenuation.

insulation displacement connector (IDC)   **1.** A wire connection device in which the insulation that is surrounding a conductor is displaced at the connection point and consequently making a connection to the conductor. **2.** A type of wire terminating connection in which the insulation jacket is cut by the connector where the wire is inserted.

interbuilding backbone cable   Cable that handles traffic between buildings.

interbuilding backbone pathway   Provides the means to connect main terminal spaces with other main terminal spaces.

interbuilding telecommunications backbone   A pathway and/or cable facility from the entrance room/space provided for interconnecting to other buildings, as in a campus environment.

intercom/paging system   A communications device used for either one-way or two-way voice broadcasts.

interconnect   A location where interconnections are made.

interconnection*   **1.** A connection scheme that employs connecting hardware for the direct connection of a cable to another cable without a patch cord or jumper. **2.** A type of connection in which single-port equipment connections (e.g., 4-pair and optical fiber connectors) attach to horizontal or backbone cabling.

intermediate cross-connect   The connection point between a backbone cable that extends from the main cross-connect and the backbone cable from the horizontal cross-connect.

intrabuilding backbone cable   Cable that handles traffic between multi-dwelling unit telecommunications rooms in a single building.

intrabuilding backbone pathway   Provides the means to connect the main terminal space and the multi-dwelling unit telecommunications room.

intrabuilding telecommunications backbone   A pathway or cable facility for interconnecting telecommunications service entrance rooms, equipment rooms, telecommunications rooms, or multiple-dwelling unit telecommunications rooms within a building.

J

jack   Common term for telecommunications outlet/connector. See modular jack.

job change order   The documents detailing changes to the original work plan, including all extra work hours and materials expended.

job site   The physical location where work is to be performed.

jumper*   An assembly of twisted-pairs without connectors, used to join telecommunications circuits/links at the cross-connect.

K

key telephone system (KTS)   Terminals and equipment in a local environment that provide immediate access from all terminals to a variety of telephone services without attendant assistance.

keyed   A jack, outlet, or connector is considered keyed when it requires a specific orientation to prevent mismating.

keypad   User interface with buttons used for arming and disarming an alarm system with numerical codes.

kilohertz (kHz)   One thousand cycles per second (hertz).

L

ladder cable tray   A prefabricated structure consisting of side rails connected at the bottom by transverse members (rungs) for supporting and routing cables or conductors within the structure.

ladder rack   A device similar to a cable tray, but more closely resembles a single section of a ladder. It is constructed of metal with two sides affixed to horizontal cross members.

legend   A list of symbols on a blueprint or other construction document.

legend sheet   A part of a drawing set, maintained by the architect and providing a list of standard symbols and abbreviations used throughout the document.

level   The absolute or relative voltage, current, or power measured at a certain point in the circuit or system.

lifting belt   Safety device designed to be worn around the abdomen to help support the stomach muscles while encouraging proper posture.

line cord   Telecommunications cord using stranded or tinsel conductor.

link*   A transmission path between two points, not including terminal equipment, work area cables, and equipment cables.

list of materials   Bidding document containing all of the items to be installed by description, catalog number, quantity, unit price, and total price.

listed*   Equipment included in a list published by an organization, acceptable to the authority having jurisdiction, that maintains periodic inspection of production of listed equipment, and whose listing states either that the equipment or material meets appropriate standards or has been tested and found suitable for use in a specified manner.

local area network (LAN)   The standard industry term for a network installation that serves a relatively small area (e.g., a structured cabling installation serving a building).

logical topology   The path taken by messages as they travel from one device to another on a network. Contrast with physical topology.

loose tube (fiber)   A protective tube loosely surrounding a cabled fiber, often filled with gel.

low-voltage bracket   A device that may be inserted in a non-fire rated finished wall allowing the mounting of a faceplate that may hold telecommunications outlet/connectors.

M

M drawing   Mechanical drawings includes drawings for heating, ventilating, and air conditioning and plumbing.

macrobend   Cable bend that will increase attenuation in optical fiber cable but is recoverable.

main cross-connect (MC)   The cross-connect normally located in the main equipment room for cross-connection and interconnection of entrance cables, first-level backbone cables, and equipment cables.

main terminal space*   The location of the cross-connect point of incoming cables from the telecommunications external network and the premises cable system.

Material Safety Data Sheet (MSDS)   Information system describing hazardous chemicals and materials.

mechanical drawing   A part of the drawing set covering installation of the plumbing, heating, ventilating, and air conditioning systems within and into a building.

media (telecommunications)*   Wire, cable, or conductors used for telecommunications.

megahertz (MHz)   A unit of frequency equal to one million cycles per second (hertz).

megahertz•kilometer (MHz•km)   The expression of optical fiber cable bandwidth where half power is realized at a specific point in frequency at 1 kilometer (i.e., 160 MHz•km @ 850 nanometer and 500 MHz•km @ 1300 nanometer).

mesh grip   See wire mesh grip.

microbend   Cable bend that will increase attenuation in optical fiber cable but is not recoverable.

micrometer   See micron.

micron   **1.** A unit of length equal to one millionth of a meter (0.000001 meter). **2.** An abbreviation for micrometer.

millimeter (mm)   One thousandth of a meter.

minimum point of entry (MPOE)*   Either the closest practicable point to where the carrier facilities cross the property line or the closest practicable point to where the cabling enters a multi-unit building or buildings.

modal dispersion   A characteristic of transmission in an optical fiber that results from different lengths of the light paths taken by the many modes of light as they travel down the fiber from source to receiver.

mode*   **1.** Light path in optical fiber. **2.** A path of light in an optical fiber.*

modem   Shortening of modulator/demodulator, this device converts digital signals to analog for transmission over analog telephone lines and then reconverts analog signals to digital for processing by computers.

modular jack*   A female telecommunications connector that may be keyed or unkeyed and may have 6 or 8 contact positions, but not all the positions need be equipped with jack contacts.

modular patch panel (MPP)*   A facility enabling the terminating of cable elements on insulation displacement connector modules and their connections by means of a patch cord.

modular plug*   A male telecommunications connector for cable or cord that may be keyed or unkeyed and may have 6 or 8 contact positions, but not all the positions need be equipped with contacts.

module   Units installed in the distribution device to support communications functionality.

mud ring (plaster ring/low-voltage ring)   An adapter that is designed to convert a standard electrical box to accept a single-gang faceplate and raise the faceplate attachment location so that it is flush with the Drywall.

multi-dwelling unit (MDU)   Category includes apartments, townhouses, condominiums, and assisted-living facilities. These facilities may be under a single roof or consist of multiple buildings in a residential campus.

multi-dwelling unit telecommunications room (MDU-TR)   The space where backbone and auxiliary disconnect outlet cables terminate.

multimedia*   **1.** An application that communicates to more than one of the human sensory receptors. **2.** Applications that communicate information by more than one means.

multimode fiber   An optical waveguide that allows many bound modes to propagate.

mutual capacitance   Effective capacitance between the two conductors of a pair.

N

N connector   A type of threaded-coaxial connector.

nanometer (nm)   1. A unit of length equal to one billionth of a meter (0.000000001 meter). **2.** The most common unit of measurement for optical fiber operating wavelengths.

near-end crosstalk (NEXT)   The signal transfer between circuits at the same (near) end of the cable.

network interface device (NID)*   The point of connection between networks.

nominal velocity of propagation (NVP)   The coefficient used to determine the speed of transmission along a cable relative to the speed of light in a vacuum.

O

ohm   Unit of measure of electrical resistance; one ohm is defined as the resistance that allows one ampere, the unit of electric current, to flow when one volt is applied.

Ohm's law   The mathematical relationship among electric current, resistance, and voltage. The voltage in volts is equal to the current in amperes multiplied by the resistance in ohms.

optical fiber   Construction of glass used to transmit light pulses in telecommunications.

optical fiber cable*   An assembly consisting of one or more optical fibers.

outlet box (telecommunications)*   A metallic or nonmetallic box mounted within a floor, wall, or ceiling and used to hold telecommunications outlets/connectors or transition device.

outlet cable*   A cable placed in a residential unit extending directly between the telecommunications outlet/connector and the distribution device.

outlet/connector (telecommunications)*   A connecting device in the work area on which horizontal cable or outlet cable terminates.

outside plant (OSP)   Telecommunications infrastructure exterior to buildings.

P

padding down the circuit   Adding attenuation to a transmission line so that the signal is not too high at the receiver end.

patch cord*   A length of cable with connectors on one or both ends used to join telecommunications links at the cross-connect.

patch panel*   A connecting hardware system that facilitates cable termination and cabling administration using patch cords. See modular patch panel.

pathway*   A facility for the placement of telecommunications cable.

personal protective equipment (PPE)   Safety equipment worn by the residential cabling installer that is designed to prevent injury (e.g., hard hat, gloves, and protective eyewear).

personnel lift   Mechanical device for lifting construction personnel when a ladder cannot be used safely because of the required working height or weight of equipment.

physical topology   The physical appearance of a network with logical topology.

plaster ring   See mud ring.

plot plan   See site drawing.

plug   See modular plug.

power (P)   The energy required to operate an electrical device (e.g., a motor, amplifier, or telephone transmitter). Composed of voltage and current.

power-line carrier (PLC)   A technology that sends electronic information (e.g., on/off commands) through a home's alternating current power lines.

power sum equal level far-end crosstalk loss (PSELFEXT)*   A computation of unwanted signal coupling from multiple transmitters at the near-end into a pair measured at the far-end, and normalized to the received signal level.

power sum near-end crosstalk loss (PSNEXT)*   A computation of the unwanted signal coupling from multiple transmitters at the near-end into a pair measured at the near-end.

preinstallation meeting   Internal meeting convened by the project manager and including the telecommunications installation team and appropriate contractors to discuss all aspects of upcoming construction project.

premises   Building or set of buildings on common property that are occupied by a single tenant or landlord.

premises wiring   Interbuilding or intrabuilding, and horizontal cabling that is owned by a single tenant or landlord.

prewiring*   **1.** Wiring installed before walls are enclosed or finished. **2.** Wiring installed in anticipation of future use or need. See rough-in.

primary protector   A protective device placed on telecommunications conductors in accordance with ANSI/NFPA 70.

private branch exchange (PBX)   A device allowing private local voice (and other voice-related services) switching over a network.

project plan   Construction plan reflecting all aspects of the work to be performed using a priority schedule.

project schedule   Scheduling document including all activities to take place in a construction project.

propagation delay   The time interval required for a signal to travel between network devices.

protective glasses   See protective goggles.

protective goggles   Protective eyewear that provides side protection and front protection when the hazards involve flying objects.

pull cord   See pull string.

pull strength   See pull tension.

pull string   A cord or wire placed within a raceway and used to pull wire and cable through the raceway.

pull tension*   The pulling force that can be applied to a cable.

Q

quad cable   A 4-conductor nontwisted-pair cable with a red, green, black, and yellow conductor.

quad-shielded   A cable consisting of a foil-braid-foil-braid shield.

R

raceway*   Any enclosed channel designed for holding wires or cables.

radio frequency interference (RFI)   See electromagnetic interference.

reflection coefficient   The amount of reflected energy in a transmission line, which is affected by the degree of mismatch between the load and the line.

residential gateway*   A device that enables communication among networks in the residence and between residential networks and service providers' networks.

residential structured cabling system (RSCS)   The complete or collective configuration of cabling and associated hardware at a given site that has been installed to provide a comprehensive telecommunications infrastructure; the RSCS begins at the point where the service provider terminates.

resistance   Force opposing the flow of electrical energy, and measured in ohms.

retrofit   **1.** To furnish a device or system with new parts or equipment not available at the time of manufacture. **2.** The installation of a structured cabling system into a preexisting residential structure.

retrofit ring   See low-voltage bracket.

return loss   When the termination (load) impedance does not match (equal) the value of the characteristic impedance of the transmission line, some of the signal energy is reflected back toward the source and is not delivered to the load; this signal loss contributes to the insertion loss of the transmission path, and is called return loss.

ribbon cable   Cable in which many fibers are embedded in a plastic material in parallel, forming a flat ribbon-like structure.

rough-in   **1.** Wiring installed before walls are enclosed or finished. **2.** Wiring installed in anticipation of future use or need. **3.** The time in which a residential cabling installer runs the cable between the various termination points, the main task during this period being to complete all the work that is difficult to do once Drywall and other wall materials are installed.

S

safety grounding wand   Electrical protective device consisting of an insulated handle with a metal tip that is connected to an insulated cable with a large metal clip on the opposite end; the clip is connected first to a ground source and the metal tip is used to short any transient voltages left on a deenergized circuit.

safety harness Full-body harness worn with attachable lanyards any time an elevating device (e.g., a personnel lift) is used.

safety lanyard A device made of nylon ropes with self-closing and locking keepers (metal safety hooks) on each end, which can incorporate a shock from an accidental fall.

safety plan A plan prepared by a company and put in place before beginning any work operation, covering all safety issues likely to arise on a particular job site.

scissor lift Personnel lift composed of a working platform mounted on a large scissor jack.

scope of work Guideline document that lists all of the elements of the installation.

secondary protector A device that provides a means to safely limit currents to less than the current-carrying capacity of listed indoor communications wire and cable, listed telephone set line cords, and listed communications terminal equipment having ports for external wire line communications circuits.

service provider (SP) An entity that provides connection to a part of the Internet, or other services (e.g., application programming interfaces). An SP, when reached by the user through an access provider (AP), becomes the AP to the service they provide. See access provider (AP).

sheath See cable sheath.

shield* **1.** The metal covering over wire pairs or other conductors, usually in the form of copper braids, metal foils, or solid tubing. **2.** A metallic layer placed around a conductor or group of conductors.

short circuit An accidentally established low-resistance connection between two points in an electric circuit.

simplex signaling A unidirectional signaling method in which data transfer can take place in only one direction, with no capabilities to change directions.

singlemode optical fiber* **1.** Optical fiber with a relatively small diameter, featuring a core of 8-9 μm (micrometers or microns) and a cladding diameter of 125 μm; light is restricted to a single path, or mode, in singlemode fiber. **2.** An optical fiber that carries only one path of light.

site drawing Architectural or construction drawing prepared by the civil engineer and indicating the location of any exterior pathways that are being installed for use by the telecommunications contractor.

site survey A process used to identify the characteristics of an environment. Required for the design of a wireless network.

sleeve* An opening, usually circular, through the wall, ceiling, or floor to allow the passage of cables.

slot* An opening through a wall, floor, or ceiling, usually rectangular, to allow the passage of cables.

small form factor (SFF) fiber-optic connector Developed to address issues raised with existing optical fiber connectors, this connector generally consists of two fibers secured in a plug about the size of an 8-position modular jack. It is easy to field terminate and keyed to maintain, transmit, and receive fibers in a consistent position.

smart appliances   Products that include some processing power within themselves and have the ability to send or receive data or control messages to/from another product. All use some digital control to automate some product function and some send control messages, but most cannot both receive and send information.

solid conductor   A material or object that permits an electric current or light wave to flow easily.

solid-state protector   Arc protection (arrestor) devices based on high-power semiconductor technology that do not deteriorate with age below a rated maximum surge current.

space (telecommunications)*   An area used for housing the installation and termination of telecommunications equipment and cable (e.g., common equipment rooms, equipment rooms, common telecommunications rooms, work areas, and maintenance holes/handholes).

speaker   A device that changes electrical signals into sounds loud enough to be heard at a distance.

splice   A joining of conductors in a splice closure, meant to be permanent.

split grip   A device similar to a mesh grip but is split, that when attached to cable facilitates the physical support of cable.

splitter   A network device that provides signals to a number of outputs, which are individually matched and isolated from each other.

staging area   An area at a job site where the bulk of the materials and tools is kept and distributed.

standard   A collection of requirements that encompasses properties of components and systems that is intended to ensure an accepted degree of functionality and longevity. Standards are intended to reflect accepted norms as typically determined through a balloting process conducted by a nationally or internationally accredited organization.

star topology   A network topology in which services are distributed from a central point.

stranded cable   A cable with individual conductors composed of groups of wires twisted together.

stress detector   A device that detects the stress in beams when walking above or near the detector.

subwoofer   A powered or nonpowered low-frequency speaker.

surface-mounted raceway   Plastic or metallic raceway that is installed on the surface of a wall, floor, or ceiling that provides a protective pathway for cables and/or power from public access.

surge protection device   May be electrical or lightning arrestors designed to protect equipment between the power source and the equipment being protected.

surround sound   A home theater application that has multiple audio channels.

surround speaker   A speaker used to reproduce surround channel information primarily to create ambience and sonic realism.

switching   Networking protocol in which a station sends a message to a hub, which then routes the message to the specified destination station.

synchronous transmission   Data is transferred at specific intervals using a common timing mechanism called a clock. Contrast with asynchronous transmission.

T

T sheet   Drawings created by electrical engineers to show the telecommunications installation in a building.

telecommunications*   Any transmission, emission, and reception of signs, signals, writings, images, and sounds, that is, information of any nature by cable, radio, optical, or other electromagnetic systems.

telecommunications entrance facility*   See entrance facility (telecommunications).

telecommunications entrance point*   See entrance point (telecommunications).

telecommunications entrance room or space   See space (telecommunications).

telecommunications equipment room*   See equipment room (telecommunications).

telecommunications infrastructure*   See infrastructure (telecommunications).

telecommunications media*   See media (telecommunications).

telecommunications outlet/connector   Provides the means for the tenant to connect premises equipment.

telecommunications room (TR)*   An enclosed space for housing telecommunications equipment, cable terminations, and cross-connect cabling that is the recognized location of the horizontal cross-connect.

telecommunications service entrance   See entrance facility (telecommunications).

telecommunications space*   See space (telecommunications).

terminal*   **1.** A point at which information may enter or leave a communications network. **2.** The input-output associated equipment. **3.** A device by means of which wires may be connected to each other.

termination   **1.** The ending of a transmission or transmission pathway. **2.** The act of connecting a cable/wire/fiber to connecting hardware.

tight-buffered optical fiber cable   Fiber is protected by supporting each strand of glass in a tight-buffed coating, which increases strand diameter to 900 μm.

topology*   **1.** The physical or logical arrangement of a telecommunications system.* **2.** The features, shape, or physical appearance of an item (e.g., what is shown by a topographical map). For a cabling plan, its topology is representative of the plan's features and physical appearance. See logical topology and physical topology.

transition point (TP)*   A location in the horizontal cabling where flat undercarpet cable connects to round cable.

transmission   The movement of information as electrical or optical signals from one point to another via a medium such as air, water, copper wire, optical fiber, or whatever else might be used to carry the signal.

transmission speed   The rate of flow (speed) of binary digit information over a transmission line, expressed in bits per second.

trim-out   Usually one of the last steps in the home construction process, this is the period when all cables are terminated to the appropriate devices and testing is completed.

trunk distribution and feeder   Rigid coaxial cable, typically used for backbone cabling.

twisted-pair cable   A multiconductor cable comprising two or more copper conductors twisted in a manner designed to cancel electrical interference.

U

underground entrance   Entrance facility using conduit or other types of mechanical pathways to provide out-of-sight service to a building.

uninterruptible power supply (UPS)   A device that is inserted between a primary power source (e.g., a commercial utility) and the primary power input of equipment to be protected (e.g., a computer system) to eliminate the effects of transient anomalies or temporary outages.

V

voltage (V)   Electrical potential or potential difference expressed in volts.

volume control   A device used to increase and decrease the audio level of a speaker or speakers.

W

wire mesh grip   A device attached to the end of a cable, which facilitates the pulling of cable. This is the same as a mesh grip.

work area*   A building space where the occupants interact with telecommunications terminal equipment.

Acronyms

A

A	ampere
ac	alternating current
ACEG	alternating current equipment ground
ACR	attenuation-to-crosstalk ratio
ADO	auxiliary disconnect outlet
ADSL	asymmetric digital subscriber line
AHJ	authority having jurisdiction
AIA	American Institute of Architects
AIA	American Insurance Association
AISG	American Insurance Service Group
ALF	assisted-living facility
AME	architectural, mechanical, electrical
AMES	architectural, mechanical, electrical, structural
amp	ampere
ANSI	American National Standards Institute
AP	access provider
ASHRAE	American Society of Heating, Refrigerating, and Air Conditioning Engineers
ASME	American Society of Mechanical Engineers
ASTM	American Society for Testing and Materials
ATIS	Alliance for Telecommunications Industry Solutions
ATR	all-threaded-rod
A/V	audio/video
AWG	American wire gauge

B

b/s	bit per second
BAS	building automation system
BCT	bonding conductor for telecommunications
BER	bit error rate
BICSI®	Building Industry Consulting Service International
bit	binary digit
BNC	Bayonet Neil-Concelman
BOCA	Building Officials and Code Administrators International, Inc.
BOMA	International Building Owners and Management Association

C

°C	degrees Celsius
C	mutual capacitance
CABA	Continental Automated Buildings Association

CableLabs	Cable Television Laboratories	
CAD	computer-aided design	
CATV	community antenna television	
CB	circuit breaker	
CBC	coupled bonding conductor	
CCBC	closely coupled bonding conductor	
CCTV	closed circuit television	
CD	compact disc	
CD-ROM	compact disc–read only memory	
CEA	Consumer Electronics Association	
CEBus®	Consumer Electronics Bus	
CEC®	*Canadian Electrical Code®*	
CEDIA	Custom Electronic Design and Installation Association	
CFR	*Code of Federal Regulations*	
CISCA	Ceilings and Interior Systems Construction Association	
CO	carbon monoxide	
coax	coaxial	
CP	consolidation point	
CPE	customer premises equipment	
CPE	customer provided equipment	
CPR	cardiopulmonary resuscitation	
CRT	cathode-ray tube	
CSA	Canadian Standards Association	
CS/BS	cable system/broadcast system	
CSC	Construction Specifications Canada	
CSI	Construction Specifications Institute	
CSMA/CD	carrier sense multiple access/collision detection	

D

dB	decibel	
dBm	decibel milliwatt	
DBS	direct broadcast satellite	
dc	direct current	
DD	distribution device	
demarc	demarcation point	
DLP	digital light processor	
DP	demarcation point	
DP	distribution point	
DSL	digital subscriber line	
DSS	digital satellite system	
DVC	desktop video conference	
DVD	digital versatile disc	

E

EDP	electrical distribution panel
EF	entrance facility
EIA	Electronic Industries Alliance
ELFEXT	equal level far-end crosstalk loss
EMC	electromagnetic compatibility
EMI	electromagnetic interference
EMP	electromagnetic pulse
EMS	Emergency Medical Services
EMT	electrical metallic tubing
EP	entrance point
ER	equipment room
ETA	Electronic Technicians Association

F

°F	degrees Fahrenheit
F	farad
FA	fire alarm
fax	facsimile
FCC	Federal Communications Commission
FDM	frequency division multiplexing
FDU	fiber distribution unit
FEP	fluorinated ethylene propylene
FEXT	far-end crosstalk
FIPS	Federal Information Processing Standards
FOTP	fiber optic test procedure
FPN	fine print note
ft	foot
FTTH	fiber-to-the-home

F

G	conductance
Gb/s	gigabit per second
GC	general contractor
GEC	grounding electrode conductor
GFCI	ground fault circuit interrupter
GFI	ground fault interrupter
GHz	gigahertz
GRC	galvanized rigid conduit
GSA	General Services Administration

H

H	henry
HAA	Home Automation Association

HANA	Home Automation and Networking Association
HC	horizontal cross-connect
HDSL	high-bit-rate DSL
HDTV	high definition television
HLAN	home-based local area network
HomePNA™	Home Phoneline Networking Alliance
HS-IR	high-speed infrared
HVAC	heating, ventilating, and air conditioning
Hz	hertz

I

I	current
IAEC	International Association of Electrical Inspectors
IC	intermediate cross-connect
ICBO	International Conference of Building Officials
ICEA	Insulated Cable Engineers Association
ID	identification
ID	inside diameter
IDC	insulation displacement connector
IEC	International Electrotechnical Commission
IEEE®	Institute of Electrical and Electronics Engineers, Inc.®
IMC	intermediate metallic conduit
in	inch
I/O	input/output
IP	Internet protocol
IR	infrared
IRC	Institute for Research in Construction
IrDa	Infrared Data Association
IROB	in range out of building
ISDN	integrated services digital network
ISO	International Organization for Standardization
ITU	International Telecommunication Union
IWTA	International Wireless Telecommunications Association
IXC	interexchange carrier

J

JTC	Joint Technical Committee

K

kb/s	kilobit per second
kHz	kilohertz
km	kilometer
kts	Key Telephone System
kVA	kilovoltampere

L

LAN	local area network
laser	light amplification by stimulated emission of radiation
lbf	pound-force
LCD	liquid crystal display
LEC	local exchange carrier
LED	light emitting diode
LEOM	low earth orbit microsatellite
LSA	link state algorithm

M

m	meter
µm	micron; one millionth of a meter (0.000001 meter); also micrometer
mA	milliampere
MAC	move, add, or change
MAN	metropolitan area network
Mb/s	megabit per second
MC	main cross-connect
MCM	thousand circular mils
MDP	main distribution panel
MDU	multi-dwelling unit
MDU-TR	multi-dwelling unit telecommunications room
MGN	multigrounded neutral
MH	maintenance hole
MHz	megahertz
MHz•km	megahertz•kilometer
mm	millimeter
MMDS	multi-channel multipoint distribution system
modem	modulator/demodulator
MPOE	minimum point of entry
MPOP	minimum point of presence
MPP	modular patch panel
MRI	magnetic resonance imaging
MSDS	Material Safety Data Sheet
MUTOA	multiuser telecommunications outlet assembly
mW	milliwatt

N

N	Newton
NBC	National Building Code
NBCC®	National Building Code—Canada
NEC®	*National Electrical Code®*

NECA	National Electrical Contractors Association
NEMA®	National Electrical Manufacturers Association®
NESC®	*National Electrical Safety Code®*
NEXT	near-end crosstalk loss
NFPA®	National Fire Protection Association®
NIC	network interface card
NID	network interface device
NIST	National Institute of Standards and Technology
nm	nanometer
NRC-IRC	National Research Council of Canada, Institute for Research in Construction
NSCA	National Systems Contractors Association
NTIS	National Technical Information Services
NVP	nominal velocity of propagation

O

Ω	ohm—the Greek capital letter omega traditionally used to represent the electrical unit
OC	outlet cable
OD	outside diameter
OPL	Omega Point Laboratory
OSHA	Occupational Safety and Health Administration
OSP	outside plant
OTDR	optical time domain reflectometer

P

P	power
PBX	private branch exchange
PC	personal computer
PDA	personal digital assistant
PE	polyethylene
pF/ft	picofarad per foot
PIR	passive infrared
PLC	power-line carrier
PO	purchase order
POF	plastic optical fiber
POTS	plain old telephone service
PPE	personal protective equipment
PPT	post, telephone, and telegraph
PSELFEXT	power sum equal level far-end crosstalk loss
PSNEXT	power sum near-end crosstalk loss
PSTN	public switched telephone network
PTC	positive temperature coefficient
PVC	polyvinyl chloride

R

R	resistance in ohm
RCD	residual current device
RCDD®	Registered Communications Distribution Designer
RF	radio frequency
RFI	radio frequency interference
RFP	request for proposal
RFQ	request for quote or quotation
RG	radio grade
RG	residential gateway
RJ	registered jack
RSCS	residential structured cabling system
RUS	Rural Utilities Services

S

SBCCI	Southern Building Code Congress International, Inc.
SC	subcommittee
SC	subscriber connector
SCTE	Society of Cable Telecommunications Engineers Inc.
ScTP	screened twisted-pair
SFF	small form factor
SOHO	small office, home office
SP	service provider
SS	spread spectrum
ST	straight tip
STB	set top box
STP	shielded twisted-pair
STP-A	shielded twisted-pair-A
SVHS	super video home system
SWAP	shared wireless access protocol
SWRi	Southwest Research, Inc.

T

TAG	Technical Advisory Group
TBB	telecommunications bonding backbone
TBBIBC	telecommunications bonding backbone interconnecting bonding conductor
TCIM	*Telecommunications Cabling Installation Manual*
TDMM	*Telecommunications Distribution Methods Manual*
TDR	time domain reflectometer
TEBC	telecommunications equipment bonding conductor
TEF	telecommunications entrance facility

TGB	telecommunications grounding busbar
TIA	Telecommunications Industry Association
TMGB	telecommunications main grounding busbar
TO	telecommunications outlet
TP	transition point
TR	telecommunications room
TSB	Telecommunications Systems Bulletin
TV	television

U

UBC	Universal Building Code
UL®	Underwriters Laboratories Inc.®
ULC	Underwriters Laboratories, Inc. of Canada
UPS	uninterruptible power supply
USOC	universal service order code
UTP	unshielded twisted-pair
UV	ultraviolet

V

V	volt
Vac	volt alternating current
VCR	video cassette recorder
VCSEL	vertical-cavity surface-emitting laser
VDSL	very high-bit-rate digital subscriber line
VDV	voice/data/video
VESA	Video Electronics Standards Association
VFL	visual fault locator
VGA	video graphics array
VoIP	voice over Internet protocol

W

W	watt
WAN	wide area network
WDF	Wireless Data Forum
WG	Working Group
W-H	Inchcape Warnock Hersey (associated with ETL)
WLL	wireless local loop
WP	waterproof

X

xDSL	x digital subscriber line

Bibliography
and Resources

Bibliography

Abruzzino, James. *Communications Cabling Technicians' Handbook*. Alexandria, VA: CNC Press, 1998.

Allen, Neal. *Network Maintenance and Troubleshooting Guide*. Everett, WA: Fluke Networks, 1997.

American National Standards Institute/Electronic Industries Alliance/Telecommunications Industry Association. ANSI/EIA/TIA-455-A, *Standard Test Procedures for Fiber Optic Fibers, Cables and Transducers, Sensors, Connecting and Terminating Devices, and other Fiber Optic Components*. Arlington, VA: Electronic Industries Alliance/Telecommunications Industry Association, 1991.

American National Standards Institute/Institute of Electrical and Electronics Engineers, Inc®. ANSI/IEEE C2. *National Electrical Safety Code®*. New York, NY: Institute of Electrical and Electronics Engineers, Inc., 1997.

American National Standards Institute/Insulated Cable Engineers Association. ANSI/ICEA S-80-576. *Communications Wire and Cable for Premises Wiring*. Yarmouth, MA: Insulated Cable Engineers Association, 1994.

———. ANSI/ICEA S-83-596. *Fiber Optic Premises Distribution Cable*. Yarmouth, MA: Insulated Cable Engineers Association, 1994.

———. ANSI/ICEA S-87-640. *Fiber Optic Outside Plant Communications Cable*. Yarmouth, MA: Insulated Cable Engineers Association, 1992.

———. ANSI/ICEA S-89-648. *Telecommunications Aerial Service Wire*. Yarmouth, MA: Insulated Cable Engineers Association, 1993.

American National Standards Institute/National Fire Protection Association, Inc. ANSI/NFPA-70. *National Electrical Code®*. Quincy, MA: National Fire Protection Association, Inc., 2002.

———. ANSI/NFPA-71. *Installation, Maintenance, and Use of Signaling Systems for Central Station Service*. Quincy, MA: National Fire Protection Association, Inc., 1989.

———. ANSI/NFPA-72. *National Fire Alarm Code*. Quincy, MA: National Fire Protection Association, Inc., 1999.

———. ANSI/NFPA-75. *Protection of Electronic Computer/Data Processing Equipment*. Quincy, MA: National Fire Protection Association, Inc., 1999.

———. ANSI/NFPA-780. *Standard for the Installation of Lightning Protection Systems*. Quincy, MA: National Fire Protection Association, Inc., 1999.

American National Standards Institute/Telecommunications Industry Association/Electronic Industries Alliance. ANSI/TIA/EIA-232-F. *Interface Between Data Terminal Equipment and Data Circuit-Terminating Equipment Employing Serial Binary Data Interchange*. Arlington, VA: Telecommunications Industry Association/Electronic Industries Alliance, 1997.

———. ANSI/TIA/EIA-422-B. *Electrical Characteristics of Balanced Voltage Digital Interface Circuits*. Arlington, VA: Telecommunications Industry Association/Electronic Industries Alliance, May 1994.

————. ANSI/TIA/EIA-455-61. *FOTP-61 Measurement of Fiber or Cable Attenuation Using an OTDR*. Arlington, VA: Telecommunications Industry Association/Electronic Industries Alliance, 2000.

————. ANSI/TIA/EIA-455-171-A. *Attenuation by Substitution Measurement for Short Length Multimode and Graded Index and Singlemode Optical Fiber Cable Assemblies*. Arlington, VA: Telecommunications Industry Association/Electronic Industries Alliance, 1999.

————. ANSI/TIA/EIA-485-A. *Electrical Characteristics of Generators and Receivers for Use in Balanced Digital Multipoint Systems*. Arlington, VA: Telecommunications Industry Association/Electronic Industries Alliance, March 1998.

————. ANSI/TIA/EIA-492-AAAA. *Detail Specification for 62.5 μm Core Diameter/125 μm Cladding Diameter Class Ia Graded Index Multimode Optical Fibers*. Arlington, VA: Telecommunications Industry Association/Electronic Industries Alliance, January, 1998.

————. ANSI/TIA/EIA-492-AAAB. *Detail Specification for 50 μm Core Diameter/125 μm Cladding Diameter Class Ia Multimode, Graded-Index Optical Waveguide Fibers*. Arlington, VA: Electronic Industries Alliance/Telecommunications Industry Association, November 1998.

————. ANSI/TIA/EIA-492-CAAA. *Detail Specification for Class IVa Dispersion—Unshifted Singlemode Optical Fibers*. Arlington, VA: Telecommunications Industry Association/Electronic Industries Alliance, May 1998.

————. ANSI/TIA/EIA-526-7. *Method 1: Measurement of Optical Power Loss of Installed Single-Mode Fiber Cable Plant—OFSTP-7*. Arlington, VA: Telecommunications Industry Association/Electronic Industries Alliance, August 1998.

————. ANSI/TIA/EIA-526-14-A. *Optical Power Loss Measurements of Installed Multimode Fiber Cable Plant OFSTP-14A*. Arlington, VA: Telecommunications Industry Association/Electronic Industries Alliance, 1998.

————. ANSI/TIA/EIA-568-A-5. *Transmission Performance Specifications for 4-Pair 100-Ohm Category 5e Cabling*. Arlington, VA: Telecommunications Industry Association/Electronic Industries Alliance, 1999.

————. ANSI/TIA/EIA-568-B.1. *Commercial Building Telecommunications Cabling Standard, Part 1: General Requirements*. Arlington, VA: Telecommunications Industry Association/Electronic Industries Alliance, 2001.

————. ANSI/TIA/EIA-568-B.2. *Commercial Building Telecommunications Cabling Standard, Part 2: Balanced Twisted-Pair Cabling Components*. Arlington, VA: Telecommunications Industry Association/Electronics Industries Alliance, 2001.

————. ANSI/TIA/EIA-568-B.3. *Optical Fiber Cabling Components Standard*. Arlington, VA: Telecommunications Industry Association/Electronics Industries Alliance, 2000.

————. ANSI/TIA/EIA-569-A. *Commercial Building Standard for Telecommunications Pathways and Spaces*. Arlington, VA: Telecommunications Industry Association/Electronic Industries Alliance, 1998.

————. ANSI/TIA/EIA-570-A. *Residential Telecommunications Cabling Standard*. Arlington, VA: Telecommunications Industry Association/Electronic Industries Alliance, 1999.

————. ANSI/TIA/EIA-598-A. *Optical Fiber Cable Color Coding*. Arlington, VA: Telecommunications Industry Association/Electronic Industries Alliance, May 1995.

————. ANSI/TIA/EIA-604-3. *FOCIS 3—Fiber Optic Interconnector Intermateability Standard*. Arlington, VA: Telecommunications Industry Association/Electronic Industries Alliance, August 1997.

————. ANSI/TIA/EIA-606. *Administration Standard for the Telecommunications Infrastructure of Commercial Buildings*. Arlington, VA: Telecommunications Industry Association/Electronic Industries Alliance, 1993.

————. ANSI/TIA/EIA-607. *Commercial Building Grounding and Bonding Requirements for Telecommunications*. Arlington, VA: Telecommunications Industry Association/Electronic Industries Alliance, August 1994.

————. ANSI/TIA/EIA-758. *Customer-Owned Outside Plant Telecommunications Cabling Standard*. Arlington, VA: Telecommunications Industry Association/Electronic Industries Alliance, 1999.

Better Homes and Gardens Step-by-Step Wiring. Editors of Better Homes and Gardens. Des Moines, IA: Better Homes and Gardens Books, 1997.

BICSI®. *BICSI Telecommunications Dictionary*. Tampa, FL: BICSI, 1999.

————. *Customer-Owned Outside Plant Design Manual*. Tampa, FL: BICSI, 1999.

————. *Telecommunications Cabling Installation Manual*, 3rd edition. Tampa, FL: BICSI, 2001.

————. *Telecommunications Distribution Methods Manual*, 9th edition. Tampa, FL: BICSI, 2000.

————. *Telecommunications Quick Reference Guide for Code Officials*. Tampa, FL: BICSI, 1999.

Bigelow, Stephen J. *Understanding Telephone Electronics*. 3rd edition. Indianapolis, IN: Sams Publishing, 1991.

Briere, Danny and Pat Hurley. *Smart Homes for Dummies*. Foster City, CA: IDG Books, 1999.

Brosch, Ernest. *How to Cable the Home Office/Small Office: A Complete Guide*. New York, NY: Flatiron Publishing, 1995.

Canadian Standards Association. CSA-C22.1. *Canadian Electrical Code®, Part 1*. Pointe Claire, Canada: Canadian Standards Association, 1998.

————. CSA-T525. *Residential Wiring for Telecommunications*. Pointe Claire, Canada: Canadian Standards Association, 1994. Reaffirmed in 1999.

————. CSA-T527. *Grounding and Bonding for Telecommunications in Commercial Buildings*. Pointe Claire, Canada: Canadian Standards Association, 1994. Harmonized with ANSI/TIA/EIA-607. Reaffirmed in 1999.

————. CSA-T528. *Design Guidelines for Administration of Telecommunications Infrastructure in Commercial Buildings*. Pointe Claire, Canada: Canadian Standards Association, 1993. Harmonized with ANSI/TIA/EIA-606. Reaffirmed in 1997.

————. CSA-T529. *Design Guidelines for Telecommunications Wiring Systems in Commercial Buildings*. Pointe Claire, Canada: Canadian Standards Association, 1995. Harmonized with ANSI/TIA/EIA-568-A.

————. CSA-T530. *Building Facilities, Design Guidelines for Telecommunications*. Pointe Claire, Canada: Canadian Standards Association, 1997. Harmonized with ANSI/TIA/EIA-569-A.

CEBus Industry Council, Inc. *CEBus Standard*. EIA-600 Series. Chevy Chase, MD: CEBus Industry Council, Inc.

Chomyez, Bob. *Fiber Optic Installations: A Practical Guide*. New York, NY: McGraw-Hill, 1996.

Codex, Motorola. *Basics Book of Information Networking*. Reading, MA: Addison-Wesley Publishing Co., 1992.

Complete Guide to Home Wiring. Editors, Black and Decker Home Improvement Library. Minneapolis, MN: Creative Publishing International, 2001.

Connor, Deni and Mark Anderson. *Networking the Desktop: Cabling, Configuration, and Communications*. Boston, MA: AP Professional, 1995.

Dodd, Annabel Z. *Essential Guide to Telecommunications*. Saddle River, NJ: Prentice Hall PTR, 2000.

Federal Communications Commission. *FCC Code of Federal Regulations, Title 47, Telecommunications, Part 68, Connection of Terminal Equipment to the Telephone Network*. Washington, DC: Federal Communications Commission. Revised 1 October 1998.

Fiber U Guide to Fiber Optic Training (booklet). Fiber U/FOTEC. Medford, MA: FOTEC, 1998.

FIPS PUB 174. *Federal Building Telecommunications Writing Standard of 1992*. Available from the U.S. Department of Commerce/National Institute of Standards and Technology. Gaithersburg, MD, 1992.

Gerhart, James. *Home Automation and Wiring*. New York, NY: McGraw-Hill, 1999.

Groth, David and Jim McBee. *Cabling: The Complete Guide to Networking*. Alameda, CA: Sybex, 2000.

Guide to the Digital Home: How to Computerize Your Home. Lincoln, NE: Sandhills Publishing, 2000.

Hayes, Jim. *Fiber Optics Technician's Manual*, 2nd edition. Albany, NY: Delmar/Thompson Learning, 2001.

Hayes, Jim and George Shinopoulos. *Fiber Optic Testing: A Practical Guide to Testing Fiber Optic Components and Networks*. Medford, MA: FOTEC, 1994.

Hayes, Jim and Paul Rosenberg. *Data, Voice, and Video Cabling*. Albany, NY: Delmar/Thompson Learning, 1999.

Held, Gilbert. *Understanding Data Communications*, 4th edition. Indianapolis, IN: Sams Publishing, 1994.

Ideas for Great Home Offices. Editors, Sunset Books. Menlo Park, CA: Sunset Publishing Corp., 1995.

Installation Manual for Residential Structured Wiring. Version 2.0. Bothell, WA: Leviton Integrated Networks, 2001.

Institute of Electrical and Electronics Engineers, Inc.® (IEEE®). IEEE 802.3. Part 3. *CSMA/CD Access Method and Physical Layer Specifications, Information Technology—Telecommunications and Information Exchange Between Systems—Local and Metropolitan Area Networks—Specific Requirements*. New York, NY: Institute of Electrical and Electronics Engineers, Inc., 1985.

———. IEEE 802.11b. *IEEE Standard for Information Technology—Telecommunications and Information Exchange Between Systems—Local and Metropolitan Area Networks—Specific Requirements—Part 11: Wireless LAN MAC and Physical Layer Specifications: Higher Speed Physical Layer Extension in the 2.4 GHz Band*. New York, NY: Institute of Electrical and Electronics Engineers, Inc., 1999.

———. IEEE Standard 1394. *High Performance Serial Bus (Fire Wire) to Provide High Speed Communications for Digital Audio, Digital Video, Signal Routing, and Home Networking*. New York, NY: Institute of Electrical and Electronics Engineers, Inc., 1995.

International Building Officials Code Administrators. *The BOCA National Building Code*. Country Club Hill, IL: Building Officials Code Administrators International, January 1999.

International Conference of Building Officials. *Uniform Building Code*. Whittier, CA: International Conference of Building Officials, 1997.

International Electrotechnical Commission. IEC 60603-7. *Connectors for Frequencies Below 3 MHz for Use with Printed Boards-Part 7: Detail Specification for Connectors, 8-Way, Including Fixed and Free Connectors with Common Mating Features, with Assessed Quality*. Geneva, Switzerland: International Electrotechnical Commission, 1996.

International Organization for Standardization/International Electrotechnical Commission. ISO/IEC 11801. *Information Technology—Generic Cabling for Customer Premises*. Geneva, Switzerland: International Organization for Standardization/International Electrotechnical Commission, July 1995.

Johnston, Mark. *Field Testing of High Performance Premise Cabling*. Phoenix, AZ: Microtest, 2000.

Lennie Lightwave's Guide to Fiber Optic Installations. Medford, MA: Fiber U/FOTEC, 1997.

Lennie Lightwave's Guide to Fiber Optic Testing. Medford, MA: Fiber U/FOTEC, 1996.

Lowe, Doug. *Networking for Dummies*. San Mateo, CA: IDG Books, 1994.

Maybin, Harry B. *Low Voltage Wiring Handbook: Design, Installation and Maintenance*. New York, NY: McGraw-Hill, 1995.

McClimans, Fred J. *Communications Wiring and Interconnection*. New York, NY: McGraw-Hill, 1992.

Miller, Brent A. and Chatschik Bisdikian. *Bluetooth Revealed*. Saddle River, NJ: Prentice Hall PTR, 2001.

National Fire Protection Association, Inc. *National Electrical Code*® (NFPA 70), 2002 edition. Quincy, MA: National Fire Protection Association, Inc., 2002.

———. *National Electrical Code®* Handbook, 8th edition. Quincy, MA: National Fire Protection Association, Inc., 1999.

National Research Council of Canada, Institute for Research in Construction. NRC/AT&T 555-400-021. *A Guide to Premises Distribution*. Ottawa, Canada: National Research Council of Canada, Institute for Research in Construction, 1988.

———. NRCC 38726. *National Building Code of Canada*. Ottawa, Canada: National Research Council of Canada, Institute for Research in Construction, 1995.

———. NRCC 38727. *National Fire Code of Canada*. Ottawa, Canada: National Research Council of Canada, Institute for Research in Construction, 1995.

Occupational Safety and Health Administration. *Code of Federal Regulations, Title 29, Part 1910. Occupational Safety and Health Standard*. Washington, DC: U.S. Department of Labor, Occupational Safety and Health Administration, 1998.

———. *Code of Federal Regulations, Title 29, Part 1910.146-Permit-Required Confined Spaces*. Washington, DC: U.S. Department of Labor, Occupational Safety and Health Administration, 1998.

———. *Code of Federal Regulations, Title 29, Part 1926. Safety and Health Regulations for Construction*. Washington, DC: U.S. Department of Labor, Occupational Safety and Health Administration, 1997.

———. Occupational Safety and Health Act of 1970. Title 29. Washington, DC: U.S. Department of Labor, 1970.

Pearson, Eric R. *Complete Guide to Fiber Optic Cable System Installation*. Albany, NY: Delmar/Thompson Learning, 1997.

Society of Cable Telecommunications Engineers, Inc. (SCTE) IPS-SP-001. *Flexible R.F. Coaxial Drop Cable*. Exton, PA: Society of Cable Telecommunications Engineers, Inc., June 1996.

———. IPS-SP-100. *Specification for Trunk, Feeder, and Distribution Coax Cable*. Exton, PA: Society of Cable Telecommunications Engineers, Inc., January 1997.

———. IPS-SP-401. *"F" Port (Male Feed Thru) Physical Dimensions*. Exton, PA: Society of Cable Telecommunications Engineers, Inc., October 1997.

———. IPS-SP-404. *"F" Connector (Male Indoor) Installation and Performance*. Exton, PA: Society of Cable Telecommunications Engineers, Inc., October 1997.

Southern Building Code Congress International, Inc. *The Standard Building Code*. Birmingham, AL: Southern Building Code Congress International, Inc.

Sterling, Donald J. *Premises Wiring for High-Performance Buildings*. Albany, NY: Delmar, 1995.

Sterling, Donald J. *Technician's Guide to Fiber Optics*. 3rd edition. Albany, NY: Delmar, 1999.

Sunset Complete Home Wiring. Editors, Sunset Books. Menlo Park, CA: Sunset Publishing Corp., 2000.

Telcordia Technologies, Inc. (Bellcore). GR-1503-CORE. B*ellcore Generic Requirements for Coaxial Connectors (Series 59, 6, 7, and 11)*. Morristown, NJ: Telcordia Technologies, Inc., March 1995.

Telecommunications Technologies Education Guide. Springfield, MA: Northeast Center for Telecommunications Technologies, 1999.

Telecommunications Industry Association/Electronic Industries Alliance. TIA/EIA TSB67. *Transmission Performance Specifications for Field Testing of Unshielded Twisted-Pair Cabling Systems*. Arlington, VA: Telecommunications Industry Association/Electronic Industries Alliance, October 1995.

———. TIA/EIA TSB75. *Additional Horizontal Cabling Practices for Open Offices*. Arlington, VA: Telecommunications Industry Association/Electronics Industries Alliance, August 1996.

Token Ring Network Introduction and Planning Guide. 4th edition. New York, NY: IBM Corporation, 1990.

Underwriters Laboratories Inc.® UL497. *Protectors for Paired Conductor Communications Circuits*. Northbrook, IL: Underwriters Laboratories Inc., 2001.

————. UL497A. *Secondary Protectors for Communications Circuits*. Northbrook, IL: Underwriters Laboratories Inc., 2001.

————. UL497B. *Protectors for Data Communications and Fire Alarm Circuits*. Northbrook, IL: Underwriters Laboratories Inc., 1999.

————. UL 1863. *Underwriters Laboratories Standard for Safety-Communication Circuit Accessories*. Northbrook, IL: Underwriters Laboratories, Inc., 1995.

Weisman Carl J. *Essential Guide to RF and Wireless*. Saddle River, NJ: Prentice Hall PTR, 2000.

Wiring Strategies for Voice and Data Systems. Bothell, WA: Leviton Telcom, 1998.

Young, Harry E. *Wireless Basics*. 2nd edition. Overland Park, KS: Telephony Books, 1996.

Resources

Periodicals

Archi-Tech. Portland, ME: DKNA Multimedia. www.architechmag.com

Building Operations Management. Milwaukee, WI: Trade Press Publishing Company.

Building Design and Construction. Newton, MA: Cahners Publishing Company.

Cabling Contractor. Nashua, NH: PennWell Corp. www.cable-install.com

Cabling Installation and Maintenance. Nashua, NH: PennWell Corp. www.cable-install.com

Cabling Product News. Nashua, NH: PennWell Corp. www.cable-install.com

CEDIA News. Indianapolis, IN: CEDIA. www.cedia.org

CE Pro – Custom Electronics Professional. Wayland, MA: EH Publishing Co. www.ce-pro.com

Comments & Connections. Cedar Rapids, IA: NSCA. www.nsca.org

Consumer Electronics Vision. Arlington, VA: CEA. www.ce.org

Design-Build Magazine. New York, NY: McGraw-Hill Construction Information Group. www.designbuildmag.com

Electronic House Buyer's Guide. Wayland, MA: EH Publishing, 2000.

Electronic House Guide to Home Planning. Wayland, MA: EH Publishing, 2001.

Electronics Servicing and Technology. Port Washington, NY: Mainly Marketing Enterprises.

H.A. Pro. Wayland, MA: EH Publishing Co. www.hapro.com

High-Tech News. Greencastle, IN: ETA. eta@indy.tdsnet.com

Home Automation. Wayland, MA: EH Publishing Co. www.HomeAutomationMag.com

Home Automation and Networking News. HANA. Washington, DC. www.HANAonline.org

Home Automator. Mebane, NC. www.HomeAutomator.com

Home and Building Automation Quarterly. Ottawa, ON: CABA. www.caba.org

Home Systems. Newton, MA: Cahners Publishing Co. www.gohomesystems.com

Home Technology Products. Dallas, TX: Stevens Publishing. www.HomeTechProducts.com

Intelligence Quarterly. Oyster Bay, NY: ABI. www.alliedworld.com

Popular Home Automation. Wayland, MA: EH Publishing Co. www.pophome.com

Professional Builder. Newton, MA: Cahners Publishing Co.

Residential Systems. Baldwin, NY: Miller Freeman Publishing Co. www.resmagonline.com

Security Dealer. Melville, NY: Cygnus Publishing.

Skylines. Washington, DC: BOMA International. www.boma.org

Specs. Louisville, CO: CableLabs. www.cablelabs.com

Sterling, Donald J. *Premises Cabling*. Albany, NY: Delmar/Thompson Learning.

Premises Networks Online. www.premisesnetworks.com

Systems Contractor News. New York, NY: Miller Freeman Publishing Co.

TecHome Builder. Wayland, MA: EH Publishing Co. www.techomebuilder.com

The Perfect Vision—High Performance Home Theater. Austin, TX: Absolute Multimedia. www.theperfectvision.com

Training resources

BICSI Residential Network Cabling Training and Registration Program. Tampa, FL: BICSI, 2002.

BICSI *Residential Telecommunications Cabling Standard: Understanding ANSI/TIA/EIA-570-A*. (online training: www.bicsi.org) Tampa, FL: BICSI, 2000.

Bowling, Roy. *Wire Running: "As Easy As That"* (7 videos). Lakewood, CO: Complete Vision Productions, 1995.

Cable Pulling Series (four videos). Nashua, NH: PennWell Corp., 1999.

Infrastructure Wiring for Existing Homes (video). CDA/The Training Dept., Tucson, AZ, 2001.

Infrastructure Wiring for New Homes (video). CDA/The Training Dept., Tucson, AZ, 2001.

Residential Infrastructure Wiring (2-video training course). The Training Dept., Tucson, AZ.

Residential Installation Basics (2-video training course). The Training Dept., Tucson, AZ.

Residential Retrofit Wiring (2-video training course). The Training Dept., Tucson, AZ.

Wiring Your Home for the Future (video). The Training Dept., Tucson, AZ.

Web sites

1394 Trade Association. www.1394ta.org

Allied Business Intelligence (ABI). www.alliedworld.com

Alliance for Telecommunications Industry Solutions (ATIS). www.atis.org

American Institute of Architects (AIA). www.aiaonline.com

American Insurance Association (AIA). www.aiadc.org

American Insurance Service Group (AISG). www.iso.com/aisg/index.html

American National Standards Institute (ANSI). www.ansi.org

American Society of Heating, Refrigerating, and Air Conditioning Engineers (ASHRAE). www.bacnet.org

American Society for Testing and Materials (ASTM). www.astm.org

BACnet Committee. www.bacnet.org

BICSI®. www.bicsi.org

Bluetooth SIG. www.bluetooth.org

Cable Television Laboratories (CableLabs). www.cablelabs.com

Cable Testing. www.cabletesting.com

Cahners In-Stat Group. www.instat.com

Canadian Alarm and Security Association (CANASA). www.canasa.org

Canadian Standards Association (CSA International). www.csa-international.org

CEBus® Industry Council (CIC). www.cebus.org

Ceilings and Interior Systems Construction Association (CISCA). www.cisca.org

Construction Marketplace. www.construction.com

Construction Specifications Institute (CSI). www.csinet.org

Consumer Electronics Association (CEA). Formerly Consumer Electronics Manufacturers Association (CEMA). www.ce.org

Continental Automated Buildings Association (CABA). www.caba.org

Custom Electronic Design and Installation Association (CEDIA). www.cedia.org

Datacomm Research Co. www.datacommresearch.com

Division 17. www.division17.net

EIBA. www.eiba.org

Electric Smarts. www.electricsmarts.com

Electrical Contracting Online. www.econline.com

ElectroniCast Corp. www.electronicast.com

Electronic House Online. www.electronichouse.com

Electronic Industries Alliance (EIA). www.eia.org

Electronics Technicians Association (ETA). eta@indy.tdsnet.com or www.eta-sda.com

Encyclopedia of Computer Security. www.itsecurity.com

Energy Conservation and Homecare Network Consortium (Echonet). www.echonet.gr.jp

ETL SEMKO. www.etlsemko.com

European Committee for Electrotechnical Standardization (CENELEC). www.cenelec.org

European Telecommunications Standards Institute. ETSI InfoCentre. www.etsi.org

Extend the Internet (ETI) Alliance. www.etialliance.com

Faulkner Information Services. www.faulkner.com

Federal Communications Commission (FCC). www.fcc.gov

Federal Information Processing Standards (FIPS). www.itl.nist.gov/fipspubs

Fiber Optic Association (FOA). www.thefoa.org

Fiber Optics Online. www.fiberopticsonline.com

General Services Administration (GSA). www.gsa.gov

GetPlugged.com. www.getplugged.com

Global Engineering Documents. www.global.ihs.com

HiperLAN/2 Global Forum (H2GF). www.hiperlan2.com

Home Audio Visual Interoperability (HAVI) Organization. www.havi.org

Home Automation and Networking Association (HANA). www.homeautomation.org

Home Automation Times. www.homeautomationtimes.com

Home Director. www.homedirector.com

Home Electronic System (HES) Working Group. www.Sc25wg1.metrolink.com

Home Phoneline Networking Alliance (HomePNA™). www.homepna.org

HomePlug Powerline Alliance. www.homeplug.org

HomeRF Working Group. www.homerf.org

Home Safety Central. www.homesafetycentral.com

HomeToys. www.hometoys.com

House Ear Institute. www.hei.org

Information Technology Laboratory Publications. Federal Information Processing Standards (FIPS). www.itl.nist.gov/fipspubs

Infrared Data Association (IrDA). www.irda.org

Insight Research Group. www.insight-group.com

Institute of Electrical and Electronics Engineers, Inc.® www.ieee.org

Insulated Cable Engineers Association (ICEA). www.icea.net

International Association of Electrical Inspectors (IAEI). www.iaei.com

International Building Officials and Code Administrators International (BOCA). www.bocai.org

International Building Owners and Managers Association (BOMA). www.boma.org

International Conference of Building Officials (ICBO). www.icbo.org

International Electrotechnical Commission (IEC). www.iec.ch

International Facility Management Association (IFMA). www.ifmaorg

International Organization for Standardization (ISO). www.iso.ch

International Telecommunication Union. www.itu.int

International Wireless Telecommunication Association (IWTA). www.iwta.org

Internet Home Alliance. www.internethomealliance.com

Intertek Testing Services (ITS). www.info@ETLSEMKO

Jini Community. www.jini.org

KMI Corp. www.kmicorp.com

Konnex Association. www.konnex-knx.com

LonMark Interoperability Association. www.lonmark.org

McGraw-Hill Construction Outlook. (F. W. Dodge Report). www.fwdodge.com

MultiMedia Telecommunications Association (MMTA). www.MMTA.org

National Association of Communications Contractors (NACC). www.cabcert.com

National Association of Home Builders (NAHB). www.nahb.com

National Building Code-Canada (NBCC®). www.nrc.ca/irc and www.ccbfc.org

National Center for Construction Education and Research. www.nccer.org

National Electrical Contractors Association (NECA). www.necanet.org

National Electrical Manufacturers Association® (NEMA®). www.nema.org

National Fire Protection Association® (NFPA®). www.nfpa.org

National Institute of Standards and Technology (NIST). www.nist.gov

National Research Council of Canada (NRC-IRC). Institute for Research in Construction. www.nrc.ca/irc and www.ccbfc.org

National Systems Contractors Association (NSCA). www.nsca.org

National Technical Information Services (NTIS). U.S. Department of Commerce. www.ntis.gov

Network Wiring.com. www.network.wiring.com

Occupational Safety and Health Administration (OSHA). www.osha.gov

Parks Associates. www.parksassociates.com

R7 Home Networking Committee. www.ce.org

Rural Utilities Services (RUS—formerly REA). www.usda.gov/rus

Salutation Consortium. www.salutation.org

Satellite Broadcasting and Communications Association (SBCA). www.sbca.com

Security Industry Association (SIA). www.siaonline.org

SmartContractor.com. www.smartcontractor.com

SmartElectrical.com. www.smartelectrical.com

SMARTHOME.COM. www.smarthome.com

Smart House. www.smart-house.com

Society of Cable Telecommunications Engineers (SCTE). www.scte.org

Southern Building Code Congress International, Inc. (SBCCI). www.sbcci.org

TechHome. www.TechHome.org

Technocopia. www.technocopia.com

Technology Futures, Inc. (TFI). www.tfi.com

Telecommunications Industry Association (TIA). www.tiaonline.org

Telcordia™ Technologies (formerly Bellcore). www.telcordia.com

Test and Measurement Online. www.testandmeasurement.com

Training Dept. www.trainingdept.com

Underwriters Laboratories Inc.® (UL®). www.ul.com

Universal Plug and Play Forum (UPnP). www.upnp.org

USB Implementers Forum. www.usb.org

Video Electronics Standards Association (VESA). www.vesa.org

Wireless Ethernet Compatibility Alliance (WECA). www.wi-fi.org

Wireless LAN Association (WLANA). www.wlana.com

Wireless LAN Interoperability Forum (WLIF). www.wlif.com

Wiring America's Homes. www.connectedhome.org

X10.ORG. www.x10.org

NOTE: For a list of recognized regulatory and reference bodies, see Chapter 3: Codes, Standards, and Regulations.

Index

NOTE: Boldface numbers indicate illustrations.

We wrote
the book!

The BICSI Approach

Recognized throughout the world as a leader in telecommunications education, BICSI specializes in low-voltage cabling. Focusing on commercial distribution design and installation, BICSI recently added residential installation to its telecommunications mix. BICSI offers a full line of educational products and services, including registration programs, courses, technical publications, and conferences.

Professional Registrations

Earn international credentials by demonstrating your proficiency in a specific area. Currently, BICSI offers professional registration programs in cabling installation (BICSI

Registered Installer, Level 1; Installer, Level 2; and Technician), residential cabling (BICSI Registered Residential Installer), distribution design (Registered Communications Distribution Designer—RCDD®), LAN design (RCDD/LAN Specialist), and customer-owned outside plant design (RCDD/OSP Specialist).

Vendor-Neutral Training

BICSI offers a variety of learning opportunities, from structured classroom instruction to flexible Internet home study. With instruction now available via the Web and at over 130 licensed training centers, BICSI presents leading-edge technical training in many parts of the world. BICSI's virtual campus features courses such as *The Residential Telecommunications Cabling Standard— Understanding ANSI/TIA/EIA-570-A* course. BICSI's *Educational Resource Catalog* outlines BICSI's 25-plus courses, and is available on request.

BiCSi®
A Telecommunications Association

Technical Publications

Since 1974, BICSI has been developing a library of technical publications, several of which have become industry standards. These reference books span the subjects of cabling installation, outside plant design, network design, and more. Many of these valuable reference tools double as study guides for BICSI courses and exams. BICSI members receive a significant discount on manuals and may enjoy even greater savings if the previous edition is owned. *Note: Most manuals available on CD-ROM at manual price.*

- BICSI's *Telecommunications Distribution Methods Manual* (*TDMM*), now the accepted guideline of the industry, is a valuable reference tool for those who design the telecommunications infrastructure. The *TDMM* also serves as a detailed study guide for those preparing to take the Registered Communications Distribution Designer (RCDD®) exam. US$179 BICSI member; US$329 nonmember. ISBN 1-928886-04-3 (manual); ISBN 1-928886-05-1 (CD).

- The newest BICSI publication, the *Residential Network Cabling Manual* provides detailed how-to information in residential voice, data, and video distribution design, installation, and systems integration. US$49 BICSI member; US$49 nonmember. ISBN 0-07-138211-9 (manual).

- In the *Telecommunications Cabling Installation Manual*, you'll find step-by-step procedures for installing telecommunications cable and useful information included in the BICSI Installer and Technician exams. Also available as a hard cover, 3-ring binder. US$99 BICSI member; US$99 nonmember (3-ring binder version). ISBN 1-928886-08-6 (manual); ISBN 1-928886-09-4 (CD).

- The *Network Design Reference Manual* describes all aspects of networking—LANs, local internetworks (backbone/campus), and wide area internetworks and associated telecommunications links. US$179 BICSI member; US$329 nonmember (3-ring binder version). ISBN 1-928886-10-8 (manual); ISBN 1-928886-11-6 (CD).

- Ideal for those with previous outside plant design experience, our *Customer-Owned Outside Plant Design Manual* offers an overview of outside plant design, including pathways and spaces, CO-OSP cabling infrastructure, and more. US$99 BICSI member; US$179 nonmember. ISBN 1-928886-06-X (manual); ISBN 1-928886-07-8 (CD).

- BICSI's *Introduction to Commercial Voice/Data Cabling Systems* Video and Workbook provides a visual tour of voice/data cabling for the modern commercial building, as well as the spaces and systems that comprise its infrastructure. US$249 BICSI member; US$349 nonmember.

Introduction to Commercial Voice/Data Cabling Systems Video and Workbook

- Learn design guidelines for the integration of emerging applications into existing LANs in the *LAN and Internetworking Applications Guide*. US$29 BICSI member; US$29 nonmember. ISBN 1-928886-00-0.

- The *BICSI Telecommunications Dictionary* is a compilation of glossaries from all of BICSI's publications. Acronyms, abbreviations, symbols, and international telecommunications standards can all be found in this convenient book. US$19 BICSI member; US$19 nonmember. ISBN 1-928886-03-5.

- BICSI also offers the *On-the-Job Training Booklet*, which provides a performance checklist of key cabling installation tasks—perfect for candidates studying for registration as a BICSI Installer or Technician, or for the supervisor looking to evaluate employee performance. US$9 BICSI member; US$9 nonmember.

- And finally, BICSI developed the *Telecommunications Quick Reference Guide for Code Officials: Summary and Excerpts from the NEC® 2002*, which outlines portions of the National Fire Protection Association (NFPA 70), *National Electrical Code®* (*NEC*), *2002 Edition*. Free to both BICSI members and nonmembers.

New Member Instant Reference Library Deal

Save over US$250! For a limited time only, BICSI is offering a New Member Instant Reference Library Deal, allowing new members to purchase all of the above publications (with their CD-ROM counterpart) at a significant savings—20% off of the discounted member price! New members who purchase the entire library within 30 days of becoming a BICSI member pay only US$1035 plus shipping!

Check out BICSI's publications in more detail. Visit the BICSI Web site (www.bicsi.org) for additional information on any of BICSI's reference books. There you'll find sample chapters of all BICSI manuals, a complete price listing, and online ordering. You may also call BICSI's Customer Service Department at 813-979-1991 or 800-242-7405 (USA/Canada toll free) to request a publications brochure (also available on the BICSI Web site). An information request form may be found on the following page.

Membership

To remain competitive in our changing environment, you need to stay abreast of telecommunications issues, standards, and technology—locally and around the globe. Fortunately, BICSI is here to keep you informed and knowledgeable in all aspects of the telecommunications profession. With more than 22,000 members, BICSI offers members substantial discounts on our quality technical publications, design courses, and conference fees. In fact, the cost of membership can more than pay for itself with the purchase of just one BICSI manual.

We recognize that you may have questions about BICSI, so we encourage you to contact BICSI for a BICSI Information Packet. BICSI membership is your key to a successful career in telecommunications, and we want you to appreciate all of the member benefits.

BICSI Information Packet Available

To request a BICSI Information Packet, complete the form below and fax it to 813-971-4311. You may also request a packet by calling 813-979-1991 or 800-242-7405 (USA/Canada toll free) or e-mailing bicsi@bicsi.org.

Yes, I want to find out more about BICSI. Please send me a BICSI Information Packet:

_____
name

_____
address

_____
city   state   zip

_____
phone   fax

_____
e-mail

For more information or to request a BICSI
Information Packet, contact BICSI today!

8610 Hidden River Parkway, Tampa, FL 33637-1000 USA
813-979-1991 or 800-242-7405 (USA/Canada toll free)
fax: 813-971-4311
e-mail: bicsi@bicsi.org
Web site: www.bicsi.org